HUMAN PATHOGENIC MICROBES

DEVELOPMENTS IN MICROBIOLOGY

HUMAN PATHOGENIC MICROBES

DISEASES AND CONCERNS

MANZOOR AHMAD MIR
Department of Bioresources, School of Biological Sciences, University of Kashmir, Srinagar, India

ELSEVIER
ACADEMIC PRESS
An imprint of Elsevier

Academic Press is an imprint of Elsevier
125 London Wall, London EC2Y 5AS, United Kingdom
525 B Street, Suite 1650, San Diego, CA 92101, United States
50 Hampshire Street, 5th Floor, Cambridge, MA 02139, United States
The Boulevard, Langford Lane, Kidlington, Oxford OX5 1GB, United Kingdom

ISBN: 978-0-323-96127-1

For information on all Academic Press publications visit our website at https://www.elsevier.com/books-and-journals

Publisher: Stacy Masucci
Acquisitions Editor: Linda Versteeg-Buschman
Editorial Project Manager: Mica Ella Ortega
Production Project Manager: Maria Bernard
Cover Designer: Christian Bilbow

Typeset by TNQ Technologies

Contents

5. Combating human fungal infections: need for new antifungal drugs and therapies

Manzoor Ahmad Mir, Hafsa Qadri, Shariqa Aisha, and Abdul Haseeb Shah

6. Significance of immunotherapy for human bacterial diseases and antibacterial drug discovery

Manzoor Ahmad Mir, Syed Suhail Hamdani, and Hafsa Qadri

7. Significance of immunotherapy for human fungal diseases and antifungal drug discovery

Manzoor Ahmad Mir, Ulfat Jan, and Hafsa Qadri

8. Combinatorial approach to combat drug resistance in human pathogenic bacteria

Manzoor Ahmad Mir, Manoj Kumawat, Bilkees Nabi, and Manoj Kumar

9. Combinatorial approach to combat drug resistance in human pathogenic fungi

Manzoor Ahmad Mir, Hafsa Qadri, Shariqa Aisha, and Abdul Haseeb Shah

10. Recent trends in the development of bacterial and fungal vaccines

Manzoor Ahmad Mir, Muhammad Usman, Hafsa Qadri, and Shariqa Aisha

List of contributors

Sushma Ahlawat Department of Biochemistry and Biochemical Engineering, Sam Higginbottom University of Agriculture, Technology and Sciences, Prayagraj, Uttar Pradesh, India

Shariqa Aisha Department of Bioresources, School of Biological Sciences, University of Kashmir, Srinagar, Jammu and Kashmir, India

Bader Alshehri Department of Medical Laboratory Sciences, College of Applied Medical Sciences, Majmaah University, Saudi Arabia

Syed Suhail Hamdani Department of Bioresources, School of Biological Sciences, University of Kashmir, Srinagar, Jammu and Kashmir, India

Nusrat Jan Department of Bioresources, School of Biological Sciences, University of Kashmir, Srinagar, Jammu and Kashmir, India

Ulfat Jan Department of Bioresources, School of Biological Sciences, University of Kashmir, Srinagar, Jammu and Kashmir, India

Manoj Kumar Department of Microbiology, ICMR-NIREH, Bhopal, Madhya Pradesh, India

Manoj Kumawat Department of Microbiology, ICMR-NIREH, Bhopal, Madhya Pradesh, India

Manzoor Ahmad Mir Department of Bioresources, School of Biological Sciences, University of Kashmir, Srinagar, Jammu and Kashmir, India

Bilkees Nabi Department of Biochemistry and Biochemical Engineering, Sam Higginbottom University of Agriculture, Technology and Sciences, Prayagraj, Uttar Pradesh, India

Hafsa Qadri Department of Bioresources, School of Biological Sciences, University of Kashmir, Srinagar, Jammu and Kashmir, India

Ubaid Rasool Department of Microbiology, School of Biological Sciences, University of Kashmir, Srinagar, Jammu and Kashmir, India

Abdul Haseeb Shah Department of Bioresources, School of Biological Sciences, University of Kashmir, Srinagar, Jammu and Kashmir, India

Muhammad Usman Department of Bioresources, School of Biological Sciences, University of Kashmir, Srinagar, Jammu and Kashmir, India

Abrar Yousuf Department of Bioresources, School of Biological Sciences, University of Kashmir, Srinagar, Jammu and Kashmir, India

Preface

Human pathogenic microorganisms represent a socioeconomic burden and a major public healthcare issue. All the leading human bacterial and fungal pathogens have been found to be associated with high rates of mortality and morbidity affecting millions of individuals around the globe. These pathogens have evolved different mechanisms to escape the host immune responses and become more virulent and resistant to different antimicrobial drugs due to continuous mutations. Drug-resistant bacterial diseases represent a major challenge in routine medical practice globally. The evolution of antibiotic drug-resistant pathogenic bacteria constitutes a severe healthcare threat worldwide. The major concern in the present times is resistance developed by the pathogenic bacteria against the antibiotics to withstand the adverse effects of drugs, resulting in their efficiency of virulence. The major classes of drugs commonly used for bacterial infections include penicillin, tetracyclines, cephalosporins, fluoroquinolones, macrolides, and sulfonamides.

Infections associated with multiple fungal pathogenic organisms generally belonging to *Candida, Cryptococcus, Aspergillus* genera are affecting vast numbers of people extensively each year. These pathogenic fungi have adapted diversified resistance mechanisms against a variety of antifungal classes to withstand the adverse drug effects. Different categories of antifungals like azoles, echinocandins, etc., represent the most commonly employed antifungal agents. The majority of these available antifungals are not entirely potent which ultimately limits their use in clinical usage and practice. Many naturally derived compounds from microorganisms, plants, marine organisms, and nano-based particles are currently being investigated for their possible antifungal properties.

Immunotherapies are disease-management approaches that specifically target or modulate immune system components. As countries continue to grapple with a slew of new and resurgent diseases, the most current worldwide health concern being the SARS-CoV2 pandemic, infectious diseases continue to pose a substantial threat to human health. While significant progress has been achieved in diagnosing and understanding the etiology of infectious diseases, current antimicrobial chemotherapy has been shown to be ineffective against a large number of current pathogens. Emerging and re-emerging microorganisms create new infectious diseases, have developed resistance to available antibacterial drugs, or are unable to be treated due to a lack of treatment options or are generally ineffective due to underlying host immunological impairment. A variety of immunotherapeutic strategies are currently being examined as viable alternative therapeutics for infectious diseases as a result, which has resulted in remarkable advances in the understanding of pathogen—host immunity interactions.

There has been a tremendous increase in the occurrence of invasive fungal diseases. In immunocompromised persons, invasive fungal diseases are found correlated with high mortality and morbidity rates. Unfortunately, there are limited treatment strategies available against the increasing fungal drug-resistance phenomenon associated with different human fungal pathogenic species. Moreover, the currently available antifungals are associated with severe issues including host toxicity, fungistatic property, and the development of fungal drug resistance in the concerned pathogens. There is an instant need for the identification, establishment, and advancement of novel treatment strategies for the eradication of increasing human fungal diseases. Drug combination therapy could be a novel successful therapeutic approach to improve the potency of the drug and alleviate the advancement of the phenomenon of drug resistance.

Antimicrobial drug resistance represents an emerging issue. Microbial infections cause many challenging diseases and hence the high death rates worldwide. Individuals with compromised immune systems have been mainly seen to be under attack. With the help of different research methodologies, novel approaches have been proposed for the establishment of bacterial and fungal vaccines. Furthermore, a more desirable knowledge about the proper functioning of the immune system against different bacterial and fungal pathogens has resulted in the further establishment of such novel bacterial and fungal vaccination strategies. Although few bacterial and fungal vaccines have developed via clinical trials, there is a greater scope for the advancement of the clinical development of such vaccines.

Acknowledgments

First, I would like to thank the Almighty for giving me strength, belief and good health. It is "HE" who is the creator of everything between Heavens and Earth and beyond. May "HE, open his clandestine treasure of knowledge upon me, Ameen".

I am bereft of expression that would describe my gratitude and respect to my teachers and mentors who stood prop all the way with zealous and enthusiastic spirit of scientific adepts. I will benefit for a long time to come from their sincerity, originality and truthfulness, which have nourished my intellectual maturity. I admire and respect them all for their sincerity, dedication, devotion, amazing memory, thorough knowledge and constructive criticism.

Apart from above, I am indebted to a number of people who have always been a constant source of encouragement and enthusiasm, directly or indirectly, in making this book, entitled *Human Pathogenic Microbes: Diseases and Concerns*. I put on record my abysmal gratitude to all those people who lent their help in every possible way with gestures whenever and wherever the same was required by me.

I feel great happiness in expressing profound thanks and venerations to my parents (Haji Mohammad Abdullah Mir and Zoona Begum) who supported me in attaining new zeniths in every field of life and knowledge. They have been a source of great strength throughout my life, whose tears and toil with zeal in planting the seed of loyalty, hard work, dedication and sincerity in me warrants recognition with laud. My father-in-law Mr Mohammad Ashraf Malik deserves a special mention for his support, care and guidance.

I would like to make a special thank-you for my entire family (Sumaira Manzoor, Aariz Manzoor, Ayzel Manzoor), my scholars Umar Mehraj, Hina Qayoom, Hafsa Qadri, Shazia Shafi, Shariqa Aisha, who deserve a special mention. At the same time, I would like to thank my other family members and everybody who were important to the successful realization of this book, as well as express my apology that I cannot mention personally one by one.

I express my intensity of emotions towards the Prime Mover Almighty in whom I have great faith.

Dr. Manzoor A Mir

Abbreviation list

5-FC	5-Fluorocytosine
ABC	ATP-binding cassette
AIDS	Acquired immunodeficiency syndrome
Amp-B	Amphotericin-B
AMR	Antimicrobial drug resistance
APC	Antigen presenting cells
APHL	Association of Public Health Laboratories
ART	Antiretroviral therapy
ATP	Adenosine 5′- triphosphate
BCG	Bacille Calmette-Guerin (a vaccine for tuberculosis (TB) disease)
BSIs	Blood stream infections
CAR	Chimeric antigen receptor
CARD9	Caspase recruitment domain-containing protein 9
CARSS	China antimicrobial resistance surveillance system
$CD4^+$	Cluster of differentiation 4
CDC	Centre for Disease Prevention and Control
CDC	Centers for Disease Control
Cdr	Candida drug resistance
CGB	Cnavanine-glycine-bromthymol blue agar
CHINET	China Antimicrobial Surveillance Network
CIDT	Culture-independent diagnostic test
CLR	C-type lectin receptor
CLSI	Clinical and Laboratory Standards Institute
COVID-19	Coronavirus disease 2019
CSF	Caspofungin
CTLA4	Cytotoxic T-lymphocyte-associated protein 4
DC	Dendritic cells
DDDs	Defined daily doses
DST	Drug susceptibility test
E. coli	Escherichia coli
ECM	Extracellular matrix
EDLB	Enteric Diseases Laboratory Branch
EEHM	Ethanolic extract of Hyptis martiusii
EID	Emerging infectious disease
ELIZA	enzyme-linked immunosorbent assay
EMB	Ethambutol
ERG	ETS-related gene
EUCAST	The European Committee on Antimicrobial Susceptibility Testing
FDA	Food and Drug Administration
FDC	Fixed-dose combination
FDOSS	Food borne disease outbreak surveillance system
FLC	Fluconazole
GAP	Global action plan
GIT	Gastrointestinal tract
GMCSF	Granulocyte macrophage colony-stimulating cells

GPI	Glycosylphosphatidylinositol
HIV	human immunodeficiency viruses
HKS	Heat-killed Saccharomyces cerevisiae
HSCT	Hematopoietic stem cell transplantation
HSP	Heat shock protein
IBD	Inflammatory bowel disease
ICU	Intensive care unit
IFIs	Invasive fungal infections
IFN	Interferon
IGRA	interferon - gamma release assays
ILCs	Innate lymphoid cells
INH	Isoniazid
IPA	Invasive pulmonary aspergillosis
iPSCs	Induced pluripotent stem cells
IV	Intravenous
LEDS	Laboratory-based enteric disease surveillance
LPS	Lipopolysaccharide
LTBI	Latent tuberculosis infection
MABs	Monoclonal antibodies
MALDI-TOF	Matrix-assisted laser desorption/ionization-time of flight
MALT	Mucosa-associated lymphoid tissue.
MAM	Multivalent adhesion molecule
MDMs	Monocyte-derived macrophages
MDR	Multidrug resistance
MDR TB	Multidrug-resistant tuberculosis
MDR	Multidrug resistance
MDR	Multidrug resistant
MFC	Minimal fungicidal concentration
MFS	Major facilitator superfamily
MHC	Major histocompatibility complex
MIC	Minimum inhibitory concentration
MLS	Macrolide-lincosamide-streptogramin
MRSA	Methicillin-resistant S. aureus
MTB	Mycobacterium tuberculosis
NAC	Non-Albicans Candida
NARMS	National antimicrobial resistance monitoring system
NK	Natural killer cells
NNDSS	National notifiable diseases surveillance system
NOD	Nucleotide oligomerization domain
NTS	Nontyphoidal salmonellosis
OCV	Oral cholera vaccination
ORS	Oral rehydration solution
PAF	Penicillium chrysogenum antifungal protein
PAMPs	Pathogen-associated molecular patterns
PCR	Polymerase chain reaction
PCV	Pneumococcal conjugate vaccine
PFGE	Pulsed-field gel electrophoresis
PH	Potential of hydrogen
POT	Pathogen-oriented therapy
PRRs	Pattern recognition receptors
PZA	Pyrazinamide
QRDR	Quinolone-resistance-determining region
RDT	Rapid diagnostic tests

RIF	Rifampin
ROS	Reactive oxygen species
RT-PCR	Real-time polymerase chain reaction
SAP	Secreted aspartic protease
SARS-CoV2	Severe acute respiratory syndrome coronavirus 2
SCV	Salmonella containing vacuoles
TB	Tuberculosis
TCBS	Thiosulfate—citrate—bile salts agar
TCR	T cell receptor
TIGIT	T cell immunoreceptor with Ig and ITIM domains
TILs	Tumor-infiltrating lymphocytes
Tim-3	T cell immunoglobulin and mucin domain-containing protein 3
TMP-SMX	Trimethoprim-sulfamethoxazole
TRB	Terbinafine
TST	Tuberculin skin test
TSTs	Tuberculosis skin tests
US	United States
US FDA	United States Food and Drug Administration
USDA	US Department of Agriculture
UTI	Urinary tract infection
VVC	Vulvovaginal candidiasis
WASH	"Water, sanitation and hygiene"
WHO	World Health Organization
XDR TB	Extensively drug-resistant tuberculosis
XDR	Extensively drug-resistant
ZnONPs	Zinc oxide nanoparticles

Glossary

Acquired immunity The type of immunity developed with the passage of time in an individual. It consists of special B and T cells.

AIDS Acquired immunodeficiency syndrome is an infection caused by HIV virus that infects $CD4^+$ T-lymphocytes and body's cellular immunity weakens day by day.

Anaerobic organism Any organism that does not require molecular oxygen to grow is known as an anaerobic organism or anaerobe. If free oxygen is present, it may respond negatively or possibly perish. An aerobic organism (aerobe), on the other hand, is one that requires oxygen to survive.

Antibiotic resistance Antibiotic resistance occurs when organisms such as bacteria and fungi develop the ability to resist drugs that are supposed to kill them.

Antibiotics Antibiotics are medications or chemicals used to treat bacterial infections in people and animals. They act by either killing germs or preventing them from growing and multiplying.

Antifungal agents A fungicidal chemical molecule or a biological microorganism utilized to destroy or suppress the growth of a pathogenic fungal organism or its spores is known as an antifungal agent.

Antigen Substance which has the ability to evoke an immune response and produce antibodies against the pathogens.

Antimicrobial agents Any of a wide range of chemical compounds and physical agents used to kill or prevent the development of microorganisms is known as an antimicrobial agent.

Antimicrobial drug resistance Capability of a disease-causing organism such as fungi, bacteria, etc., to reproduce despite the availability of various antimicrobial drugs that would normally destroy them.

Antimicrobial resistance Antimicrobial resistance refers to the resistance of bacteria, viruses, parasites, and fungi to antibacterial, antiviral, antiparasitic, and antifungal treatments. Antimicrobial resistance occurs naturally, although it is facilitated by inappropriate pharmaceutical usage, such as the use of antibiotics for bacterial illnesses like salmonellosis.

Antimicrobials Class of drugs which are used to treat bacterial infections.

Antiretroviral Class of drugs that inhibits the activity of retroviruses such as HIV.

Aspergillosis It is a fungal disease caused by various pathogenic Aspergillus species.

Bacteremia It refers to bacteria present in blood.

Bacteria Bacteria are prokaryotic, unicellular microorganisms which are not visible to the naked eye and lack nucleus and various other cell organelles. Some bacteria species are infectious while others are noninfectious to humans, plants, and animals.

Biofilms	Biofilms are the most advanced and highly useful microbial lifestyle and prevail in both the natural and human host systems. Even though biofilms are useful to the microorganism itself, they can also exert both positive and negative impacts on the surrounding milieu.
Bispecific antibodies	Bispecific antibodies (BsAbs) have two binding sites that target two different antigens or two different epitopes on the same antigen.
Cancer	Cancer is a group of diseases that are involved in uncontrolled cell division and possess the property of metastasis and loss of programmed cell death.
Candidiasis	Candidiasis represents a fungal infection caused by pathogenic Candida species, e.g., Candida albicans.
Checkpoint inhibition	Checkpoint inhibitors operate by relaxing a natural brake on our immune system, allowing T cells, which are immunological cells, to recognize and fight cancers.
Combination therapy	The application of more than one drug agent (in combination) to cure the same condition is referred to as combination therapy.
Communicable diseases	A communicable disease spreads from person to person or animal to animal. Communicable diseases are caused by pathogens such as bacteria, viruses, fungi, and protists.
Cosmopolitan	Cosmopolitan means those which are adaptive and can occur in any geographical condition throughout the world.
COVID-19	Coronavirus disease 2019 (COVID-19) is an ailment caused by a novel Coronavirus known as severe acute respiratory syndrome Coronavirus 2. (SARS-CoV-2).
Cryptococcal meningitis	It is a fungal infection of the tissues covering the brain and the spinal cord. This disease is caused by cryptococcus species.
Cryptococcosis	It is a fungal infection/disease caused by the species of Aspergillus genus.
Cytokines	Cytokines are small proteins that regulate the growth and function of other immune system and blood cells.
Dendritic cells	Dendritic cells also called as DCs or accessory cells are antigen presenting cells (APCs) that act as connecting link between innate and adaptive immunity. The main function of such type of cells is to process antigen material and present it on the cell surface to the T-cells for further immune responses.
Drug resistance	Drug resistance is the ability of microorganisms such as bacteria, fungi, etc., to withstand drug therapy and reduce the effectiveness of the drug. E.g., Azole resistant C.auris.
Enteric fever	The term used to describe typhoid or paratyphoid.
Ergosterol	A sterol found on the cellular membrane of fungal organisms that functions similarly to mammalian cholesterol in maintaining the stability of the cellular membrane.
Evolution	It is defined as the process by which new types of living organisms develop from other existing organisms.
Extensively drug-resistant tuberculosis	This is a rare form of MDR TB that is resistant to isoniazid, rifampin, any fluoroquinolone, and at least one of three injectable second-line drugs, namely amikacin, kanamycin, or capreomycin.
Facultative aerobe	The microorganism that can live in presence of oxygen but does not require it to survive.

Flavonoids A class of natural compounds having different phenolic structural forms. These compounds are mainly present in flowers, fruits, roots, vegetables, barks, etc. Such natural compounds are considered very significant for their useful health impacts.

Food and Drug Administration The United States Food and Drug Administration is a government body within the Department of Health and Human Services. The FDA regulates and manages food safety, tobacco products, dietary supplements, prescription and over-the-counter pharmaceutical drugs, vaccines, biopharmaceuticals, blood transfusions, medical devices, and other products.

Foodborne pathogens Foodborne pathogens are biological organisms that can cause food poisoning. They include viruses, bacteria, and parasites.

Food chain Food chain is the flow of energy in the form of food from one organism to another.

Human fungal pathogens These are the pathogenic fungi which invade the humans like C. albicans, etc.

Immune response It is a reaction that occurs in a living organism for the purpose of defending itself in the case of foreign invaders.

Immunization Immunization or immunisation is the process of strengthening an individual's immune system against an infectious pathogen (known as the immunogen).

Immunocompromised individuals Those individuals who are unable to elicit strong immune responses as their immune system is not working properly or is weakened.

Immunocompromised They are host organisms having a weakened immune system. They have a low capacity to fight against infections.

Immunogenicity It is the ability of a foreign substance to incite an immune response in the body of human.

Immunoglobulins These are glycoprotein molecules produced by plasma cells. They act as a critical part of the immune response by specifically recognizing and binding to particular antigens, such as bacteria or viruses, and aiding in their destruction.

Immunosuppressive drugs Immunosuppressive drugs are medications that suppress or prevent immune system activity.

Immunotherapy Immunotherapy is a treatment that employs specific components of a person's immune system to combat diseases such as cancer.

Innate immunity It is the inborn immunity which protects an individual against all pathogens.

Interferon Interferons are a type of signaling protein that is produced and released by host cells in response to the presence of certain viruses.

Invasive aspergillosis Invasive aspergillosis is a severe infection caused by aspergillus species, e.g., A. fumigatus.

Invasive fungal infections Infections in which the pathogenic fungal organisms have infiltrated deep into the tissues and maintained permanently, causing long-term sickness. These infections are most commonly found in the elderly and immunocompromised persons.

Major histocompatibility complex The major histocompatibility complex (MHC) is a group of genes that code for proteins located on cell surfaces that aid the immune system in recognizing foreign substances. MHC proteins are found in all higher vertebrates.

Medicinal plants Those plants which are utilized to treat a specific disease condition. Capsules, infusions, ointments, tablets, and other forms of such plants parts or extracts are generally utilized.

Metagenomics Metagenomics is the study of genetic material extracted directly from environmental materials. Metagenomics is both a research tool and a diagnostic strategy for analyzing the impact of many microorganisms on human health without disturbing their natural environment.

Microbes Organisms which are very small, cannot be seen without a naked eye. They include bacteria, archae, fungi, protozoa, and algae

Microbial pathogen A microbe usually (bacteria, fungi, virus, etc.) able to cause different infectious diseases in its respective hosts.

Minimum inhibitory concentration The minimal antibiotic concentration that suppresses the observable proliferation of the concerned microorganism is known as the minimum inhibitory concentration (MIC).

Monoclonal antibodies An antibody composed of identical antibody molecules produced by a single clone of cells or cell line.

Monotherapy A single-drug therapy in which a single drug is employed for treating a specific illness.

Multidrug resistance A phenomenon or process enabling a pathogenic microorganism to withstand multiple drugs/chemicals of a broad range of structure and activity targeted at eradicating the concerned organism.

Mycobacterium tuberculosis It is a species of a lethal bacteria belonging to the family Mycobacteriaceae that causes tuberculosis a highly resistant bacterial infection. This disease is commonly called as Tuberculosis.

Natural killer cells Natural killer (NK) cells are innate immune system effector lymphocytes that control many types of tumors and microbial infections by limiting their spread and consequent tissue damage.

Pathogen-associated molecular patterns These are molecules with conserved motifs that are associated with pathogen infection that serve as ligands for host pattern recognition molecules such as Toll-like receptors.

Pathogens Any of the microorganisms whether bacteria, fungi, virus, protozoa, etc., that can cause illness or disease are known as pathogens.

Pathogenicity it is the potential ability of some microbial organisms/species or viruses to cause disease. The process is marked by complex pathogenic characteristics that develop in the course of their fight for survival.

Pathophsiology The altered physiological conditions because of disease or infection or injury.

Peritrichous flagella Arrangement of flagella all over the body and all of which are directed in different ways.

Plant essential oil It represents a highly concentrated hydrophobic liquid comprising volatile chemical components isolated from different parts of the plants.

Pneumococcal conjugate vaccine The pneumococcal vaccine protects against severe life-threatening pneumococcal infections. It's also known as the pneumonia vaccine.

Probiotics The use of microorganisms for its beneficial qualities by introducing them into the body.

Reactive oxygen species (ROS) Reactive oxygen species represents the highly reactive chemicals developed from O_2. Some of the important examples of ROS involve hydroxyl radical, peroxides, superoxide, singlet oxygen, and alpha-oxygen.

Salmonellosis It is a Salmonella-caused infectious disease which effects the intestinal tract.

Sepsis Sepsis is a potentially deadly disease that arises when the body's response to an infection causes tissue damage.

Serotype A serotype, also known as a serovar, is an unique variant within a bacteria or virus species or among immune cells of different people.

Surveillance It is a close observation of public health data for decision making.

Synergy A mechanism in which two or more entities, chemicals, or other agents interact or cooperate to yield a cumulative impact higher in contrast to the sum of their single impacts.

Systemic disease The disease which affects the whole body rather than a single organ or part of body such as high blood pressure.

T cell T cells are immune cells that develop from stem cells in the bone marrow. They assist to defend the body from infection and may aid in the fight against cancer.

T-lymphocytes T-lymphocytes are white blood cells that develop from stem cells in the bone marrow. They protect our body from infection and may help fight cancer.

Terpenes Terpenes represent the widest category of naturally occurring products. Depending upon the number of isoprene units they comprise, these compounds are generally categorized as monoterpenes, di-terpenes, tri-terpenes, tetra-terpenes, and sesquiterpenes. These compounds are mainly present in different plant parts and form the crucial part of plant essential oils.

Transgenic animals Transgenic animals are those that have had a foreign gene (gene of interest) purposely put into their genome (most typically mice).

Tuberculosis Tuberculosis (TB) is caused by bacteria (Mycobacterium tuberculosis), which primarily affects the lungs.

Vaccine Vaccine is the substance used to stimulate the production of antibodies and provide immunity against one or several diseases, prepared from the causative agent of a disease, its products, or a synthetic substitute, treated to act as an antigen without inducing the disease. E.g., measles, mumps, and rubella (MMR) vaccine.

Virulence The intrinsic capability of an infectious organism to produce disease/illness/infection is referred to as virulence.

Zoonotic pathogen Those pathogens which can naturally transmit between animals and humans.

CHAPTER

1

Human pathogenic microbes (bacterial and fungal) and associated diseases

Manzoor Ahmad Mir[1], *Ubaid Rasool*[2], *Shariqa Aisha*[1], *Bader Alshehri*[3] *and Syed Suhail Hamadani*[1]

[1]Department of Bioresources, School of Biological Sciences, University of Kashmir, Srinagar, Jammu and Kashmir, India; [2]Department of Microbiology, School of Biological Sciences, University of Kashmir, Srinagar, Jammu and Kashmir, India; [3]Department of Medical Laboratory Sciences, College of Applied Medical Sciences, Majmaah University, Saudi Arabia

1.1 Introduction

Humans and microorganisms have shared a unique relation for centuries; and this relation has evolved accordingly to the conditions, from which most of the non-pathogenic microorganisms have become pathogenic.

Normal Microbial Flora (nonpathogenic), including Resident and Transient flora, are the permanent flora (fixed); resident flora are present in a given area while flora which are present for hours, weeks on the skin or mucous membrane for some time are transient flora. Normal flora plays an important beneficial role in the human body. Normal flora prevents attachment and penetration of pathogenic microorganisms. They start competing for habitat and nutrition with pathogenic microorganisms. Some strains of *Escherichia coli* also produce antimicrobial chemicals that can neutralize pathogenic microorganisms, such as *Bacteriocin*. *Bacteriocin* binds to the outer-membrane receptors, using them to translocate to cytoplasm were they execute their cytotoxic effect, DNA and RNA activity, inhibition of membrane synthesis. Intestinal normal flora produces enzymes such as cellulose, galactosidase, and glycosidase to help in digestion of food. They also help to carry out oxidation and hydrolysis of steroids rings of bile salt.

Normal flora can also have harmful effects, i.e., they become a source of unscrupulous contagion once immunity of the congregation becomes feeble or when normal microflora

Human Pathogenic Microbes
https://doi.org/10.1016/B978-0-323-96127-1.00002-4

of individual tissue migrate to other habitat. Urine tract infection is caused by *E. coli* spreading from the intestinal system to the urinary system. *Staphylococcus aureus* is an opportunistic pathogen of the skin; in normal conditions *S. aureus* is common in nasal region, but it causes secondary pneumonia in immune-compromised host. Patients infected with influenza virus are at very high risk of *S. aureus* because viral infection damages the mucous lining of the respiratory tract, and the deeper soft tissue is exposed to the common flora. Normal flora of skin include *Staphylococcus epidermidis, S. aureus,* and *Propionibacterium acnes; P. acnes* is a type of bacteria that causes acne (acnes is the relatively slow-growing, typically aerotolerant, anaerobic, Gram-positive bacterium linked to the skin condition of acne, it can also cause chronic blepharitis and endophthalmitis, the latter particularly following intraocular operation) (Scholz and Kilian, 2016). Normal flora of gut include *Lactobacillus, Helicobacter pylori* (is a spiral Gram-negative, microaerophilic bacterium that is commonly found in the stomach. Its helical shape is assumed to have evolved in order for it to pierce the stomach's mucoid lining and consequently start contagion) (Alfarouk et al., 2019), *Enterococcus*, and Bacteroides, which are Gram-negative, anaerobic bacteria. Bacteroides are nonendospore-forming bacilli that can be motile or nonmotile depending on the species. The DNA base content is 40%–48% G-C, which can lead to pleural effusion in a variety of body sites such as the abdomen, brain, liver, pelvis, and lungs. *E. coli, S. faecalis, E. faecalis* can originate in vigorous individuals besides from jumble-sale as a probiotic. Probiotics such as Symbio flor1 and EF-2001 are distinguished by means of the absence of precise genetic factor involved in drug confrontation and pathogenesis (de Almeida et al., 2018). As an unscrupulous pathogen, *E. faecalis* is able to persuade lethal contagions, specifically in the nosocomial area, where antibiotic resistance is unsurprisingly abundant, and enhances pathogenicity (Panthee et al., 2021). *Bifidobacterium* probiotics are microorganisms that generally live in human intestines and stomach. They assist our body in vital tasks such as metabolism and the prevention of dangerous germs. *Streptococcus* spp. is a normal flora of the respirational tract. These Gram-positive, sphere-shaped bacteria cause a variety of diseases, including strep throat, pneumonia, wound skin, heart valve, and circulatory diseases. *Corynebacterium* are Gram-positive bacteria that live in aerobic environments. They are bacilli, and at certain stages of development, they are club-shaped. Diphtheria is a dangerous illness caused by *Corynebacterium diphtheria* strains, which generate the diphtheria toxin. It can cause breathing problems, cardiac arrest, paralysis, and even death (Hoskisson, 2018).

Fungi have a wide range of interactions with humans. While some of these interactions can be favorable, others can be dangerous for health. Some fungi, like bacteria and viruses, can be pathogens. Infections or fungal toxins can cause human fungal diseases. Fungi are also found in our natural environment. Candida species are common fungi that live in our skin, respiratory, genital, and gastrointestinal system. Candida is usually kept at bay by our immune response and other microflora. In immunocompromised humans, they can cause infections ranging from minor to severe oral thrush, esophagitis, and deadly systemic illnesses. Candida is a fungus which can generate hyphae. Multiple mycobiomes may interact with one another in different places of the body. *Candida albicans* disrupts the gastrointestinal microbiome and promotes allergic respiratory infection caused by *A. fumigatus* in the respiratory microbiome, the well-studied gastrointestinal-respiratory interaction. Systemic immunological responses can be triggered by the transfer of fungal molecules hooked on the blood and following circulation, such as RNA, DNA, or peptidoglycans, resulting in infection distant beyond the original infection site (Noverr et al., 2005; Morris et al., 2012).

Some pathogens can effectively undermine the congregation immune response to inaugurate intracellular preservation through the use of strategies such as immunosuppression, molecular imitation, antigen costume or confiscation, complement and cytokine cascade circumvention, antigen exhibition obstruction, escape from apoptosis and autophagy, and costimulatory signal modulation (Mir, 2015a,b,c and d). Interactions here between microbiota and the bacterium microbiome might be harmful to one's health and perhaps cause disease. In other circumstances, the presence of bacteria is linked to the presence of fungus; for example, *Mycobacterium* superinfection can occur simultaneously with aspergillosis in some cases (Darling, 1976). Bacteria and fungi compete in various situations; for example, when *Pseudomonas aeruginosa* dominates in cystic fibrosis, *Candida* species, as well as possibly other fungus, are being repressed. Several theories have been presented to explain this interaction's dual character. Commensalism between bacteria and fungi may strengthen the susceptibility of *Mycobacterium* and Aspergillosis to the environment in a synergistic way. The composition of the microbiome is influenced by the immunological response of the host. TLR-2, TLR-4, dectin-1, dectin-2, and galectin-3 are phagocyte pattern-recognition receptors (PRRs) that identify fungus pathogen-associated molecular patterns (PAMPs) such as α-mannans, β-mannans, and β-glucans (Iliev et al., 2012; Jawhara et al., 2012). Through an antigen-presenting mechanism, macrophages and dendritic cells develop and activate T-cells in response to this pattern recognition. Activated -cells can develop into Th-1, which promotes fungus phagocytosis, or Th-2, which stimulates B-cells to release fungi-specific antibodies, depending on which cytokines are triggered (Saijo et al., 2010; van der Velden et al., 2013). The activation of this immune system is dependent on both innate and adaptive responses to environmental threats such as infectious illness. Antigen-specific T- and B-cell receptor signals must be activated for adaptive immune responses to occur. This antigen-specific signal, however, is inadequate to activate naive T cells, necessitating the addition of a costimulatory signal, which is an antigen-independent nonspecific signal (Mir, 2015a,b,c and d).

1.1.1 Modes of transmission

Different pathogens spread in different ways. Respiratory pathogens, for example, are frequently disseminated by the air, while intestinal microbes usually spread through food or water. There are several stages to transmitting: escape from the host or reservoir of infections, transfer to the new-fangled host, entry into the new host, and flee from the new host.

1. **Airborne Transmission:** The pathogens remain suspended in air and are transmitted through droplet nuclei, i.e., small particles left from evaporation of large particles. The droplet nuclei remain in air for hours or days and carried to individuals because the pathogens cannot grow in air. Shaking the hand of an infected individual who used their hand to wipe their runny nose can give you a cold. A bioaerosol is a biological substance suspended in the atmosphere. Bioaerosols contain bacterial cells and fragments, as well as fungus spores and hyphae. *E. coli, Pantoea (Enterobacter) agglomerans, Pseudomonas* spp., and *Acinetobacter* spp., which are widely extracted in bovine barns, pig houses, and poultry barns, are among the airborne Gram-negative bacteria linked to animal facilities that might cause human sickness. Sewage treatment

plants, vegetable and herb processing industries, and recycling centers all emit disease-causing Gram-negative bacteria into the environment. These microorganisms can cause a variety of health issues, especially in youngsters, the elderly, and those who have weakened immune systems. Tuberculosis (TB) is a contagious infection caused by *M. tuberculosis*, a bacillus. Although pulmonary tuberculosis primarily affects the lungs, it can also affect other regions of the body. Tuberculosis (TB) is the world's second biggest cause of death, after HIV, and the major cause of death among HIV patients. Pulmonary tuberculosis is very contagious spreadable through the air. Over 80% of tuberculosis infections are pulmonary, and if remain unattended, an active tuberculosis patient can transfer the disease to up to 10–15 other people over the course of a year through direct contact.

Histoplasmosis, coccidiomycosis, blastomycosis, cryptococcosis, and aspergillosis are among the fungal infections. Histoplasmosis can occur after inhaling spores of the fungus *Histoplasma capsulatum*. This disease is spread through airborne transport from environmental surfaces rather than from infected people or animals to others. Histoplasmosis mainly affects the respiratory system and causes a wide range of symptoms. People who are infected are usually asymptomatic or have cold or flu symptoms that do not necessitate immediate treatment. Chronic lung disease caused by *H. capsulatum* infection embodies tuberculosis and can deteriorate over months or years. When histoplasmosis becomes disseminated and spreads outside the lungs, it becomes the most severe and rare form. Disseminated histoplasmosis is fatal if left untreated.

2. **Contact Transmission:** Some of the pathogens spread when contact of host is done with the reservoir of pathogen. Contact refers to person to person contact through touching; cold can be caused by shaking a hand of a person who has cold. Contaminated blood can or body fluids can be source of transmission. *Bacillus anthracis*, a Gram-positive, rod-shaped bacterium, causes anthrax, a fatal disease. When people come into contact with an infectious animal or breathe spores, they become infected. Symptoms differ depending on how the infection was spread. They can range from a dark scabbed skin ulcer to breathing problems. *Burkholderia mallei*, a Gram-negative, bipolar, aerobic bacterium causes glanders in humans and animals.

   Fungi reproduce by producing spores, which can be picked up or breathed through interacting directly. As a result, diseases are more capable of harming the skin, nails, or respiratory tract. Disease can also enter the body through the skin, causing organ damage and a chronic illness. Athlete's foot, jock itch, ringworm (*Trichophyton, Microsporum,* and *Epidermophyton*), yeast infection, onychomycosis—or a fungal infection of the nail are all examples of common fungal contagions.

1.2 Human–pathogen interaction

The host–pathogen interaction defines how bacteria or fungus maintain themselves within host species on a genetic, cellular, biological, or population level (Fig. 1.1). Pathogens produce an extensive variety of chemicals that trigger a variety of reactions in the host. Bacteria, fungus, protozoa, helminths, and viruses are examples of pathogens. Based on their spread, each of these diverse sorts of organisms can therefore be categorized as a pathogen.

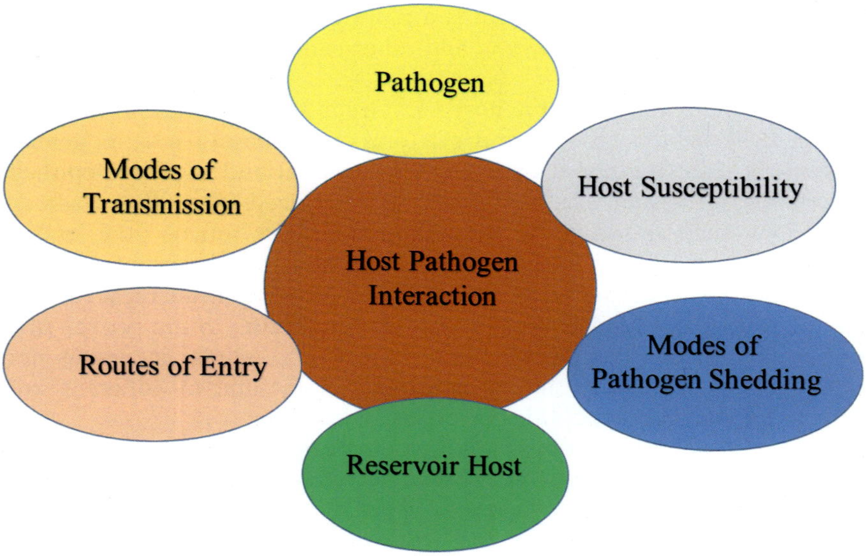

FIGURE 1.1 The host–pathogen interaction depends on the host's susceptibility, pathogen virulence factors, route of entry, mode of pathogen transmission, and a variety of environmental factors.

This involves foodborne illness, airborne illness, waterborne illness, bloodborne illness, and vectorborne illness. Toxins are secreted into the host by a variety of pathogenic bacteria, including food-borne *Staph. aureus* and Botulinum. Aspergillus niger, the most common pathogenic fungus, produces aflatoxin, a carcinogenic agent that affects many foods, particularly those grown underground (San-Blas and Calderone, 2008). Pathogens can do lots of activities within the host to cause illness and elicit an immune response. Because of their fast rate of reproduction and tissue permeation, pathogens cause sickness. As phagocytic cells start breaking down the pathogen within the body, an immune response occurs, resulting in frequent signs. Some bacteria, such as *Helicobacter pylori*, can release chemicals that cause cellular damage or impair normal tissue function. Infectious diseases remain the foremost source of bereavement, not only are newfangled contagious illnesses ascending, but the recurrence of fatal communicable infections, as well as the widespread availability of antimicrobial resistance strains, pose a fearsome menace to community wellbeing. The pathogen can participate in one of three host–pathogen interactions, based on how well it interacts with the host. Pathogen benefits from a contact but the host does not, this is referred to as *commensalism*. *Bacteroides thetaiotaomicron*, for example, found in the human digestive system still have no known benefits. Mutualistic relationship arises when pathogen and host benefit from interaction, as seen in the gut of humans. Various bacteria help host by breaking down nutrients, and our bodies serve as their home in return (Backhed et al., 2005). When a pathogen gains from a relationship while the host suffers, this is known as *parasitism*. Despite the fact that infections have the power to cause disease, this is not always the case in humans, this bacterium normally survives in the intestines as part of the normal, healthy microbiota. It can, though, produce severe diarrhea whether it spreads toward other human digestive system or the body. Although *E. coli* is categorized as a pathogen, it does not necessarily act as one. This illustration also applies for *S. aureus* and other commonly occurring in microbial flora (Clermont et al., 2000).

The adhesion of the pathogen to the host's surface is a critical stage in the host—pathogen interaction. The host surfaces are skin, mucosa (oral cavity, nasal passages, genital area), and deeper tissues (Fig. 1.2) (lymphoid tissue, gastrointestinal and digestive epithelia, alveolar lining, and endothelial tissue). The host produces motorized forces that wash down germs off these exteriors, such as saliva efflux, choking, sniffling, mucus flow, and blood flow. Microbial pathogens frequently produce factors that attach to fragments on the shallow of diverse microbial tissue cells and render the microbe prone to mechanical rinsing. The pathogen begins its specialized biochemical process after adhering to the surface of a particular human host, which includes replication, toxin production, cellular incursion, and initiation of host cell signaling pathways, all of which result in illness. Adherens are microbial adherence factors that are polypeptides or polysaccharides.

After adhering to a host, around pathogens increase in-depth entree into the host to continue the infectious disease sequence. Pathogenic incursion can occur both extracellularly and intracellularly. Extracellular invasion happens when a type of bacteria penetrates a tissue's barriers and spreads within the host while trying to remain external of the host cells. This is the strategy used by *Streptococcus AB* and *Streptococcus aureus*. Coagulases are enzymes that clot plasma and protect the bacterial cell, presumably by inhibiting lysis. Hyaluronidase also referred as *spreading factor* helps hyaluronic acid spread. Deoxyribonuclease, a DNA-breaking enzyme, protects *S. aureus* from neutrophil extracellular trap-mediated death (Berends et al., 2010; Monteith et al., 2021). Lipase, which digests lipids, staphylokinase, which breaks fibrin and facilitates spread, and beta-lactamase, which

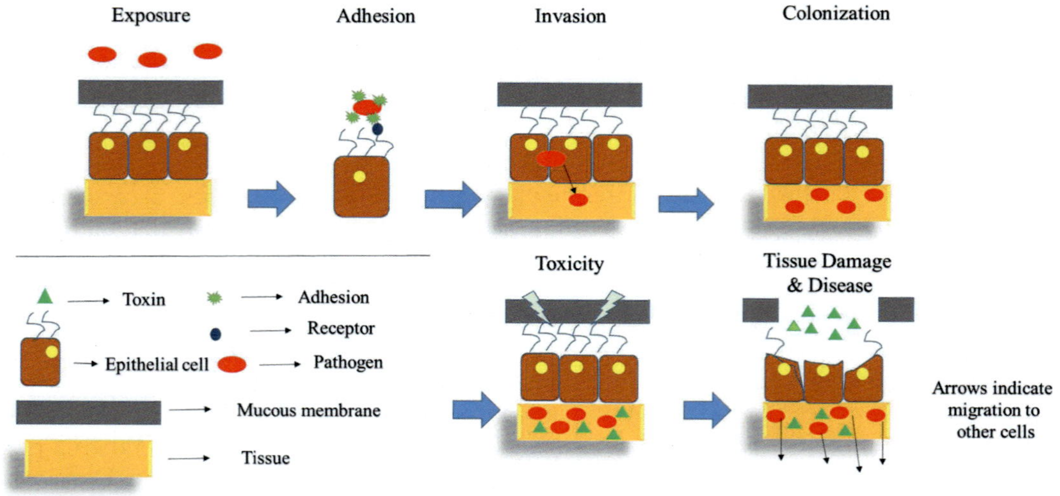

FIGURE 1.2 Mechanism of pathogenesis by microbes to the host cell depend on the strategies for interacting with the host, firstly exposure, adhesion, invasion, colonization, toxicity and then leads to tissue damage. Pathogenic microbes express a diverse set of molecules that bind to host cell targets to facilitate a wide range of host responses.

enhances drug resistance, are also produced by *S. aureus*. When a microorganism enters the cells of the host tissue and survives in the environment, this is known as intracellular invasion. A wide range of gram-negative, gram-positive, and mtb pathogens have been found to invade host cells, and both phagocytic and nonphagocytic cells can become invasion targets.

Toxins are proteinaceous and nonproteinaceous chemicals created by bacteria to kill the host cell. They are similar to biological weapons. Gram-negative bacteria, LPS (endotoxins), and Teichoic acid (for Gram-positive bacteria) are nonproteinaceous toxins. Proteinaceous toxins are important enzymes that are secreted into the environment and inserted into the cytoplasm of eukaryotic cells through the type III secretion system or other methods. The amino acid content and characteristic of microbe exotoxins are used to classify them. AB toxins are found in many bacteria species, including *E. coli, V. cholerae,* and *C. diphtheria*. The A-subunit is responsible for enzymatic commotion, while the B-subunit is accomplished of binding and delivering the toxic substance into the host cell. Proteolytic toxins degrade specific host proteins.

Each fungus species has its own set of glycan polymers and proteins inside the cell wall. Cell wall proteins' electrostatic charge is heavily influenced by glycosylation and the existence of adversely charged phosphate groups in their carbohydrate side chains. They may also contribute to an increase in the hydrophobic nature of the cell surface, which is important for biomaterial binding. Furthermore, fungus growth in the host necessitates ongoing cell wall manufacturing and remodeling, and several fungal diseases may infect the host based on morphological stages, each with a distinct cell wall organization and biochemical composition (Tronchin et al., 2008). In the host cell invasion process, adhesion comes before invasion. The two main processes by which host cells invade are by causing endocytosis and active penetration. Fungi can also pass through epithelial and endothelial cell barriers via hydrolytic enzymes breakdown of intercellular tight connections or a Trojan horse process in which leukocytes transport them (Sheppard and Filler, 2014).

1.3 Human pathogenic fungal microorganisms

Fungi are fascinating organisms that belong to their own kingdom. Eukaryotic microorganisms have a nucleus that is surrounded by a membrane, larger cells than bacteria, and molecular mechanisms that are similar to the plants and animals. In contrast to mammalian cells, fungi have a resilient cell membrane formed of chitin molecules that almost invariably surrounds their plasma membrane. Because fungi do not produce chlorophyll, they are not plants.

The shapes, structures, and behavioral characteristics of fungi are used to recognize and identify them. Yeasts are fungi that exist mostly as single cells, whereas molds are fungi that are formed from hyphal threads (i.e., hyphal fungi). A mycelium is a collection of hyphal threads, branches, and any spore-bearing structures. Yeast and hyphae both grow vegetatively through mitotic, asexual cell division, which is commonly caused by a burst from a parental fungal cell. In most fungi, meiotic, sexual reproduction is also conceivable. Matings have been recorded not even just between two fungal strains, but also inside a hypha between distinct cell units or among mother and descendant yeast cells; fungal parthenogenesis has also been observed. In fungus, meiosis leads to the production of a sexual spore. A crosswall or septum separates a cell unit of a fungal hypha from its neighbor, though not completely. In the hyphae of Zygomycetes, there seem to be a very few septa. Certain zygomycetes species deposit large numbers of asexual spores within a spherical sac called a *sporangium*, and these spores are referred as *sporangiospores*. A spore that might not be enclosed by a sporangium is known as a *conidium*. Conidia develop on or within a conidiophore, which can range in size from a single hyphal cell to a visible tissue complex. The majority of yeasts reproduce through budding, which results in the creation of blastospores. Any disease caused by a darkly pigmented mold is referred to as *phaeohyphomycosis*. The correct term *is hyalohyphomycosis,* which refers to any infection caused by a colourless mold. A fungal infection that affects at least two deep organs and/or the skin is known as *dispersed mycosis*. *Dermatomycosis* is a term that refers to almost any yeast infection that affects and penetrates the skin. To enable penetration, several fungal infections alter the structure of their cells. In the natural environment, these dimorphic diseases typically change from a mold to a blooming, round-celled form in tissue. *Blastomyces dermatitidis, Histoplasma capsulatum, Paracoccidioides brasiliensis,* and *Sporothrix schenckii* are some of the most well-known instances of the dimorphic form. Minor morphological changes between tissue forms and shapes observed in cultures are seen in many fungi, and other, less frequent fungal diseases act similarly. Infected tissues undergo changes. *Coccidioides immitis* changes from a mold to a sporangia-bearing unicellular spherule.

Only some few human pathogenic fungi are vigorous enough to infect a healthy host. Unless they come into contact with an immunocompromised patient, whose immune system has been impaired, allowing them to enter the body, most are rather harmless. The undamaged epithelial coverings of the intestinal tract and the mucus fence of the respirational tract prevent fungal cell and spore aspiration under normal conditions, whereas dead or injured tissue can serve as a breeding site for infectious disease. Glucose, the most advantageous carbon source, aids in the important survival and niche colonization processes (Qadri et al., 2021a). As a result, invasive illnesses caused by fungi must be classified as opportunistic infections in most cases. Recently, it was discovered that biological variability within critical inherent or adaptive immune response genes influences susceptibility to invasive fungal infections, resulting in IL-10 production failures, Tolllike receptor polymorphism, and plasminogen gene polymorphism (Carvalho et al., 2008). Endemic or dimorphic mycoses, which are caused by real pathogenic fungus, and opportunistic mold and yeast infections, which are saprobes that exclusively attack immunocompromised hosts, are the two forms of fungal diseases. Endemic mycoses such as histoplasmosis, coccidioidomycosis, blastomycosis, and paracoccidioidomycosis were originally limited to specific areas of the American continent. *Penicillium marneffei*, on the other hand, was mostly found in Southeast Asia. However, these illnesses can now spread to any location. Candidiasis and aspergillosis are prevalent diseases characterized by opportunistic fungi (Pfaller and Diekema, 2007).

1.3.1 Candida species

Candida albicans is a member of the Kingdom Fungi, Division Ascomycota, Class Saccharomycetes, Order Saccharomycetales, Family Saccharomycetaceae, Genus *Candida,* and Species *Candida albicans*. *C. albicans* is highly pathogenic opportunistic yeast (Gow and Yadav, 2017). It is a communal associate of human intestinal bacteria and can sustain outside the humanoid body (Bensasson et al., 2019). It is normally a commensal, but it can become an opportunistic pathogen in immunocompromised people under a range of circumstances. Candida can be found all around the world, although it is more common in immunocompromised persons with serious conditions like HIV and cancer. Candida is one of the most common organisms that cause infections in hospitals. Patients who have undergone surgery, received a transplant, or are in intensive care units are extremely vulnerable (Erdogan and Rao, 2015). *Candida albicans'* genome is the foremost eukaryotic pathogen genomes to be sequenced. Clinical isolates have a 16-megabyte genetic code, 33.3% GC content, 132 non-coding RNAs, many transposons, eight diploid chromosomes, and 6735 (haploid) ORFs. The genome of *Candida albicans* is extremely flexible, with considerable heterozygocity, intra-chromosome recombination, and aneuploidy seen in different strains. Stress adaptation is produced by heterozygozity loss, chromosomal rearrangements, and whole-chromosome or segmental aneuploidies, which can change the copy number of important drug-resistance genes, allowing effective sensitivity (Jones et al., 2004). In vivo, *C. albicans* switches between distinct vegetative growth forms in response to changing circumstances such as food availability, temperature, pH, CO_2, and serum presence (Noble et al., 2017). This virulence characteristic facilitates epithelia invasion, host dispersion, and survival in a range of host habitats, as well as regulating the host immune response and counteracting immune monitoring. Due to posttranscriptional rewiring and constitutive development of other metabolic processes, it is metabolically flexible and can use several nutrition sources at the same time.

1.3.2 Aspergillus species

Kingdom Fungi, Division Ascomycota, Class Eurotiomycetes, Order Eurotiales, Family Trichocomaceae, and Genus Aspergillus are the classifications for *Aspergillus* species. Clinically relevant species can be found in the *Aspergillus*: Fumigati, Circumdati, Terrei, Nidulantes, Ornati, Warcupi, Candidi, Restricti, Usti, Flavipedes, and Versicolores (Peterson, 2008). *Aspergillus* species are saprophytic filamentous fungi that flourish on soil, rotting plants, seeds, and grains. Humans are occasionally poisoned by Aspergillus species (Seyedmousavi et al., 2015). The majority of Aspergillus species can be found throughout the year in a wide range of habitats and substrates (Kwon-Chung and Sugui, 2013). Only a few well-known opportunistic infections in humans are considered relevant. Some *Aspergillus* species cause significant sickness in people. *A. fumigatus* and *A. flavus* are the most frequent pathogenic species, both of which produce aflatoxin, a toxin and cancer-causing agent that can harm foods like almonds. *A. fumigatus* and *A. clavatus* are the two most prevalent species that cause allergic disease. Aspergillus can be dangerous to neonates (Cloherty, 2012). *A. fumigatus* cause primary lung infections that can quickly progress to necrotizing pneumonia with the potential to spread. Unlike *C. albicans*, which is a dimorphic mold in the environment and a yeast in the body, the organism takes on the form of a mold in both the environment and the host, separating it from other frequent mold infections. *Aspergillus* species creates a variety of microbial components that allow it to cause disease in immunocompromised persons. These are being researched the most for *A. fumigatus*, which is the most common cause of sickness (>90% in most locations). Multiple poisons and metabolites are secreted by the organism to protect it from host defences, and it has unique cell wall components that can help it survive in the host.

1.3.3 *Cryptococcus* species

Cryptococcus belongs to the Kingdom Fungi, Division Basidiomycota, Class Tremellomycetes, Order Tremellales, Family Tremellaceae, and Genus *Cryptococcus*. The *Cryptococcus* species that frequently cause sickness in people include *Cryptococcus neoformans* and *Cryptococcus gattii*. C. neoformans has serotypes A, D, and an AD hybrid. Serotypes B and C are the most common *C. gattii* strains (Springer et al., 2014). Pathogenic *Cryptococcus* spp. are classified into nine molecular types, VNI, VNII, VNB, VNIII, and VNIV for *C. neoformans* isolates and VGI, VGII, VGIII, and VGIV for *C. gattii* isolates. Breathing of communicable propagules from environmental reservoirs, such as poorly encapsulated yeast cells or basidiospores, causes cryptococcal infection, which is then deposited in the pulmonary alveoli. Tissue inoculation after trauma has been described and is thought to be uncommon (Christianson et al., 2003). The yeast could enter the body through the gastrointestinal tract, though this is a less reliable route. In spite of high rates of serologic reactivity in progenies, particularly in urban areas, primary lung infection is commonly assumed to be asymptomatic or mildly symptomatic. The most common and medically significant species is Cryptococcus neoformans. It is finest recognized for initiating severe meningitis and aseptic meningitis in HIV/AIDS patients. It has the potential to infect organ transplant recipients as well as cancer patients receiving particular therapy (Goldman et al., 2001). *C. gattii* is exclusively found in Africa and Australia's tropical regions. In persons who are not immunocompromised, it can cause disease (cryptococcosis). It was discovered in eucalyptus trees in Australia. Air, dry moss, grasshoppers, and tubercular lungs have all been found to contain *C. albidus* (Fonseca et al., 2000). On a macroscopic level, the colonies range in color from cream to pale pink, with the preponderance of colonies appearing soft and mucoid. It has been revealed that some of the colonies appear rough and wrinkled, however this is a rare occurrence.

1.4 Human pathogenic fungal diseases

Prior to infecting people, fungi are required to meet four aspects: growth at normal body temperatures, circumvention or penetration of surface defenses, tissue lysis and absorption, and immunity, especially resistance to increased body temperatures. The process by which fungi overcome locomotion challenges around or through recipient obstacles is morphogenesis with small spherical, detachable cells and long, linked cells. Whenever a fungus wants to nutritionally exploit human tissue, it must produce lytic enzymes and have absorption mechanisms for the released nutrients. Finally, the robust body's immune system evolves in reaction to contacts with potential fungal infections; yet, few fungi fit all four criteria for posing a health danger (Kohler et al., 2017).

1.4.1 Candidiasis

Candida albicans causes candidiasis, which is an infection caused by the yeast *Candida albicans*. The *C. albicans* species is perhaps the most prevalent *Candida* species that can infect humans. *C. albicans* is a yeast that may live on the skin's surface as well as inside the body, in places like the mouth, throat, intestine, and vaginal canal, without causing any difficulties. Candida can cause disease if it grows in an uncontrolled manner or if it enters the body deeply Fig. 1.3, such as the bloodstream or internal organs such as the kidney, heart, or brain. Candida infection is an opportunistic infection. Fungi are eukaryotic organisms that can be yeasts, molds, or dimorphic fungi (Arya, 2021). When the conditions are right, they can become pathogenic. It can affect the mouth, vaginal area, penis, and other body regions. The common word for candidiasis of the aperture is thrush. It manifests itself as white reinforcements on the tongue, throat, and other mouth areas. Other symptoms of thrush include soreness and difficulty swallowing. Some *Candida* strains may be transmitted sexually. It is an acute or chronic superficial or disseminated mycosis. It has a cosmopolitan distribution. *C. albicans* grows rapidly in the presence of a diet rich in fruits and carbohydrates. The frequency of its occurrence in the human intestinal tract varies according to climatic zone and socioeconomic environment. Pseudomembranous candidiasis, erythematous candidiasis, and chronic hyperplastic candidiasis are all types of oral candidiasis. Chronically unwell people and babies are both susceptible to pseudomembranous candidiasis. On the tongue and oral mucosa, it usually appears as white, soft, slightly raised plaques. Plaques resembles froth because they are tangled collections of fungal hyphae, desquamated epithelium, necrotic debris, keratin, leukocyte, fibrin, and bacteria. An erythematous patch is left behind when this white plaque is removed. *Antibiotic mouth* is a term used to describe erythematous candida infection. It happens after broadspectrum antibiotics or corticosteroids have been administered. Painful erythematous spots on the tongue, as well as central papillary atrophy, are symptoms of the lesions. Genital discomfort, burning, and a white discharge from the vaginal canal are all signs of vaginal candidiasis. An itching rash is caused by a yeast infection of the penis.

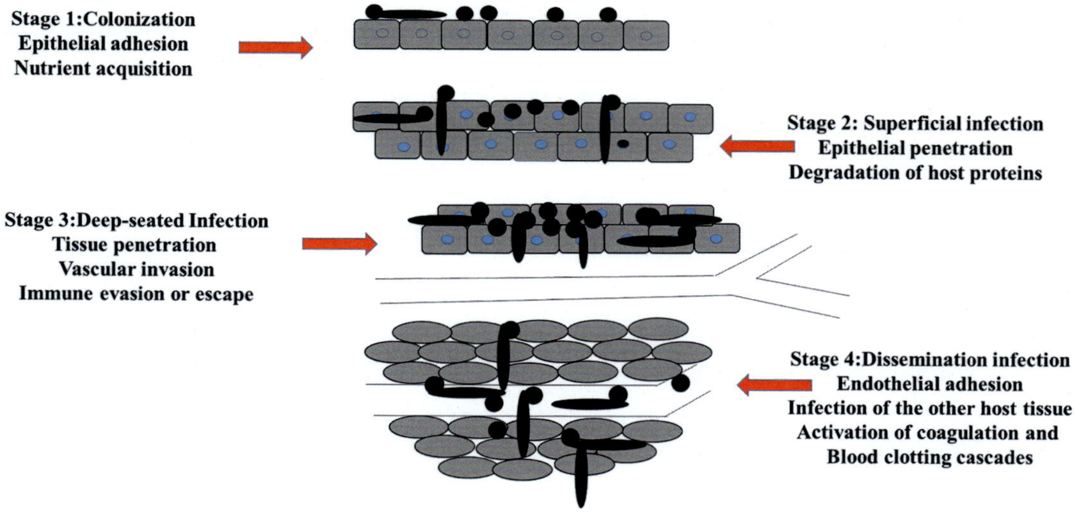

FIGURE 1.3 Candidiasis pathogenesis starts with colonization, epithelial adhesion, epithelial penetration, tissue penetration, vascular invasion, and then to endothelial adhesion Infection of the other host tissue activation of coagulation and blood clotting cascades.

Fungal infections can spread to other parts of the body, causing fevers and other symptoms, and in rare circumstances, they can become invasive. Oral *Candida* species include *C. albicans, C. glabrata, C. guillermondii, C. krusei, C. parapsilosis, C. pseudotropicalis, C. stellatoidea, C. tropicali. Candida* in both healthy and immunocompromised patients, oral candidiasis, can manifest as a number of diseases. Candida infections include hyperplastic or atrophic (denture) candidiasis, pseudomembranous candidiasis (thrush), linear gingival erythema, median rhomboid glossitis, and angular cheilitis, to name a few. It can cause a variety of clinical signs, ranging from minor surface infections to catastrophic disseminated illness. Disseminated candidiasis is almost always associated with acquired or inherited immunodeficiency.

Candida endocarditis is characterized by fever, heart murmur, congestive heart failure, and anemia. These characteristics are similar to bacterial endocarditis, but differ from bacterial disease in the high frequency of large vegetation on the valves and emboli to the splenic, renal, and iliac arteries. This disease is found in drug addicts, and the source of infection is contamination through blood in an oxygenator. *C. parapsilosis, C. guilliermondii, C. krusei*, and *C. pseudotropicalis* are the etiological agents. *C. albicans* causes meningeal candidiasis. *C. albicans* is a small, oval, budding yeastlike fungus that measures $2.5 \times 4 \times 6$ m in diameter. The cell grows longer and develops pseudomycelium. *C. albicans* contains typical chlamydospores but not others.

This generates short filaments in 1 to 2 h on egg albumin or human serum. On Sabouraud's glucose agar medium, these grow rapidly (24–48 h) at room temperature. The colonies are medium in size, smooth and pasty, and smell like yeast. Older colonies develop radial furrows and a honeycomblike appearance in the center. Glucose and maltose are converted to acid and gas as a result of fermentation. The types of colonies formed on blood agar plates after 10 days at 37°C, the types of growth in Sabouraud's glucose broth after 48 h at 37°C, the morphology and development of blastopores, chlamydospores, and pseudomycelia on corn meal agar at room temperature, and the fermentation interactions in glucose, maltose, sucrose, and lactose after 10 days at 37°C can all be used to differentiate between species.

Antifungal medications such as nystatin, clotrimazole, amphotericin B, and miconazole are used to treat *Candida* infections. Antifungal vaginal cream can be used to treat mild to moderate genital *Candida* infections. The antifungal creams are available in 1, 3, or 7-day treatments. econazole or fluconazole 150 mg tablets are also available as one-time doses (Fang et al., 2021). Although oral and topical therapies are equally effective, oral drugs are more costly. During the first trimester of pregnancy, doctors avoid prescribing fluconazole (Zhang et al., 2019). Fluconazole is taken weekly for 6 months on days 1, 4, and 7 for repetitive vaginal *Candida* infections. Oral thrush can be treated in a similar way, with oral chewable tablets as an alternative dosing form. Antifungal drugs such as caspofungin, fluconazole, and amphotericin B must be taken orally or intravenously to treat systemic candidiasis.

1.4.2 Aspergillosis

Aspergillosis is a fungus that can produce an infection, an allergic reaction, or a fungal growth. People who have tuberculosis or chronic obstructive lung disease are more likely to be affected. The principal pathogen is *Aspergillosis fumigatus*, which is very harmful for birds and can cause invasive illnesses in people on occasion. *Aspergillus niger* and *Aspergillus flavus* are two more species that can cause infection. Soil and dust are their natural habitats, and spores abound. *Aspergillus* is a common saprophyte, and *Aspergillus fumigatus* is the pathogen that causes human infection. Infection begins with the release of conidia into the air, which are inhaled and consumed by alveolar macrophages. Conidia swell and germinate in the lungs, producing hyphae that have a proclivity for invading preexisting cavities or blood vessels. In allergic illness, an activated immune response causes a persistent proinflammatory state marked by antibody production, immunoglobulin E, and T-cell activation by normally well-tolerated antigens. T cells play a critical role in allergic response etiology. Antigenic peptide engagement of T-cell receptors is inadequate to activate naive T cells; a second costimulatory signal is required. Costimulatory molecules are the second type of signal that cause T-cell activation and proliferation. CD80/CD86 interacting with CD28/cytotoxic T-lymphocyte-associated antigen-4 is one of the best-studied costimulatory pathways (CTLA-4) (Mir, 2015a,b,c and d). Allergic bronchopulmonary aspergillosis inhaled spores create a hypersensitivity reaction, which may result in Type I hypersensitivity asthma, which occurs in atopic patients after exposure to inhaled aspergillus spores; Types III extrinsic hypersensitivity alveolitis; Combined type I and type III hypersensitivity reaction.

Due to old tuberculosis or bronchiectasis, *aspergilloma* is a disorder in which a fungal ball forms within and is usually restricted to an existing lung cavity. Surgical removal is required in this kind of colonization aspergillosis. since the disease frequently produces significant hemoptysis (Lee et al., 2004). Invasive aspergillosis is caused by a fungus that causes pneumonia before spreading to other organs like the brain, kidneys, and heart. Patients who develop this potentially fatal disease are frequently immunocompromised or weakened as a result of long-term antibiotic, steroid, and cytotoxic medication use. *Aspergillus* infects persons with compromised immune systems through a skin puncture after surgery or a burn wound. When invasive aspergillosis spreads to the skin from another region of the body, such as the lungs, cancerous aspergillosis develops (van Burik et al., 1998). Infection most commonly enters the body through the respiratory tract. *Aspergillus*, on the other hand, can penetrate and is distributed throughout the body, including the epidermis, nasal cavity, neurological system, eyes, and nails. *A. fumigatus* is perhaps the most prevalent species that infects people. Sinus infections are more likely to be caused by *A. flavus*. When it's impossible to determine the exact species, the organism is classified as an *Aspergillus* species (Davda et al., 2018).

Because aspergillosis conidia are breathed continually, the first analysis and assessment for aspergillosis is to develop a high clinical suspicion in order to identify patients who are suffering from intrusive aspergillosis. To identify if an individual has invasive aspergillosis, a fungal staining of the sputum should be conducted initially. The occurrence of *Aspergillus* does not always imply a severe infection in healthy persons; nonetheless, the presence of the fungus in immunocompromised people should warn the doctor to treat the patient as if it were an acute infection.

The worldwide standard for confirming that the contagion is caused by *Aspergillus* besides not alternative mold or fungus is to culture the *Aspergillus* species in phlegm or bronchoalveolar lavage and identify the hyphae. A tissue sample for aspergilloma can help confirm the diagnosis and rule out alternative possibilities for lung cancer (Jenks and Hoenigl, 2018). Galactomannan and beta-D-glucan tests are examples of serum biomarkers that can be useful. Galactomannan levels in bronchoalveolarlavage sputum can also be measured.

In the absence of a tuberculosis test, a positive Aspergillus Immunoglobulin g can aid in the detection of chronic aspergillosis. Chest radiographs may show pulmonary aspergilloma parenchymal opacities: nodules with adjacent diminution, aspergilloma (fungal ball in a pre-existing lung hollow), cavitations, and fibrosis are all common findings on CT imaging of the lungs. A CT scan may be conducted if an aspergillosis symptom is suspected to search aimed at sinus involvement, such as tumors, opacification, or damage of the sinus walls.

Treatment for suspected invasive aspergillosis should begin as early as possible because the patient's condition can rapidly deteriorate from onset to death in 1–2 weeks. Voriconazole (4 mg/kg, twice daily), posaconazole (300 mg IV, daily), micafungin (150 mg IV, daily), and amphotericin B (1 mg/kg, daily) are all intravenous antibiotics that can be given to critically ill patients for a 6- to 12-week course. Voriconazole is the first-line therapy. Despite its effectiveness, amphotericin is considered a second-line treatment due to its negative effects. Taking into account the patient's comorbid disorders, it's also important to try to fix their immunocompromised state as much as feasible (van de Peppel et al., 2018; Alastruey-Izquierdo et al., 2018). To determine how well the treatment is working, symptoms are evaluated and an aspergillus titer is taken. Repeat CT imaging may reveal that the fungus ball is no longer present and the cavitary lesions have shrunk in size. Repeat imaging is advised after at least 2 weeks of therapy.

1.4.3 Cryptococcosis

Cryptococcosis is a fungal infection brought on by the *Cryptococcus* fungus (*C. neoformans* or *C. gattii*). The lungs and central nervous system are still the most prevalent sites for cryptococcosis, but it could also impact different regions of the body. The fungus *Cryptococcus* causes cryptococcal meningitis, which is a brain illness. *Cryptococcus neoformans* is a fungus that is available across the world in the surroundings. After inhaling the microscopic fungus, people can become infected with *C. neoformans*, but most people who are exposed to it never become ill. The symptoms of the infection are determined by the body areas affected (Sabiiti and May, 2012). Pneumonia-like illnesses can be caused by *C. neoformans* infection inside the lungs. Sniffle, difficulty breathing, chest pain, and fever are all common symptoms that are similar to those of a variety of other diseases. The fungus *Cryptococcus* causes cryptococcal meningitis when it spreads from the respiratory tract to the brain. Cryptococcal meningitis causes a wide range of symptoms. Headache, fever, neck discomfort, nausea and vomiting, sensitivity to light, disorientation, and behavioral abnormalities are all possible symptoms. Infections with *C. neoformans* are uncommon among normally healthy persons. Patients with impaired immune systems, such as those with severe HIV/AIDS, those who have had an organ transplant, or those who are on corticosteroids, rheumatoid arthritis drugs, or other immune-suppressing treatments, account for the great majority of *C. neoformans* infectious illnesses (Bratton et al., 2012). *C. neoformans* disease is diagnosed using medical history, symptoms, clinical examination, and diagnostic tests. Send a sample of tissue or biological fluids (such as blood, cerebrospinal fluid, or sputum) to a laboratory for microscopic examination, antigen testing, or culture. Also, have tests like a chest X-ray or a CT scan of your lungs, brain, or other parts of your body performed.

Infected individuals with *C. neoformans* must take antifungal suppository for a minimum of 6 months, and usually extensive. This variety of management is commonly determined by the sternness of the infection as well as the body portions affected. Fluconazole is usually used to treat asymptomatic infections (such as those discovered by targeted screening) and mild-to-moderate lung infections. Amphotericin B, in amalgamation with flu cytosine, is commended as an initial cure for people suffering from severe respiratory infections or nervous system infections (brain and spinal cord). Fluconazole is frequently prescribed to patients for a prolonged retro of time in order to eradicate the infection. Definite sets of individuals, such as prenatal women, progenies, and persons that live in resource-constrained environments, may require different antifungal treatments, doses, and durations. In some cases, surgical removal of fungal growths may be required (cryptococcomas).

Cryptococcomas (fungal growths) in the respiratory tract, skin, central nervous system, or other organs can form as a result of *C. gattii* infection, leading to symptoms within afflicted parts of the body. The gestation period for *C. gattii* disease is not well characterized. *C. gattii* disease symptoms might occur anywhere from 2 to 13 months following breathing the fungus, with a 6 to 7-month average (MacDougall and Fyfe, 2006).

People can become infected as early as 2 weeks after breathing in the fungus or as late as 3 years afterward (Tsunemi et al., 2001; Johannson et al., 2012). Cryptococcal antigen screening in serum or cerebrospinal fluid is an important first step in detecting cryptococcal infection, however it does not differentiate between *C. neoformans* and *C. gattii*. Culture has historically been used to determine if a cryptococcal infection is caused by the *C. neoformans* species complex or the *C. gattii* species complex. On canavanine-glycine-bromthymol blue (CGB) agar, *C. gattii* will turn the culture medium blue, whereas *C. neoformans* will keep the medium's color (yellow to green). Matrix-assisted laser desorption/ionization time of flight (MALDI-TOF), a recent innovation in identification systems, may also distinguish between them.

1.5 Human pathogenic bacterial microorganisms

Pathogenic bacteria have been highly adapted and endowed with mechanisms to overcome the body's usual defenses, allowing them to penetrate areas of the body where bacteria are not normally found, such as the blood. Some infections only infect the epithelium, skin, or mucous membranes on the surface, while many others penetrate deeper, spreading into the tissues and distributing through the lymphatic and blood systems. In rare cases, a pathogenic microbe can invade a healthy individual, but infection normally happens when the body's defense mechanisms are weakened by local trauma or an underlying debilitating state such as injury, drunkenness, cold, tiredness, or hunger. Pathogenic bacteria create sickness symptoms by causing harm to or interfering with the function of host tissues, by eliciting an immune response that destroys the host cells inadvertently (Greenwood, 2012), or by releasing toxins, bacteria can cause direct or indirect damage to host cells (Rudkin et al., 2017).

Pathogens are classified as facultative or obligatory microbes based on how closely their life cycle is connected to the patient's. Facultative pathogens are organisms that can replicate outside of the host in a variety of environments. Bacteria and fungi found in the environment that can cause infection are known as facultative pathogens. Antibiotics are powerful drugs that combat infections and can save lives when taken properly. Bacteria are either prevented from multiplying or are killed. Before bacteria may multiply and create symptoms, the immune system usually kills them. Even if symptoms appear, the immune system is usually capable of tolerating and combating the infection. There are instances, however, when the quantity of hazardous bacteria is excessive, and the immune system is unable to resist all of them. Antibiotics (Fig. 1.4) come in handy in this situation (Table 1.1). The antibiotic resistance pandemic includes a number of the most dangerous hospital-acquired bacteria. The difference among facultative and incidental microorganisms is occasionally established, with the latter referring to those that only infect weaker or immunocompromised hosts on a rare basis. "Unplanned" pathogenic microorganisms include pathogens such as *Neisseria meningitides* and *E. coli*. Some illnesses can exclusively infect a single type of host, whereas the others can contaminate an extensive variety of animals. Host ranges can be strange, if not perplexing. *Mycobacterium leprae* and *Mycobacterium lepromatosis*, two closely related internal bacteria that are primarily found in humans, armadillos in the Americas, and red squirrels in Scotland, cause leprosy in humans.

Antimicrobial resistance (AMR) is a complicated global problem that must be adequately handled. Antimicrobial treatment for multidrug-resistant bacterial (e.g., tuberculosis, cholera) and fungal (e.g., Candidiasis) illnesses is severely restricted, and there are a variety of causes and explanations for this progression. Due to the severe challenges of cumulative AMR, there remains an urgent need to recognize, innovate, validate, and develop fresh tactics and methodologies that may be used to quickly overcome this serious issue. Immunotherapy is a powerful strategy for boosting the immune system and countering antimicrobial drug

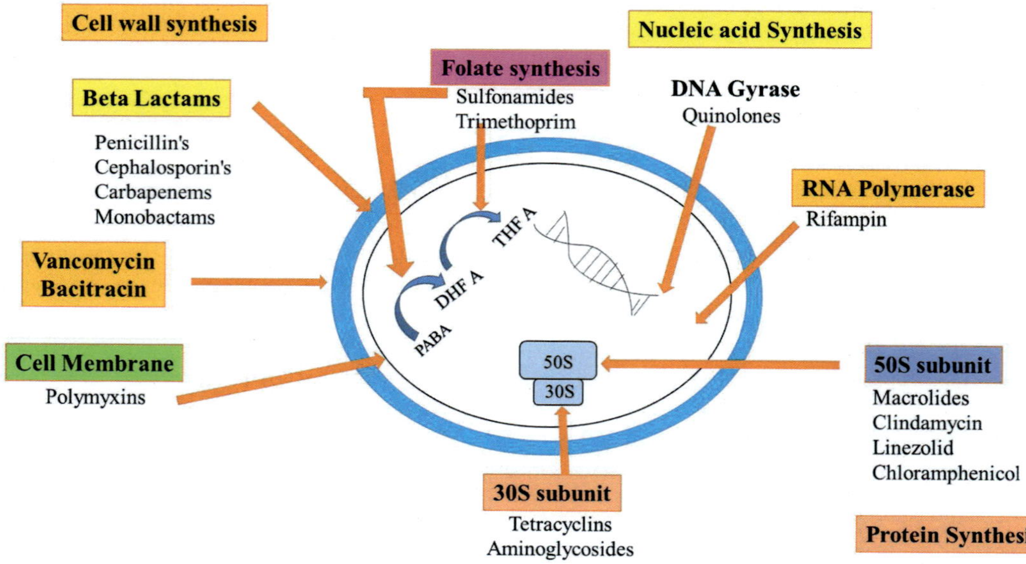

FIGURE 1.4 Antibiotics have different mechanisms: some act on cell wall synthesis, nucleic acid synthesis, and protein synthesis, for example, penicillin interferes with bacterial cell wall synthesis, cephalosporin disrupts the synthesis of the peptidoglycan layer of bacterial cell walls, fluoroquinolone blocks DNA replication via inhibition of DNA gyrase, aminoglycosides inhibit bacterial protein synthesis, linezolid binds to the 23S portion of the 50S subunit and acts by preventing the formation of the initiation complex between the 30S and 50S subunits of the ribosome.

TABLE 1.1 Mechanism of antibiotics.

S.no	Antibiotic	Inhibits	Mode of action
1.	Penicillin	Inhibits cell wall synthesis	Act as bactericidal: blocks cross-linking via competitive inhibition of the transpeptidase enzyme
2.	Cephalosporin	Inhibits cell wall synthesis	Act as bactericidal: inhibits bacterial cell wall synthesis via competitive inhibition of the transpeptidase enzyme
3.	Vancomycin	Cell wall inhibitor	Act as bactericidal: disrupts peptidoglycan cross-linkage
4.	Aminoglycosides	Protein synthesis inhibition Anti-30S ribosomal subunit	Act as bactericidal: irreversible binding to 30S
5.	Tetracycline	Protein synthesis inhibition Anti-50S ribosomal subunit	Act as bacteriostatic: reversibly binds 50S
6.	Fluoroquinolones	DNA synthesis inhibitors	Act as bactericidal: inhibits DNA gyrase enzyme, inhibiting DNA synthesis
7.	Rifampin	RNA synthesis inhibitors	Act as bactericidal: inhibits RNA transcription by inhibiting RNA polymerase
8.	Isoniazid	Mycolic acids synthesis inhibitors	inhibit mycolic acid synthesis, which interferes with cell wall synthesis, thereby producing a bactericidal effect
9.	Trimethoprim/ Sulfonamides	Folic acid synthesis inhibitors	Act as bacteriostatic: inhibition with PABA

confrontation. Drug combination treatment, on the other hand, appears to be a viable strategy for delaying the emergence of resistance and extending the life of antimicrobial medicines. Bacteriophage treatment is also a novel therapeutic strategy for preventing multidrug resistance from developing. (MDR) (Qadri et al., 2021b).

1.5.1 Mycobacterium tuberculosis

The Bacteria Domain, Phylum Actinobacteria, Class Actinobacteria, Order Mycobacteriales, Family Mycobacteriaceae, Genus *Mycobacterium*, and Species *Mycobacterium tuberculosis* are all members of the Mycobacteria. *M. tuberculosis* is a nonmotile rod-shaped bacteria associated to the Actinomycetes very distantly. Mycobacteria which are not pathogenic can be seen in both dry and oily environments, and they are all part of the human flora. The rods are 0.2–0.5 um in diameter and 2–4 μm long.

M. tuberculosis is an anaerobic bacterium, which means it cannot survive without oxygen. In the simplest version of tuberculosis, MTB complexes are invariably found in the lungs' well-aerated upper lobes. The bacterium is an internal facultative parasitic bacterium that infects macrophages. Its aggressiveness stems from its low production rate of 15–20 h. Mycolic acids are lipids present in the cell membranes of *Mycobacterium* and *Corynebacterium*.

They make up the mycobacterial cell envelope's dry mass. Mycolic acids are hydrophobic hydrocarbons that form a lipid shell around an organism and regulate cell permeability. MTB virulence is considered to be influenced by mycolic acids. Mycobacteria are hypothesized to be protected by cationic proteins, lysozyme, and oxygen radicals in the phagocytic granule. Extracellular mycobacteria are similarly immune to the buildup of serum complement. The high gratified lipids in *Mycobacterium* TB' cell wall has been linked to the following bacterial characteristics; stain and dye impermeability, antibiotic resistance is common, acid and alkaline compound resistance, complement deposition provides osmotic resistance, survival within macrophages and resistance to fatal oxidations.

M. tuberculosis sensu stricto, M. africanum, M. canetti, M. bovis, M. caprae, M. microti, M. pinniped, M. mungi, and *M. orygis* are members of the *M. tuberculosis* complex that cause human pathogenesis (van Ingen et al., 2012). Mycobacteria species produce a number of diseases; including pulmonary tuberculosis and extrapulmonary tuberculosis, which have serious consequences for both individuals and the general society. Tuberculosis (TB) is contagious and spreads through the air.

1.5.2 Vibrio cholera

Vibrio cholerae is a very motile Gram-negative rod with a comma-shaped polar flagellum. *V. cholerae* is a member of the Order Vibrionales, Domain Bacteria, Family Vibrionaceae, and Phylum Proteobacteria. It lives in water and infects the small intestine, causing cholera toxin to be produced. These bacteria possess distinct adhesion elements that allow them to adhere to the antagonistic microvilli shallow. *V. cholerae* binds to intestinal epithelial cells and produces one of two antigenically related but distinct cholera enterotoxins (CT-1 or CT-2). The cholera toxin turns on adenylate cyclase, causing an increase in cAMP and consequent hypersecretion of chloride and bicarbonate, followed by water. Despite the fact that this bacterium contains about 200 serogroups, the pandemic illness is caused by only O1 and O139 (Albert, 1996). Although *V. cholerae* can grow in lower salinity water if it is warmer and comprehends adequate carbon-based ingredients, it thrives in the presence of salt (Colwell, 1996). *V. cholerae* can use chitin as a carbon and nitrogen source and is frequently found in water with zooplankton and shellfish (Kirn et al., 2005). Chitin enhances normal proficiency in *V. cholerae,* demonstrating that adjacent inheritable transfer might happen in aquatic environment, especially during zooplankton blooms (Meibom et al., 2005). In aquatics, *V. cholerae* transforms into active but nonculturable environmental cells, correspondingly known as *conditionally viable environmental cells* (Faruque et al., 2005; Nelson et al., 2009).

1.5.3 Staphylococcus aureus

Positive bacterium, round-shaped belonging to the Firmicutes Phylum, Genus Staphylococcus, Class Bacilli, Domain Bacteria, Order Bacillales, Family Staphylococcaceae. It's found in the upper respiratory tract and on the skin, and it's a common attribute of the body's microbiota. About 30% of the total healthy adults have *Staphylococcus aureus* in their nostrils, and about 20% of the total have it on their skin. A greater proportion is found among those who are hospitalized or medical centers. Germs can be spread through direct contact, contaminated things such as fitness equipment, telephones, door handles, television remote controls, or elevators, or, less commonly, inhaling of infected droplets spread by sneezing and coughing. Infections caused by *S. aureus* might range from minor to fatal. Skin infections are the most prevalent staphylococcal infections, which typically progress to abscesses. Bacteria, on the other hand, can enter the bloodstream which is said to be bacteremia and infect virtually any part of the body, including the heart valves (endocarditis) and bones (osteomyelitis). Germs can also accumulate on hospital equipment inside the body, such as artificial heart valves or joints, implanted heart devices, and catheterization inserted through the epidermis into blood arteries. A number of strains of *S. aureus* exist. Foodborne illness, septicemia, and skin scorching are all possible side effects of some strains. Streptococci that produce toxins are responsible for toxic shock syndrome. This illness, which develops quickly and aggressively, is marked by fever, recklessness, dangerously low blood pressure, and organ failure.

1.6 Human pathogenic bacterial diseases

Bacteria that can cause disease are known as pathogenic bacteria (Ryan, 2014). While the majority of bacteria species are innocuous and frequently beneficial, others can cause infectious diseases. Only a few hundred of these pathogenic species are known to exist in humans (McFall-Ngai, 2007). The digestive tract's gut microbiome, on the other hand, contains thousands of species. Advantageous commensals that live on the outer layer of skin mucous membrane, as well as saprophytes that live mostly in soil and decomposing debris, are always present in the human body. Many bacteria can grow in blood and tissue fluids because there are enough nutrients. Pathogenic bacteria have evolved and are equipped with methods to overcome the body's natural defenses, allowing them to penetrate areas of the body where bacteria are not ordinarily found, such as the bloodstream. Some infections simply infect the epithelium, skin, or mucous membranes on the surface, whereas others penetrate deeper, spreading into the tissues and distributing through the lymphatic and blood systems. In rare situations, a pathogenic microorganism can infect someone who is otherwise healthy, but transmission usually happens when the body's natural defense mechanisms are impaired by local trauma or an underlying debilitating state like injury, drunkenness, cold, tiredness, or hunger.

1.6.1 Tuberculosis (TB)

Tuberculosis is an extremely transmittable bacterial disease caused by *Mycobacterium tuberculosis* (Mtb). Tuberculosis is spread through the airways and most usually affects the respiratory tract, but it can affect any tissue. Only around 10% of those infected with *M. tuberculosis* acquire active tuberculosis during their lifetime; the rest are able to keep their illness under control. One of the challenges of tuberculosis is that the pathogen can lay latent in many infected people for many years before becoming active and potentially pathogenic (Mir and Al-baradie, 2013). The chance of developing active TB after infection is highest shortly after infection and rises dramatically in people who have previously had the disease (Holmes et al., 2017). Tuberculosis is prevalent throughout the world (Elisa Terracciano et al., 2020). In contrast, developing countries bear a greater share of the tuberculosis disease burden. The tuberculosis burden in Asia, Africa, Eastern Europe, Latin America, and Central

TABLE 1.2 Difference between active and latent TB.

S. No.	Active tuberculosis bacteria	Latent tuberculosis bacteria
1.	TB bacilli are reproducing & spreading in the body, causing tissue damage	TB bacilli are inactive in the body. This phase can last for very long even decades
2.	Generally feel sick. Typicalsymptoms. Chest X-ray & other tests for diagnosis	Don't look or feel sick. Yourchest X-ray is usually normal. Nosymptoms
3.	TB may spread by coughing, sneezing, or singing	can't spread TB to other people
4.	Positive sputum microscopy	Negative sputum microscopy
5.	Identified by serology	Difficult to identify by serologysince bacteria in dormant state

America remains a concern. In more industrialized countries, recent arrivals from tuberculosis-endemic zones, medical employees, and HIV-positive persons have a high tuberculosis load. Immunosuppressive drugs, such as long-standing corticosteroid treatment, have also been related to a higher risk. The bacteria that causes tuberculosis can dwell in the body without causing symptoms. The infection is known as latent tuberculosis. Individuals with dormant tuberculosis don't show any symptoms and don't sense illness. The microorganisms that cause tuberculosis cannot transmit to others. Have a tuberculosis skin test reaction or a tuberculosis blood test that is positive. Many persons with latent tuberculosis do not get sick. The tuberculosis bacteria can exist for a lifetime in these persons without generating sickness (Table 1.2).

TB bacteria become active when the immune system is unable to prevent them from proliferating. When the tuberculosis bacteria become active, the disease develops. They might correspondingly be competent to blowout the bacteria to those with whom they interrelate on an everyday basis (Mir, 2013). Many people who have latent tuberculosis do not develop the disease. Gradually TB progresses in the individual within weeks of being infected (Fig. 1.5), before their immune system is able to combat the TB bacteria. Others may become ill years later if their immune response weakens as a result of a number of factors.

The immune system is unable to identify tuberculosis bacteria. In the cell wall, there is a lot of lipid. Antimicrobial resistance and permeability resistance to the toxicity of acidic and alkaline chemicals complement deposition, and lysozyme attack provides resistance to osmotic lysis. The maturation of phagosomes is halted. Phagolysosome fusion is blocked. Tubercle bacillus does not persist in phagolysosome production. Live by interfering with phagolysosome formation, Mycobacterium TB survives in macrophage phagosomes (Vergne et al., 2005).

Two examinations are used to determine tuberculosis microbes in the figure, TSTs, and blood examinations. A positive tuberculosis skin test or tuberculosis blood test merely means that the bacterium that causes tuberculosis is present in the person's system. It does not specify if the individual has a latent tuberculosis infection (LTBI) or TB illness. To determine whether the individual has tuberculosis, additional tests such as a chest X-ray and a sputum

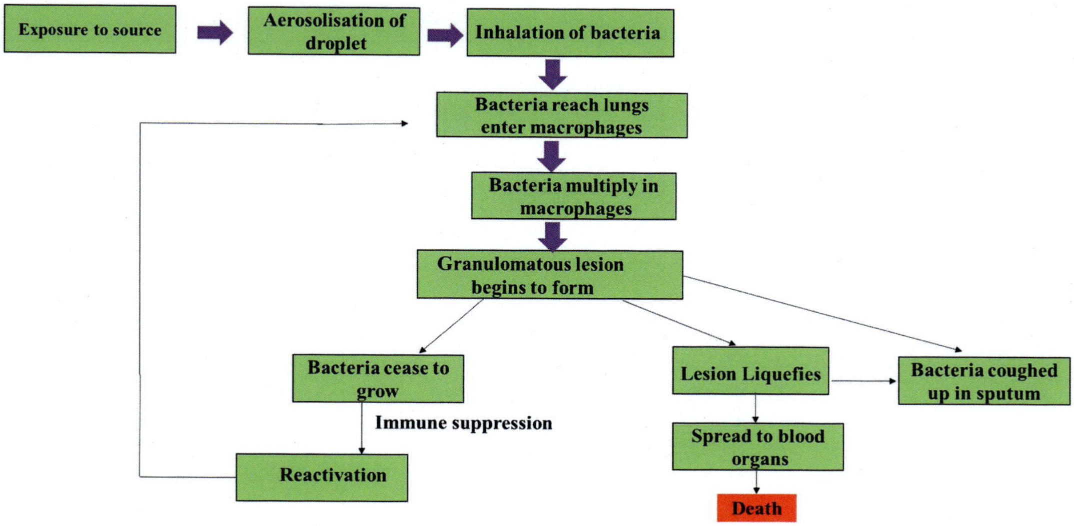

FIGURE 1.5 *Mycobacterium tuberculosis* pathogenesis occurs with Inhalation of the mycobacteria; this is followed by its interaction with resident macrophages through cellular receptors and its internalization. Macrophage bactericidal mechanisms are then activated and the initial infection results in a compact, self-limiting granulomatous lesion, which is an effective means of containing the bacteria's spread.

sample for RT-PCR, microscopy, and culturing are required (Tables 1.2 and 1.3). Line Probe assay is the molecular technique for rapid identification of pathogen along with detection of mutation PCR amplification of target gene. Subsequent solid phase reverse hybridization of PCR product to probes complementary to target genes immobilized on a nitrocellulose strip and results are analyzed by chromogenic technique, biotinylated primers for bacterial species-specific target genes, biotinylated primers specific for gene whose mutation needs to be studied, nitrocellulose strip immobilized with probes specific for target gene and gene whose mutation needs to be studied (wild type)

Tuberculosis individuals must undergo treatment, complete their prescriptions, and take their medications exactly as prescribed. The TB bacteria that are still alive may become susceptible to all of the medications if patients stop taking the drugs too soon or do not take them correctly. The treatment of drug-resistant tuberculosis is more complex and costly. A combination of drugs can be used to treat tuberculosis for 6–9 months. The Food and Drug Administration (FDA) has approved 11 drugs to treat tuberculosis in the United States. Isoniazid (INH), rifampin (RIF), ethambutol (EMB), and pyrazinamide are the first anti-TB medications approved, and they constitute the basis of therapy regimens (PZA).

Drug-resistance develops when the drugs used to treat tuberculosis are misused or mismanaged, it can result in tuberculosis. Misuse or mismanagement can be demonstrated by the following people who do not finish their TB treatment course. The wrong treatment is prescribed by medical professionals (the wrong dose or length of time). There are no medications available for proper treatment. The quality of the drugs is deplorable. MDR TB develops when tuberculosis bacteria develop resistance to at least isoniazid and rifampin, the two most powerful anti-TB drugs. All tuberculosis patients are treated with these

TABLE 1.3 Chip-based real-time PCR test for *Mycobacterium tuberculosis*.

Intended use	**Quantitative detection and diagnosis of *Mycobacterium tuberculosis* (MTB) in human pulmonary (sputum/no sputum)and EPTB specimen and aids in the diagnosis of infection with MTB**
Key features	• High primer sensitivity and specificity • Microchip based real-time PCR assay • Minimal sample requirement @ 6 μL • Smart chip with preset data for quantitation of results • Chip reuse lock • Reaction port with • Contamination/evaporation resistant design
Contents in the kit	Individually sealed pouches, each containing **1.** Truenat MTB micro PCR chip. **2.** DNase & RNase free pipette tip. **3.** Desiccant pouch. Package insert.
Testing duration	1 h

medications. Isoniazid, rifampin, any fluoroquinolone, and at least one of three injectable second-line medicines, notably amikacin, kanamycin, or capreomycin, are all resistant to XDR TB. Patients with XDR TB have considerably fewer treatment options because it is resistant to the strongest TB medications. XDR TB is extremely dangerous for people who have HIV or other immune-suppressive disorders. These individuals toned to be extra careful as they can develop tuberculosis when septic; besides the TB infection can even lead to death.

Traditional treatment has ended in failure in a large number of cases. Drugs used to treat extensively drug-resistant tuberculosis are costly, ineffective, have more long-term side effects, and are typically associated with poor prognosis. Researchers are exploring the genetic/molecular foundation of target drug delivery and drug resistance mechanisms in order to combat the spread of multidrug-resistant tuberculosis, extensively drug-resistant tuberculosis, and totally drug-resistant tuberculosis. This method can be utilized as a stand-alone or in conjunction with the medications. Through complete genome sequencing, researchers want to learn more about the molecular methods and pathology of *M. tuberculosis*, which will aid in the development of new medicines to address the present global health crisis. Understanding the biological mechanisms that cause resistance to *Mycobacterium* TB infection could lead to the discovery of immunological associates of protection, which could lead to the creation of vaccines, diagnostics, and innovative host-directed therapy techniques (Sheikh et al., 2021).

1.6.2 Cholera

A toxigenic *Vibrio cholerae* serogroup O1 or O139 bacterial infection in the colon causes cholera, which is an acute diarrheal sickness. Although the condition is usually asymptomatic or minor, it can be dangerous. One out of every 10 patients infected with cholera will experience watery diarrhea, vomiting, and leg cramps. Dehydration and shock occur as a result of the rapid loss of bodily fluids. If not addressed, death might happen in hours. Cholera can be contracted through the consumption of contaminated water or the consumption of contaminated food. Typically, the source of contamination is an infected person's feces that contaminate water or food. In regions with poor sewage and drinking water treatment, the disease can spread swiftly. Because the virus is unlikely to be transmitted from one person to the other, interaction with an infectious person is not a risk factor for becoming unwell. Severe cholera can cause severe dehydration, leading to kidney failure and can result in shock, coma, and death within hours if left untreated. The infectious *V. cholerae* germ is present in large quantities in the diarrhea of cholera patients, and if swallowed, it can infect others. This can happen if bacteria contaminate food or water. To keep the bacteria from spreading, all feces from sick people should be disposed of carefully to avoid contaminating anything nearby.

Separation and characterization of *V. cholerae* serogroup O1 or O139 from a stool sample by culture remains the gold standard for cholera diagnostic testing. Cary Blair medium is appropriate for transportation, while selective thiosulfate—citrate—bile salts agar (TCBS) is ideal for isolation and characterization. Rapid test kits are beneficial in epidemic situations, but because they do not produce an isolate for antimicrobial susceptibility testing and subtyping, they will not be used for primary screening. *V. cholerae* O1 or O139 is found in persons with symptoms suggestive of cholera by culture or PCR, and the *V. cholerae* O1 or O139 strain is aimed at boosting in places where cholera is not common or has been eliminated. In locations wherein cholera detection is limited or nonexistent, the Crystal VC dipstick rapid test can warn health professionals of an outbreak. The sensitivity and specificity of this test, however, are modest. As a result, fecal samples that test positive for *V. cholerae* O1 and/or O139 when tested with the Crystal VC dipstick should always be treated using normal *V. cholerae* isolation and characterization protocols.

If it's not properly treated, cholera can lead to dehydration, which can be potentially lethal. If cholera patients experience timely rehydration treatment, more than 99% will sustain. As an outcome, the most significant cholera treatment is electrolyte drinks. Rehydration remedy for cholera patients include sufficient capacities of an oral rehydration salt solution, intravenous (IV) fluids, and electrolytes as needed. When cholera patients are not managed with rehydration therapy, one in every four to probably half of them could die from the disease. Tetracycline has remained substantiated to be a proficient cholera cure, outperforming furazolidone, cholamphenicol, and sulfaguanidine in terms of cholera indisposition decrease. An unaccompanied 300-mg quantity of doxycycline is equitable to tetracycline and is nowadays advised as first-line cure in grownups, comprising prenatal females, and children. Tetracycline antibiotics were originally not commended for pregnant women because of the teratogenic effects, as well as for children because of teeth discoloration. There is no link between doxycycline usage as well as teratogenic consequences or dental stain in youngsters, according to a met analysis. Erythromycin is a safe and effective cholera cure for grownups and children, including heavy with child females. Ofloxacin, trimethoprim-sulfamethoxazole (TMP-SMX), and ciprofloxacin are all operative antibiotics; nonetheless, doxycycline gives analogous or better efficacy. Azithromycin has recently been discovered to be more operative than erythromycin and ciprofloxacin. Tetracycline and additional antimicrobial agent confrontation has been witnessed in both prevalent and widespread cholera setups.

Confrontation can evolve over time as a result of the effects of specific modifications or the attainment of genetic elements such as plasmids, introns, or conjugative components that allow for the increasing prevalence of resistance. Antibiotic use is widespread, including extensive dissemination for prophylaxis in asymptomatic persons, which might be a threat to antimicrobial confrontation.

1.6.3 *Streptococcus pneumoniae* and other associated infections

Pneumonia is a respiratory infectious disease that can occur in people and cause mild to severe illness. Viruses, bacteria, and fungi can all lead to pneumonia. Pneumonia symptoms include coughing, fever, and breathing difficulty. The bacteria and viruses that cause pneumonia in the community are not the same as those found in hospitals. *Streptococcus pneumoniae* (pneumococcus) causes the common bacterial pneumonia. *S. pneumoniae* is a member of the microflora of the higher respirational tract. It can be infectious, like other types of natural microflora, in certain conditions, most notably when the patient's immune response is weakened. Pneumolysin, an antiphagocytic capsule, numerous adhesins, and immunogenic cell wall elements are all important virulence factors. When *S. pneumoniae* colonises the lungs' air sacs, the body responds by triggering an inflammatory reaction, which causes the alveoli to fill with plasma, blood, and white blood cells. *Pneumonia* is the medical term for this condition. *S. pneumoniae* is the most prevalent cause of community-acquired pneumonia and meningitis in children and the elderly, and also mortality in HIV patients (van de Beek et al., 2006). In addition to pneumonia, the pathogen results in a wide range of pneumococcal contagions, including chest infections, rhinitis, acute sinusitis, otitis media, eye infections, meningitis, sepsis, osteomyelitis, septic arthritis, endocarditis, peritonitis, pericarditis, cellulitis, and brain abscess (Siemieniuk et al., 2011). When infected with *S. pneumoniae*, polymorph nuclear leukocytes (granulocytes) generate an oxidative stress and inflammation that can be deadly to the bacteria. *S. pneumonia's* potential to overhaul oxidative DNA damage in its genome induced through host immune response most probably donates to the pathogen's pathogenicity. According to this hypothesis, nasal invasion fitness and virulence (respiratory infectivity) in diverse vastly transformable *S. pneumoniae* isolates are dependent on an intact competence system (Li et al., 2016). *Haemophilus influenzae* is another species that causes pneumonia; both *H. influenzae* and *S. pneumoniae* can be identified in the humanoid higher breathing tract. By attacking *H. influenzae* with hydrogen peroxide, *S. pneumoniae* outcompeted it in an in vitro competition study. In a study where both pathogens have been placed inside the adenoidal crater of a mouse for 2 weeks, only H. influenzae survived, more investigation revealed that neutrophils visible to deceased *H. influenzae* stayed abundant supplementary confrontational in attempting to attack S. pneumoniae (Pericone et al., 2000).

Diagnosis can be confirmed based on the clinical presumption and a positive culture from almost any portion of the body. *S. pneumoniae* is commonly optochin subtle, but optochin confrontation has indeed been observed (Pikis et al., 2001). Rapid innovations in next-generation sequencing and relative genomics have allowed for the expansion of consistent molecular approaches for detecting and identifying *S. pneumoniae*. For example, the Xisco gene has been newly reported as a biomarker for *S. pneumoniae* recognition and discrepancy from meticulously associated species using PCR (Salva-Serra et al., 2018).

Broad-spectrum' antibiotics are primarily used to treat severe pneumoniae infections until antibiotics susceptibility testing reports are obtained. Antibiotics susceptibility testing involves determining which antibiotics will be most efficacious in curing a bacterial infection. Broad-spectrum medications are available in contradiction of a widespread diversity of bacteria. Once the sensitivity of the bacteria has been ascertained, health professionals can choose a more specific target antibiotic. Pneumococcal conjugate vaccine (PCV) aids in the prevention and control of infection and the spread of pneumococcal resistance. PCV has diminished pneumococcal infections caused by vaccine strains, the large majority of which were resistant in children by more than 90%. Since immunized individuals do not host the microbes, immunization has similarly reduced the spreading of resilient *S. pneumoniae* strains.

1.7 Summary and future perspectives

Despite considerable advances in trying to uncover vulnerability to fungal infections in recent years, there remain still a significant amount of *Candida* contagions for which the ecological or heritable risk subtleties have not yet been identified. Even more relevantly, despite current treatment drug therapies, mortality rates related to chronic infections remain very high, and future R&D should aim toward deeper understanding of the anti-*Candida* protective immune system for improving diagnostic and potential treatments. New findings may lead to the identification of new immunotherapy targets. Numerous studies have found association between low IFN-levels and an elevated chance of universal candidiasis (Woehrle et al., 2008). Another double, randomized, study of adjuvant IFN-treatment in sepsis is currently underway. It is very important to try to start reversing the immune paralysis (Leentjens et al., 2012). This indicates that IFN-treatment may have the potential to deal with the sepsis-induced invulnerable system (Woehrle et al., 2008).

There's still convincing anti-Aspergillus firearms on the frontier. One such fight may in the future encompass the use of new NK CAR technology, a method that could be used as an allogeneic "off-the-shelf" product (Rezvani et al., 2017). Checkpoint inhibitors can also be included, and such molecules constrain inhibiting receptors on immune cells, increasing their activity. They have been shown to be effective in cancer research (Mir, 2015a,b,c and d). Immunotherapy strategies have the potential to improve antifungal therapy and reduce high death rates. In overall, immunotherapy procedures for *Aspergillus* illnesses are still in the early stages, are cost-intensive, may cause serious side effects, and require complicated as well as time-consuming genetic and cellular manipulations before being used. Various immunotherapeutic techniques are being studied for their effectiveness, stability, and ability to overcome these obstacles.

M. tuberculosis have high lipid concentration in cell wall, leading to impenetrability of antimicrobials, resistance to carnage by acidic/alkaline amalgams, opposition to osmotic lysis by lysozyme. Host immunity is critical in the development of tuberculosis disease. Adjunct immunotherapy that also contains host-directed treatments with immune adjuvants (Efremenko et al., 2012), recombinant cytokines, or therapeutic vaccinations (Mir, 2020; Umar et al., 2020), must be investigated further to enhance the therapeutic outcomes of drug-resistant TB, extremely challenging forms of MDR/XDR-TB. There are some newly developed anti-TB drugs that can significantly improve the current treatment of

MDR/XDR-TB. Even though the effectiveness of recently developed anti-TB agents is unequivocal, the safety and tolerability of these novel drugs (Tiberi et al., 2014) must be further assessed through additional studies, and the resources required have yet to be determined. Immunotherapy is done either using humanized antibodies against CD80, CD86 and CD40 or CD28 fusogenic proteins for the treatment of intracellular pathogens like *M. tuberculosis*. This strategy can be used as an alternative strategy or in combination with the drugs (Mir et al., 2013). Natural antibodies that exclusively have the heavy chain, and their ensuing evolution into small transgenic nanobodies, are a promising new medication delivery, diagnostics, and imaging option. Nanobody descendants have unique refolding capabilities, minimal aggregation propensity, and strong target binding abilities, and are soluble, stable, and adaptable. They can be modified to produce to target enzymes, transmembrane proteins, or molecular interactions in specific ways (Mir et al., 2020).

The pneumococcus is a serious menace, having a myriad of virulence characteristics that it uses not only to hide but also to elude and frustrate host defences. As a result, in the lower airways, a proinflammatory environment emerges, which, rather than being protective, predisposes to inflammation-mediated tissue damage, favoring extrapulmonary pneumococcus dispersion. The pathogen's principal pathogenic role in the determination, the anti-phagocytic polysaccharide capsule, is the primary focus of pneumococcal vaccine research. The prevalence of 94 immunogenic capsular serotypes, on the other hand, has hindered vaccine development. Serotype pathogenicity and prevalence determine the number of serotypes encapsulated by polysaccharide-based vaccines. Aspects that may be linked to a poorer infectious illness outcome, such as cardiac problems, are being found, as well as characteristics that influence more effective therapy, both with antibiotics and with adjuvant medicines. With the discovery of effective vaccines, further advancements in illness prevention are being made in both children and adults.

References

Alastruey-Izquierdo, A., Cadranel, J., Flick, H., Godet, C., Hennequin, C., Hoenigl, M., Kosmidis, C., Lange, C., Munteanu, O., Page, I., Salzer, H.J.F., 2018. Treatment of chronic pulmonary aspergillosis: current standards and future perspectives. C. on behalf of Respiration 96 (2), 159–170.

Albert, M.J., 1996. Epidemiology & Molecular Biology of *Vibrio cholerae* O139 Bengal from:https://pubmed.ncbi.nlm.nih.gov/8783504/.

Alfarouk, K.O., Bashir, A.H.H., Aljarbou, A.N., Ramadan, A.M., Muddathir, A.K., AlHoufie, S.T.S., Hifny, A., Elhassan, G.O., Ibrahim, M.E., Alqahtani, S.S., AlSharari, S.D., Supuran, C.T., Rauch, C., Cardone, R.A., Reshkin, S.J., Fais, S., Harguindey, S., 2019. The possible role of *Helicobacter pylori* in gastric cancer and its management. Front. Oncol. 9, 75.

Arya, N.R., 2021. Candidiasis, 1, N. B. R.

Backhed, F., Ley, R.E., Sonnenburg, J.L., Peterson, D.A., Gordon, J.I., 2005. Host-bacterial mutualism in the human intestine. Science 307 (5717), 1915–1920.

Bensasson, D., Dicks, J., Ludwig, J.M., Bond, C.J., Elliston, A., Roberts, I.N., James, S.A., 2019. Diverse lineages of Candida albicans live on old oaks. Genetics 211 (1), 277–288.

Berends, E.T., Horswill, A.R., Haste, N.M., Monestier, M., Nizet, V., von Kockritz-Blickwede, M., 2010. Nuclease expression by *Staphylococcus aureus* facilitates escape from neutrophil extracellular traps. J. Innate. Immun. 2 (6), 576–586.

Bratton, E.W., El Husseini, N., Chastain, C.A., Lee, M.S., Poole, C., Sturmer, T., Juliano, J.J., Weber, D.J., Perfect, J.R., 2012. Comparison and temporal trends of three groups with cryptococcosis: HIV-infected, solid organ transplant, and HIV-negative/non-transplant. PLoS One 7 (8), e43582.

Carvalho, A., Pasqualotto, A.C., Pitzurra, L., Romani, L., Denning, D.W., Rodrigues, F., 2008. Polymorphisms in toll-like receptor genes and susceptibility to pulmonary aspergillosis. J. Infect. Dis. 197 (4), 618–621.

Christianson, J.C., Engber, W., Andes, D., 2003. Primary cutaneous cryptococcosis in immunocompetent and immunocompromised hosts. Med. Mycol. 41 (3), 177–188.

Clermont, O., Bonacorsi, S., Bingen, E., 2000. Rapid and Simple Determination of the *Escherichia coli* Phylogenetic Group.

Cloherty, J., 2012. Manual of Neonatal Care. Philadelphia. W. K. H. L. W. W.

Colwell, R.R., 1996. Global climate and infectious disease: the cholera paradigm. Science 274 (5295), 2025–2031.

Darling, W.M., 1976. Co-cultivation of mycobacteria and fungus. Lancet 308 (7988).

Davda, S., Kowa, X.Y., Aziz, Z., Ellis, S., Cheasty, E., Cappocci, S., Balan, A., 2018. The development of pulmonary aspergillosis and its histologic, clinical, and radiologic manifestations. Clin. Radiol. 73 (11), 913–921.

de Almeida, C.V., Taddei, A., Amedei, A., 2018. The controversial role of *Enterococcus faecalis* in colorectal cancer. Therap. Adv. Gastroenterol. 11, 1756284818783606.

Efremenko, Y.V., Arjanova, O.V., Prihoda, N.D., Yurchenko, L.V., Sokolenko, N.I., Mospan, I.V., Pylypchuk, V.S., Rowe, J., Jirathitikal, V., Bourinbaiar, A.S., Kutsyna, G.A., 2012. Clinical validation of sublingual formulations of Immunoxel (Dzherelo) as an adjuvant immunotherapy in treatment of TB patients. Immunotherapy 4 (3), 273–282.

Elisa Terracciano, F., Laura Zaratti, A., Franco, E., 2020. Tuberculosis: An Ever Present Disease but Difficult to Prevent.

Erdogan, A., Rao, S.S., 2015. Small intestinal fungal overgrowth. Curr. Gastroenterol. Rep. 17 (4), 16.

Fang, J., Huang, B., Ding, Z., 2021. Efficacy of antifungal drugs in the treatment of oral candidiasis: a Bayesian network meta-analysis. J. Prosthet. Dent. 125 (2), 257–265.

Faruque, S.M., Islam, M.J., Ahmad, Q.S., Faruque, A.S., Sack, D.A., Nair, G.B., Mekalanos, J.J., 2005. Self-limiting nature of seasonal cholera epidemics: role of host-mediated amplification of phage. Proc. Natl. Acad. Sci. U. S. A. 102 (17), 6119–6124.

Fonseca, A., Scorzetti, G., Fell, J.W., 2000. Diversity in the Yeast Cryptococcus Albidus and Related Species as Revealed by Ribosomal DNA Sequence Analysis.

Goldman, D.L., Khine, H., Abadi, J., Lindenberg, D.J., Pirofski, L., Niang, R., Casadevall, A., 2001. Serologic evidence for Cryptococcus neoformans infection in early childhood. Pediatrics 107 (5), E66.

Gow, N.A.R., Yadav, B., 2017. Microbe profile: Candida albicans: a shape-changing, opportunistic pathogenic fungus of humans. Microbiology 163 (8), 1145–1147.

Greenwood, D.B., Slack, M., Irving, R., 2012. Bacterial Pathogenicity.

Holmes, K.K., B, S., Bloom, B.R., et al., 2017. Major Infectious Diseases.

Hoskisson, P.A., 2018. Microbe profile: corynebacterium diphtheriae - an old foe always ready to seize opportunity. Microbiology 164 (6), 865–867.

Iliev, I.D., Funari, V.A., Taylor, K.D., Nguyen, Q., Reyes, C.N., Strom, S.P., Brown, J., Becker, C.A., Fleshner, P.R., Dubinsky, M., Rotter, J.I., Wang, H.L., McGovern, D.P., Brown, G.D., Underhill, D.M., 2012. Interactions between commensal fungi and the C-type lectin receptor Dectin-1 influence colitis. Science 336 (6086), 1314–1317.

Jawhara, S., Mogensen, E., Maggiotto, F., Fradin, C., Sarazin, A., Dubuquoy, L., Maes, E., Guerardel, Y., Janbon, G., Poulain, D., 2012. Murine model of dextran sulfate sodium-induced colitis reveals Candida glabrata virulence and contribution of beta-mannosyltransferases. J. Biol. Chem. 287 (14), 11313–11324.

Jenks, J.D., Hoenigl, M., 2018. Treatment of aspergillosis. J. Fungi. 4 (3).

Johannson, K.A., Huston, S.M., Mody, C.H., Davidson, W., 2012. Cryptococcus gattii pneumonia. Can. Med. Assoc. J. 184 (12), 1387–1390.

Jones, T., Federspiel, N.A., Chibana, H., Dungan, J., Kalman, S., Magee, B.B., Newport, G., Thorstenson, Y.R., Agabian, N., Magee, P.T., Davis, R.W., Scherer, S., 2004. The diploid genome sequence of Candida albicans. Proc. Natl. Acad. Sci. U. S. A. 101 (19), 7329–7334.

Kirn, T.J., Jude, B.A., Taylor, R.K., 2005. A colonization factor links *Vibrio cholerae* environmental survival and human infection. Nature 438 (7069), 863–866.

Kohler, J.R., Hube, B., Puccia, R., Casadevall, A., Perfect, J.R., 2017. Fungi that infect humans. Microbiol. Spectr. 5 (3).

Kwon-Chung, K.J., Sugui, J.A., 2013. Aspergillus fumigatus—what makes the species a ubiquitous human fungal pathogen? PLoS Pathog. 9 (12), e1003743.

Lee, S.H., Lee, B.J., Jung, D.Y., Kim, J.H., Sohn, D.S., Shin, J.W., Kim, J.Y., Park, I.W., Choi, B.W., 2004. Clinical manifestations and treatment outcomes of pulmonary aspergilloma. Kor. J. Intern. Med. 19 (1), 38–42.

Leentjens, J., Kox, M., Koch, R.M., Preijers, F., Joosten, L.A., van der Hoeven, J.G., Netea, M.G., Pickkers, P., 2012. Reversal of immunoparalysis in humans in vivo: a double-blind, placebo-controlled, randomized pilot study. Am. J. Respir. Crit. Care Med. 186 (9), 838–845.

Li, G., Liang, Z., Wang, X., Yang, Y., Shao, Z., Li, M., Ma, Y., Qu, F., Morrison, D.A., Zhang, J.R., 2016. Addiction of hypertransformable pneumococcal isolates to natural transformation for in vivo fitness and virulence. Infect. Immun. 84 (6), 1887–1901.

MacDougall, L., Fyfe, M., 2006. Emergence of Cryptococcus gattii in a novel environment provides clues to its incubation period. J. Clin. Microbiol. 44 (5), 1851–1852.

McFall-Ngai, M., 2007. Adaptive immunity: care for the community. Nature 445 (7124), 153.

Meibom, K.L., Blokesch, M., Dolganov, N.A., Wu, C.Y., Schoolnik, G.K., 2005. Chitin induces natural competence in *Vibrio cholerae*. Science 310 (5755), 1824–1827.

Mir, M.A., Al-baradie, R., 2013. Tuberculosis Time Bomb - A Global Emergency: Need for Alternative Vaccines, pp. 87–93.

Mir, M.A., A, R., A, J., 2013. Innate-effector Immune Response Elicitation against Tuberculosis through Anti-B7-1 (CD80) and Anti-B7-2 (CD86) Signaling in Macrophages.

Mir, M.A., Mehraj, U., Sheikh, B.A., Hamdani, S.S., 2020. Nanobodies: the "Magic Bullets" in therapeutics, drug delivery and diagnostics. Hum. Antibodies 28 (1), 29–51.

Mir, M.A., 2013. Costimulation and Costimulatory Molecules in Cancer and Tuberculosis.

Mir, M.A., 2015a. Concept of Reverse Costimulation and its Role in Diseases.

Mir, M.A., 2015b. Developing Costimulatory Molecules for Immunotherapy of Diseases Preview. Academic Press.

Mir, M.A., 2015c. Costimulation in Lymphomas and Cancers.

Mir, M.A., 2015d. Costimulation Immunotherapy in Allergies and Asthma, pp. 131–184.

Mir, M.A., 2020. Cytokines and Their Therapeutic Potential.

Monteith, A.J., Miller, J.M., Maxwell, C.N., Chazin, W.J., Skaar, E.P., 2021. Neutrophil extracellular traps enhance macrophage killing of bacterial pathogens. Sci. Adv. 7 (37), eabj2101.

Morris, A., Hillenbrand, M., Finkelman, M., George, M.P., Singh, V., Kessinger, C., Lucht, L., Busch, M., McMahon, D., Weinman, R., Steele, C., Norris, K.A., Gingo, M.R., 2012. Serum (1–>3)-beta-D-glucan levels in HIV-infected individuals are associated with immunosuppression, inflammation, and cardiopulmonary function. J. Acquir. Immune Defic. Syndr. 61 (4), 462–468.

Nelson, E.J., Harris, J.B., Morris Jr., J.G., Calderwood, S.B., Camilli, A., 2009. Cholera transmission: the host, pathogen and bacteriophage dynamic. Nat. Rev. Microbiol. 7 (10), 693–702.

Noble, S.M., Gianetti, B.A., Witchley, J.N., 2017. Candida albicans cell-type switching and functional plasticity in the mammalian host. Nat. Rev. Microbiol. 15 (2), 96–108.

Noverr, M.C., Falkowski, N.R., McDonald, R.A., McKenzie, A.N., Huffnagle, G.B., 2005. Development of allergic airway disease in mice following antibiotic therapy and fungal microbiota increase: role of host genetics, antigen, and interleukin-13. Infect. Immun. 73 (1), 30–38.

Panthee, S., Paudel, A., Hamamoto, H., Ogasawara, A.A., Iwasa, T., Blom, J., Sekimizu, K., 2021. Complete genome sequence and comparative genomic analysis of *Enterococcus faecalis* EF-2001, a probiotic bacterium. Genomics 113 (3), 1534–1542.

Pericone, C.D., Overweg, K., Hermans, P.W., Weiser, J.N., 2000. Inhibitory and bactericidal effects of hydrogen peroxide production by Streptococcus pneumoniae on other inhabitants of the upper respiratory tract. Infect. Immun. 68 (7), 3990–3997.

Peterson, S.W., V., J., Frisvad, J.C., Samson, R.A., 2008. Phylogeny and Subgeneric Taxonomy of Aspergillus.

Pfaller, M.A., Diekema, D.J., 2007. Epidemiology of invasive candidiasis: a persistent public health problem. Clin. Microbiol. Rev. 20 (1), 133–163.

Pikis, A., Campos, J.M., Rodriguez, W.J., Keith, J.M., 2001. Optochin resistance in Streptococcus pneumoniae: mechanism, significance, and clinical implications. J. Infect. Dis. 184 (5), 582–590.

Qadri, H., Qureshi, M.F., Mir, M.A., Shah, A.H., 2021a. Glucose - the X factor for the survival of human fungal pathogens and disease progression in the host. Microbiol. Res. 247, 126725.

Qadri, H., Shah, A.H., Mir, M., 2021b. Novel strategies to combat the emerging drug resistance in human pathogenic microbes. Curr. Drug Targets 22 (12), 1424–1436.

Rezvani, K., Rouce, R., Liu, E., Shpall, E., 2017. Engineering natural killer cells for cancer immunotherapy. Mol. Ther. 25 (8), 1769–1781.

Rudkin, J.K., McLoughlin, R.M., Preston, A., Massey, R.C., 2017. Bacterial toxins: offensive, defensive, or something else altogether? PLoS Pathog. 13 (9), e1006452.

Ryan, K.J.R., George, C., Ahmad, N., Drew, W.L., Lagunoff, M., Pottinger, P., Reller, L.B., Sterling, C.R., 2014. Pathogenesis of Bacterial Infections.

Sabiiti, W., May, R.C., 2012. Mechanisms of infection by the human fungal pathogen Cryptococcus neoformans. Fut. Microbiol. 7 (11), 1297–1313.

Saijo, S., Ikeda, S., Yamabe, K., Kakuta, S., Ishigame, H., Akitsu, A., Fujikado, N., Kusaka, T., Kubo, S., Chung, S.H., Komatsu, R., Miura, N., Adachi, Y., Ohno, N., Shibuya, K., Yamamoto, N., Kawakami, K., Yamasaki, S., Saito, T., Akira, S., Iwakura, Y., 2010. Dectin-2 recognition of alpha-mannans and induction of Th17 cell differentiation is essential for host defense against Candida albicans. Immunity 32 (5), 681–691.

Salva-Serra, F., Connolly, G., Moore, E.R.B., Gonzales-Siles, L., 2018. Detection of "Xisco" gene for identification of Streptococcus pneumoniae isolates. Diagn. Microbiol. Infect. Dis. 90 (4), 248–250.

San-Blas, G., Calderone, RA, e, 2008. Pathogenic Fungi: Insights in Molecular Biology.

Scholz, C.F.P., Kilian, M., 2016. The natural history of cutaneous propionibacteria, and reclassification of selected species within the genus Propionibacterium to the proposed novel genera Acidipropionibacterium gen. nov., Cutibacterium gen. nov. and Pseudopropionibacterium gen. nov. Int. J. Syst. Evol. Microbiol. 66 (11), 4422–4432.

Seyedmousavi, S., Guillot, J., Arne, P., de Hoog, G.S., Mouton, J.W., Melchers, W.J., Verweij, P.E., 2015. Aspergillus and aspergilloses in wild and domestic animals: a global health concern with parallels to human disease. Med. Mycol. 53 (8), 765–797.

Sheikh, B.A., Bhat, B.A., Mehraj, U., Mir, W., Hamadani, S., Mir, M.A., 2021. Development of new therapeutics to meet the current challenge of drug resistant tuberculosis. Curr. Pharm. Biotechnol. 22 (4), 480–500.

Sheppard, D.C., Filler, S.G., 2014. Host cell invasion by medically important fungi. Cold Spring Harb. Perspect. Med. 5 (1), a019687.

Siemieniuk, R.A., Gregson, D.B., Gill, M.J., 2011. The persisting burden of invasive pneumococcal disease in HIV patients: an observational cohort study. BMC Infect. Dis. 11, 314.

Springer, D.J., Billmyre, R.B., Filler, E.E., Voelz, K., Pursall, R., Mieczkowski, P.A., Larsen, R.A., Dietrich, F.S., May, R.C., Filler, S.G., Heitman, J., 2014. Cryptococcus gattii VGIII isolates causing infections in HIV/AIDS patients in Southern California: identification of the local environmental source as arboreal. PLoS Pathog. 10 (8), e1004285.

Tiberi, S., De Lorenzo, S., Centis, R., Viggiani, P., D'Ambrosio, L., Migliori, G.B., 2014. Bedaquiline in MDR/XDR-TB cases: first experience on compassionate use. Eur. Respir. J. 43 (1), 289–292.

Tronchin, G., Pihet, M., Lopes-Bezerra, L.M., Bouchara, J.P., 2008. Adherence mechanisms in human pathogenic fungi. Med. Mycol. 46 (8), 749–772.

Tsunemi, T., Kamata, T., Fumimura, Y., Watanabe, M., Yamawaki, M., Saito, Y., Kanda, T., Ohashi, K., Suegara, N., Murayama, S., Makimura, K., Yamaguchi, H., Mizusawa, H., 2001. Immunohistochemical diagnosis of Cryptococcus neoformans var. gattii infection in chronic meningoencephalitis: the first case in Japan. Intern. Med. 40 (12), 1241–1244.

Umar, M., S. N., Sheikh, B.A., Suhail, S., Qayoom, H., Mir, M.A., 2020. Chemokines and Cytokines in Infectious Diseases.

van Burik, J. A., Colven, R., Spach, D.H., 1998. Cutaneous aspergillosis. J. Clin. Microbiol. 36 (11), 3115–3121.

van de Beek, D., de Gans, J., Tunkel, A.R., Wijdicks, E.F., 2006. Community-acquired bacterial meningitis in adults. N. Engl. J. Med. 354 (1), 44–53.

van de Peppel, R.J., Visser, L.G., Dekkers, O.M., de Boer, M.G.J., 2018. The burden of Invasive Aspergillosis in patients with haematological malignancy: a meta-analysis and systematic review. J. Infect. 76 (6), 550–562.

van der Velden, W.J., Netea, M.G., de Haan, A.F., Huls, G.A., Donnelly, J.P., Blijlevens, N.M., 2013. Role of the mycobiome in human acute graft-versus-host disease. Biol. Blood Marrow Transplant. 19 (2), 329–332.

van Ingen, J., Rahim, Z., Mulder, A., Boeree, M.J., Simeone, R., Brosch, R., van Soolingen, D., 2012. Characterization of Mycobacterium orygis as *M. tuberculosis* complex subspecies. Emerg. Infect. Dis. 18 (4), 653–655.

Vergne, I., Chua, J., Lee, H.H., Lucas, M., Belisle, J., Deretic, V., 2005. Mechanism of phagolysosome biogenesis block by viable *Mycobacterium tuberculosis*. Proc. Natl. Acad. Sci. U. S. A. 102 (11), 4033–4038.

Woehrle, T., Du, W., Goetz, A., Hsu, H.Y., Joos, T.O., Weiss, M., Bauer, U., Brueckner, U.B., Marion Schneider, E., 2008. Pathogen specific cytokine release reveals an effect of TLR2 Arg753Gln during Candida sepsis in humans. Cytokine 41 (3), 322–329.

Zhang, Z., Zhang, X., Zhou, Y.Y., Jiang, C.M., Jiang, H.Y., 2019. The safety of oral fluconazole during the first trimester of pregnancy: a systematic review and meta-analysis. BJOG 126 (13), 1546–1552.

CHAPTER

2

Evolution of antimicrobial drug resistance in human pathogenic bacteria

Manzoor Ahmad Mir, Shariqa Aisha, Hafsa Qadri, Ulfat Jan, Abrar Yousuf and Nusrat Jan

Department of Bioresources, School of Biological Sciences, University of Kashmir, Srinagar, Jammu and Kashmir, India

2.1 Introduction

Antimicrobial therapy is one of the most widely employed infection-fighting methods in modern era. There was a popularity gain of antibiotics from 1930s to 1960s (Nathan and Cars, 2014). Antibiotics perform a critical part in decreasing microbial disease progression and death rates both in animals as well as human beings. On the other hand, extensive use of various kinds of antibiotics is a prime motive force behind the establishment and dissemination of drug resistance between the commensal and amensal microorganisms (Aarestrup et al., 2008). Drug resistance is not only seen in bacteria but it can also be seen in viruses, fungi, and parasites (Levy and Marshall, 2004). Moreover, antibacterial tolerance is rising continuously, after the introduction of main antibacterial agents with times ranging from penicillin to varomycin (Walsh, 2003; Levy and Marshall, 2004). Antibiotic-resistant bacteria strains first developed in hospitals, wherein majority of antibacterial agents were utilized (Levy and Marshall, 2004). The main problems of antimicrobial resistance (AMR) in healthcare systems and human societies are:

- Time spent in the hospital
- Failure of antibiotics
- Delay in treatment
- Immunocompromisation and many more

Human Pathogenic Microbes
https://doi.org/10.1016/B978-0-323-96127-1.00013-9

Although AMR is not a novel concept, the number of resistant strains and the afflicted areas are increasing at an alarming rate (Levy and Marshall, 2004). In 1930s sulfonamide-resistant *S. pyrogens* was observed for the first time in military healthcare systems (Levy, 1982). Streptomycin-resistant *Mycobacterium TB* appeared in the population shortly after the finding of streptomycin drug (Crofton and Mitchison, 1948; Sheikh et al., 2021a). Similarly, in 1940s, *S. aureus* emerged as penicillin-resistant bacteria shortly after penicillin was introduced, mainly in London civilian hospitals (Levy and Marshall, 2004). Along with single drug resistance, multidrug resistance has also been seen, and it was first of all discovered in enteric bacteria (Levy and Marshall, 2004). There are various aspects that increase the frequency of AMR including extensive use of antibiotics, utilization of antibiotics in agriculture, biotic elements such as mutation, etc. (Dadgostar and resistance, 2019). The rise in the number of antibiotic-resistant bacteria is directly proportional to the rise in MDR bacterial pathogens, and in current scenario, it is a major hazard to community health. It has been reported by United States CDC that antimicrobial-resistant bacteria population is responsible for thousands of deaths and millions of infections per year in United States of America. Furthermore, in the past decade the large and continuing drop in the quantity of licensed antibacterial drugs has led toward a dangerous situation that can be only tackled with novel antimicrobial therapies (Li et al., 2014). Nevertheless, some effective antibacterial drugs have progressed to phase 3rd clinical trials and many more are in phase 2 clinical trials, though there is an urge for the development of promising antimicrobial drugs against such resistant bacteria (Mandal et al., 2014) (Fig. 2.1).

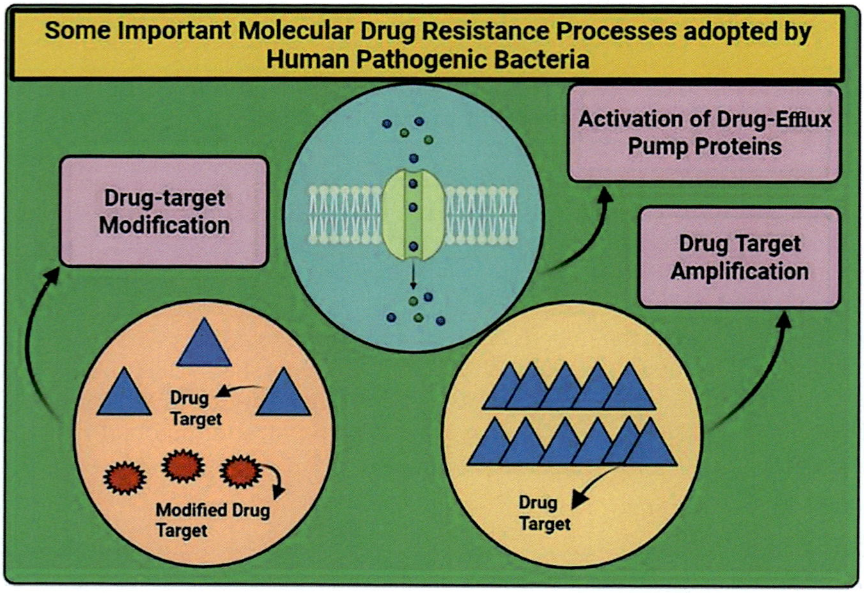

FIGURE 2.1 Diagrammatic representation of some of the significant molecular drug resistance processes including modification of the drug target, amplification of the drug target, activation of drug efflux pump proteins, etc., adopted by various bacterial pathogenic organisms to overcome the effect of different antibiotic agents.

2.2 Current status of resistance to antibacterial drugs

AMR represents the main healthcare issue impacting every part of the world to some extent. This is because of the improper utilization of numerous antimicrobial compounds both in human beings as well as animals (Fig. 2.2). It has been seen that in Europe and Northern America, AMR in case of *S. aureus, S. pneumoniae, Enterococci*, and *Enterobacteriaceae* has appeared and proliferated into healthcare systems and local populations at an alarming rate (Hsueh et al., 2002). It has been observed in Taiwan that the extensive use of antibiotics as a first line of treatment is a major cause of the appearance of numerous drug-tolerant bacterial organisms. From the last 20 years, major antibiotics that are used against these resistant bacteria in Taiwan are carbapenems, cephalosporins, aminoglycosides, and fluoroquinolones, but these bacterial infections are still rising at an alarming rate (Chang and Hsieh, 1996; Ho et al., 1999). It has been seen that AMR exists in every part of the world, with USA alone accounting for >2.8 million drug-tolerant bacterial diseases and 35,000 deaths per year. Every individual on Earth faces the risk of slow-growing bacterial infections that lead to the death of 7 lakh persons approximately every year due to the increase in the number of unchecked spread of antibiotic-resistant superbugs. This slow growing menace is still growing, and it is thought that by 2050, there will be death of 10,000,000 individuals per year, unless and until a considerable antibacterial immunotherapy is executed in response to such resistant superbugs (Arya, 2002; de Kraker et al., 2016; Pérez et al., 2020; Kasimanickam et al., 2021). From the last 10 years, there is a continuous growth of MDR Gram-negative bacteria organisms including *Enterobacteriaceae, Acinetobacter* species, and *Pseudomonas* spp. mainly

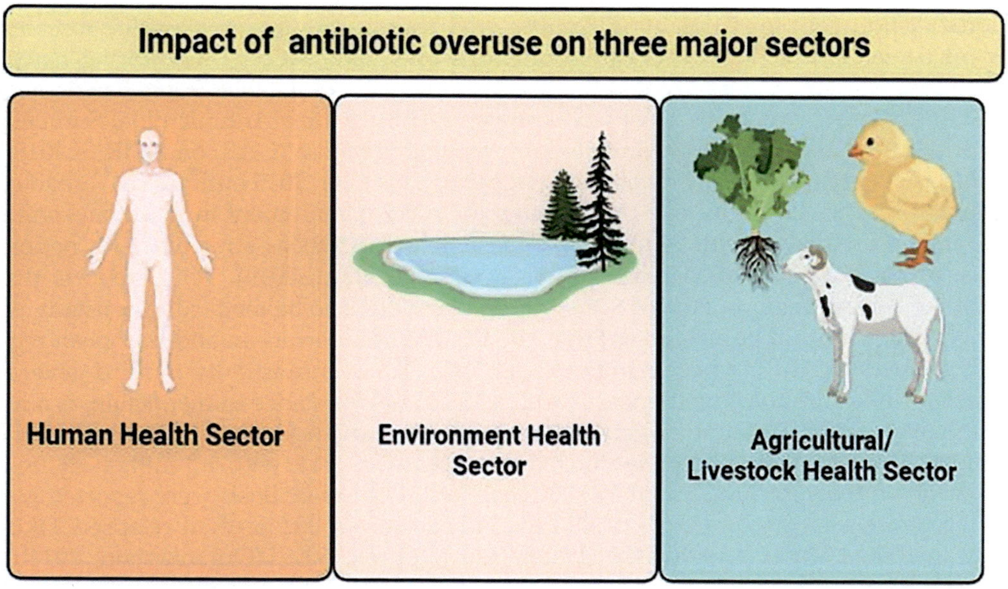

FIGURE 2.2 Representation of the impact of the overutilization of various antibiotic agents on the three major sectors viz. Human health sector, Environment health sector, and Agricultural/Livestock health sector. Antibiotic overuse represents a serious issue adding to the establishment of growing antibacterial drug resistance.

tolerant to frequently utilized carbapenems; this has encouraged the reuse of polymyxin-E as a final treatment line against such resistant infections. Bacteria producing beta-lactamase and bacteria that are resistant to methicillin are increasing day by day across the healthcare systems and populations worldwide (Lupande-Mwenebitu et al., 2020). Carbapenem-tolerant bacterial infections are touching the heights day by day mainly in healthcare systems and human populations, as there is no such antibiotic drug present that can completely combat this infection; due to this reason there are recent outbreaks in overpopulated developing countries with poor hygiene practices (Kariuki and Dougan, 2014). It has been observed that the mortality rate in the USA, the European Union, and India due to antibiotic-resistant bacterial infections are 23,000, 25,000, and 58,000, respectively (Alam et al., 2019). In 2017, Center for Disease Control and Prevention, World Health Organization, declared a list of 12 families of bacterial organisms tolerant to antibiotic compounds and pose a greatest health risk. These families are further divided into three groups—medium, high, and critical on the basis of treatment emergencies (Alam et al., 2019).

There are various factors that lead toward AMR:

- Use of excessive antibiotics and other elements in healthcare units.
- Utilization of antimicrobial drugs in agriculture and farming.
- Utilization of antimicrobials in the local factories.
- Resistant strain in the local environment (Prestinaci et al., 2015).

2.3 Surveillance of antibacterial resistance/tolerance

Antibiotic tolerance/resistance monitoring or surveillance is one of the key goals of National Action Plan for Containing Antibacterial resistance for combating life threatening resistant bacterial infections. There are numerous operating national monitoring networks for bacterial resistance, with CARSS (China Antimicrobial Resistance Surveillance System) and CHINET being two of the most well-known in China (China Antimicrobial Surveillance Network) (Hu et al., 2018). WHO initiated its global action plan (GAP) on AMR in 2015 as a model for developing local AMR confinement plans. This GAP 2015 outlines five broad organizational plans anchored by five components for AMR containment in a human—animal—environment, or "One Health," approach. This action plan acts as a reference for nations to build up their national action plans on AMR (Organization, 2019). WHO reports that in most of the African regions, 80% of *S. aureus* has been found to be methicillin resistant, while in the America, it could be as high as 90%. The Eastern Mediterranean and European regions have high rates of AMR, whereas in Southeast Asia, *E. coli* insensitivity to third-generation cephalosporins and fluoroquinolones, as well as *S. aureus* resistance to methicillin, is a major issue (Shankar and Balasubramanium, 2014). Despite the advancements in TB prevention and treatment, 8,700,000 persons still contracted the disease in 2012, with 1.3 million dying as a result of TB. In the year 2012, about 450,000 cases of MDR-tuberculosis were reported worldwide, accounting for around 3.6% of all new infections and 20.2% of all relapsed TB cases mainly in India, China, and Russian Federation (Approx. 50%) (Organization, 2014b). In pediatrics, bacterial infections have become incredibly impossible or hard to treat. Staphylococcal species, mainly *S. epidermidis* and *S. aureus*, develop around 60% to 70% of diseases, and methicillin-resistant *S. aureus* epidemics were documented in such systems on several occasions (Patel and Saiman, 2010). The CDC assessed the overall value of AMR in the USA to be $55 billion annually, and $20 billion in excess for direct medical expenditures, plus an extra $35 billion in community costs for productivity loss (Control and pdf, 2013).

2.4 Molecular mechanisms of antimicrobial drug resistance in human pathogenic bacteria

Antibiotics are the commonly used antibacterial medications interfering with bacterial growth and abidance without harming the host. Nowadays, antimicrobial drug resistance is increasing day by day at an alarming rate. Many of the antimicrobial drugs can be obtained from nature and can act as bacteriostatic and bacteriocidal drugs including cephalosporins, penicillins, erythromycin, streptomycin, vancomycin, etc. Table 2.1 depicts the list of some important antibiotic agents along with their mode of activity (Reygaert, 2018).

Main targets of the antibiotic drugs are DNA synthesis of bacteria, bacteria protein synthesis processes, and bacteria cell wall synthesis processes (Walsh, 2003). Due to antimicrobial drug resistance, these drugs are unable to combat bacterial infections. These resistances are usually associated with various categories of drug efflux pumps and chromosomal mutations, mainly mutations in extrachromosomal elements from other microorganisms including transposons and integrons. Both of these resistances lead toward healthcare emergencies worldwide (Alekshun and Levy, 2007). Antibacterial drug tolerance could be inherent or innate. Innate or fundamental pathways include Gram-negative bacteria's AmpC-lactamase and several MDR efflux pumps, defined by their natural occurrence on the host genome. Alteration in antibiotic-targeted genes, as well as transfer of resistant genes present on extra-chromosomal elements including plasmids, integrons, etc., via conjugation, transformation, and transduction, are also included in innate resistance methods (Levy and Marshall, 2004). The lack of a susceptible target of a particular antibiotic compound is the most basic example of intrinsic resistance in an individual species, e.g., the biocide triclosan has wide-spectrum efficiency in the case of Gram-positive and numerous Gram-negative bacterial organisms, but it is unable to restrict the development of representatives of the Gram-negative genus *Pseudomonas*. Though it was previously considered to be owing to active efflux, it has since been discovered because of the presence of an insensitive allele of fabI,

TABLE 2.1 List of some important antibiotic agents along with their mode of activity (Reygaert et al., 2018).

S.No	Antibiotic agent	Mode of activity
1.	Glycopeptides, cephalosporins, penicillins, etc.	Inhibition of cell wall biosynthesis.
2.	Tetracyclines, macrolides, etc.	Inhibition of protein biosynthesis.
3.	Sulfonamides, etc.	Inhibition of metabolic processes.
4.	Quinolones, etc.	Inhibition of DNA/RNA biosynthesis.

which encodes an extra enoyl-ACP reductase enzyme, which is the triclosan target in sensitive species (Chuanchuen et al., 2003; Zhu et al., 2010). Other illustration is the lipopeptide daptomycin that is effectual toward Gram-positive bacteria but not Gram-negative bacteria and was initially licensed for human use in 2003. This is because, Gram-negative bacteria possess a thinner cytoplasmic membrane compared to Gram-positive bacteria, which lowers the efficacy of Ca^{2+}-regulated transport. Daptomycin's antibacterial activity is dependent on its entry into the cytoplasmic membrane (Randall et al., 2013). Multiple genes that encode for intrinsic bacterial resistance to antibiotics have been recently identified through recent research including beta-lactams, aminoglycosides, and fluoroquinolones in *S.aureus, E. coli, P. aeruginosa* (Blake and O'Neill, 2013). Fluoroquinolone resistance-causing mutations are mostly identified in the quinolone-resistance-determining-region (QRDR) of GyrA and ParC/GrlA. Mutations are more common in certain amino acids than in others. To "train" resistance in a bacterium, numerous mutational occurences can be selected in a gradual way; subsequent mutations have additive effects. Multiple QRDR mutations and additional fluoroquinolone tolerance mechanisms, e.g., drug extrusion system, are found in highly fluoroquinolone resistant isolates (Alekshun and Levy, 2007). Rifamycins are antimicrobial medicines that stop translation by binding to RpoB, an RNA polymerase subunit (RNAP). Despite the fact that fusion treatment regimen containing rifampin as well as isoniazid, pyrazinamide, ethambutol, or gentamycin remain the first-line treatment for mTB infection, missense mutations in the rifampin-binding region of rpoB take place at a rate of 1−108 and are widely spread, resistance via point mutations in the rifampin-binding area MDR-mycobacteria, tolerant to rifampin and isoniazid and inhibits successful therapy in patients from all over the globe (Sharma and Mohan, 2006). There are various antimicrobial drugs including macrolide-lincosamide-streptogramin (MLS), tetracycline, and aminoglycosides that target ribosomes by inhibiting translation of proteins. As a result resistance through chromosomal mutations has appeared, that makes these drugs ineffective against such mutants (Alekshun and Levy, 2007). It has been shown that molecular mutation in the rplD decreases MLS susceptibility (Tait-Kamradt et al., 2000). Mycobacterium TB is responsible for one of the fatal contagious disease called as tuberculosis (TB). TB usually spreads through infectious droplets from TB patients through coughing and this disease mainly affects lungs. There are various causes that are associated with the development of new resistant Mycobacterium strains including excessive antibiotic usage, lower number of effective antimicrobials, etc. (Qadri et al., 2021). Multiple reports were documented the role and significance of transporters in the spread of antibiotic resistance in distinct mycobacterial species. *M. tuberculosis* has a genotype that contains a number of ATP/MFS transporters, and related extrusion pumps. Viveirosa et al. found at least 18 facilitators in mycobacterial organisms that play a role in antimicrobial sensitivity (Almeida Da Silva and Palomino, 2011; Viveiros et al., 2012; Paulsen et al., 2001; Rossi et al., 2006). There is a variety of mechanisms found in Mycobacterium TB through which it poses resistance to toxic effects of antibiotics. Most antituberculosis medications, including streptomycin, isoniazid, and others, are extruded by mycobacterial efflux pumps, indicating that efflux pumps perform a substantial part in the establishment of drug tolerance. Another intriguing resistance strategy used by *M. tuberculosis* to negate the activity of various medicines is molecular mimicry of the corresponding drug targets (Malinga et al., 2016; Duncan and Barry III, 2004; Smith et al., 2012).

Antimicrobial resistance can be acquired or developed by bacteria in spite of intrinsic resistance. This could be accomplished through a variety of pathways, divided into three categories: first, the ones which decrease antibiotic subcellular levels as a consequence of poor bacterial infiltration or antibiotic outflow; second, those few who modify the antibacterial drug target via gene mutation or PTMs; and third, that deactivates the antibiotic through solubilization or reconfiguration (Blair et al., 2015). A number of thoroughly studied multidrug resistant efflux transporters found in all bacteria, and novel pumps transporting drugs are still being located. MdeA in *Streptococcus mutans*, FuaABC in *Stenotrophomonas maltophilia*, KexD in *K. pneumoniae*, and LmrS in S. aureus have all been discovered in the last 2 years (Floyd et al., 2010; Hu et al., 2012; Kim et al., 2013; Ogawa et al., 2012). There are numerous enzymes that have been discovered since today, that are able to deteriorate or alter various antibacterial drugs like beta-lactamases, ESBLs, carbapenemases, etc. (Blair et al., 2015).

2.5 Resistance to antibacterial drugs in selected bacteria of international concern

Antibiotic resistance is identified as a major healthcare challenge in medical settings and is increasing dangerously from the past years. Bacteria that are responsible for nosocomial infections are also showing antibiotic resistance in human communities along with public-health settings (Theuretzbacher, 2013). The findings and analytical execution of antibacterial drugs made these drugs very attractive and useful in the medicinal era, as these antibiotics treat bacterial diseases that range from minute infections to deadly diseases. It has been reported that antibiotic-resistant bacteria cause approximately two million infections each year in the United States and 23,000 deaths per year. There is also a high risk of hospitalization and complications from antibiotic-resistant bacteria (Guyomard-Rabenirina et al., 2016). Sulfonamide, introduced in 1937, was the first effective antimicrobial agent and within 2 years, resistance to sulfonamide was reported. The basic mechanism of antibiotic resistance to this drug was horizontal transfer of foreign folP or its parts that results in the modification of drug targets due to enzyme modifications, bygenomic alterations. Historically it was thought that bacterial infections and bacterial resistance are wholly and solely associated with healthcare systems, but it has been observed that the surrounding environment containing antibiotic wastes and resistant strains is also responsible for such resistant infections (Guyomard-Rabenirina et al., 2016).

2.5.1 *Escherichia coli*—resistance to third-generation cephalosporins and fluoroquinolones

Escherichia coli, a Gram-negative bacteria occurring singly or in pairs, are motile, facultative anaerobes. *E.coli*, one of the leading causes of foodborne infections worldwide occurs generally as a normal member of microbiota in animal as well as human species. It usually causes UTI infections under abnormal conditions that can sometimes lead toward kidney diseases. It can also cause infections in peritoneum as well as skin and various other tissues. Meningitis in neonates is also caused by *E. coli*. Bacterial resistance to antibiotics represents as a serious healthcare issue in the current era. *E. coli* is resistant to several antifungal drugs including beta-lactams, quinolones, aminoglycosides, fluoroquinolones and many more. *E. coli* resistance to fluoroquinolones is frequently observed in conjunction with extended spectrum beta-lactamases. In *E. coli*, resistance toward fluoroquinolones is associated with both self-chromosomal as well as extra-chromosomal genes (Allocati et al., 2013). *E. coli* resistant to fluoroquinolones develop this type of resistance either through gene mutations or through extrachromosomal elements including plasmids, integrons, etc. Extended spectrum beta-lactamases are contagious between bacteria. These enzymes also destroy many ESBLs. Recently carbapenem resistance in *E. coli* is reported, which is an emerging threat and this type of resistance is acquired via the presence of metallo-beta-lactamases, showing resistance to nearly all useable beta-lactam antibiotics (Organization, 2014b). In Enterobacteriaceae carbapenem resistance is a newly emerging concern which is accompanied by the presence of special enzymes including carbapenemases. It was reported that the carbapenemases are usually observed in clinical isolates of *K. pneumoniae* and *E. coli*, and are usually found around the (Allocati et al., 2013) Extended Spectrum Beta Lactamase's (ESBLs), bacterial enzymes are responsible for damaging special class of antimicrobial drugs called as Beta-lactams. These enzymes provide a third-generation resistance to cephalosporins, and these enzymes can be transmitted from one bacterium to another via horizontal transfer (Organization, 2014a,b).

2.5.2 *Klebsiella pneumoniae*—resistance to third-generation cephalosporins and to carbapenems

K. pneumoniae is a Gram—negative, nonmotile, rod-shaped facultative anaerobe that usually colonizes the normal flora of the nose, mouth, skin, gastrointestinal tract (GI tract), and intestines in humans and other vertebrates. *K. pneumoniae* are the cause of common infections in hospitals among individuals who are at risk including preterm neonates, immunocompromised individuals, diabetic patients, and patients in intensive care units. In neonates, *Klebsiella pneumoniae* causes several infections including UTI infections, GIT infections, deep seated infections, etc. *K. pneumoniae* communicates from one patient to another causing outburst infection mainly in ICUs and local communities (Organization, 2014a,b). Broadspectrum antibiotics such as beta-lactams including carbapenems, penicillins, etc., are the primary line treatments against *K. pneumoniae* infections. But *K. pneumoniae* can naturally produce broadspectrum beta-lactamases that confer resistance toward beta-lactams. *K. pneumoniae* is considered as a familiar carbapenem-resistant bacteria, as it produces carbapenemases, versatile beta-lactamases, that act against carbapenem antibacterial drugs and lower its action via antibiotic hydrolysis. It has been seen that these resistant strains escalate very quickly in the United States and rest of the world. Though it could be predicted that these resistant *K. pneumoniae* strains pose a serious threat worldwide (Chen et al., 2012), bla_{KPC},a specific gene in *K. pneumoniae* encodes an enzyme namely Klebisiella pneumoniae carbapenemase that provides resistance properties in *K. pneumoniae* against various carbapenems and results in serious epidemic *K. pneumoniae* infections (Netikul and Kiratisin, 2015). It has been observed that the carbapenem-resistant *K. pneumoniae* strains possess global distribution. This observation was formulated on the basis of MLST (Kitchel et al., 2009). The resistance in *K. pneumoniae* also occurs due to the horizontal gene transfer via plasmids or transposons such as E. coli. It has been studied that *K. pneumoniae* is resistant to oral antibiotics including fluoroquinolones and cotrimazole. This study shows that there is lesser availability of US FDA approved oral drugs that can curb this resistant infection. Extended-spectrum beta-lactamases were first of all identified in 1982, when nosocomial epidemic *K. pneumoniae* infection occurred in Germany. From 1982, around 200 ESBL *K. pneumoniae* strains have been identified from different parts of the world (Organization, 2014a,b). *K. pneumoniae* is not only resistant to carbapenems but it also shows resistance toward cephalosporins, another class of beta-lactams. These drugs are administered intravenously against acute *K. pneumoniae* infections. Natural production of beta-lactamases that confer resistance in *K. pneumoniae* toward third generation cephalosporins has been identified in many parts of the world (Organization, 2014a,b).

2.5.3 *Staphylococcus aureus*—resistance to methicillin

S. aureus is a leading infective bacterium that causes acute hospital-associated and local community infection. In UK, this bacterium is the leading infective agent after *E. coli*, as well as a leading cause of nosocomial infections. During the earlier periods penicillin was administered against this bacterium, but due to production of penicillinase the drug became ineffective against such organisms. Penicillin-methicillin was used as a treatment agent, but this also failed because some *S. aureus* strains were found to be resistant to this artificial drug popularly known as MRSA. In 1960s and 1970s, these MRSA strains quickly appeared resulting in the healthcare associated and community-associated hazards mainly in Europe, United States, and rest of the parts of the world (Enright et al., 2000). These bacteria are most common cause of suppurative lesions in humans and also cause severe diseases, which are fatal. *S. aureus* is a potential pathogen due to its ability to develop antibiotic resistance toward various classes of antimicrobial agents (Table 2.2). These resistant characteristics are many a times located on extrachromosomal or mobile gene elements (Stefani et al., 2012). Methicillin, a penicillin derivative, was introduced for the first time in 1959 to combat penicillin-resistant strains soon after penicillin resistance occurs. But unfortunately, MRSA strains were observed soon after its use. MRSA was thought to be connected with penicillin-binding proteins (PBPs). The altered PBP of MRSA, PBP2a, is specified by *mecA* gene, present in staphylococci cassette chromosome. Along with *mec A*, *mecR1* is also responsible for methicillin and other antibiotic resistance. These PBPs are used in cell wall formation in this bacterium under

TABLE 2.2 List of some antimicrobial drug resistance processes in the case of *Staphylococcus aureus* (Reygaert et al., 2018).

S.No	Drug resistance process	Concerned antimicrobial compound
1.	Drug uptake alteration	Example: Tetracyclines, fluoroquinolones, glycopeptides, etc.
2.	Drug uptake reduction	Example: Glycopeptides
3.	Drug efflux pump activation	Example: Tetracyclines, etc.
4.	Drug inactivation	Example: Beta-lactams, etc.

optimal conditions. When MRSA is subjected to methicillin, MRSA strains use these PBPs to confer resistance to methicillin like PBP2 acting as a transglycolase and PBP2a confers transpeptidase activity. Due to these mechanisms, *S. aureus* gains resistance toward different classes of antimicrobials (Alekshun and Levy, 2007).

Based on the types of resistance the MRSA strains are divided as hospital-acquired and community-acquired. Information regarding the MRSA is commonly available and is epidemiologically documented. The MRSA rate varies widely from less than 1% in Europe and 50%–80% in Asia and African countries (Theuretzbacher, 2013). The MRSA strains that emerged outside the healthcare units possess characteristics common with nosocomial infectious strains (Stefani et al., 2012).

2.5.4 *Streptococcus pneumoniae*—resistance (nonsusceptibility) to penicillin

Streptococcus pneumoniae is a Gram-positive, diplococcus commonly known as pneumococcus that usually colonizes upper respiratory tract and is a leading cause of pneumonia. It is an acute infection seen commonly in children under the age of 5 years. It is the cause of three to five million deaths per year. In the United States, *S. pneumoniae* causes half a million local infections, 50,000 circulatory infections, 56,000 pneumococcal meningitis, and approximately 6 million ear infections (Tomasz, 1997). A number of bacteria such as *S. pneumoniae, S. aureus,* etc., are able to produce penicillin-binding proteins (PBPs) that naturally confer beta-lactam resistance to these bacteria. It has been seen that there are six PBPs in *S. pneumoniae* including 1a,1b,2x,2a,2b,3 that donate penicillin resistance to resistant strains (Reinert and Infection, 2009). It has been thought that this type of resistance is acquired between *Streptococcus* genus by horizontal gene transfers. It was reported that the ubiquity of penicillin-resistant *S. pneumoniae* varies from 25% to 50% percent in Spain, Greece, Israel, and France: 10% to 25% percent in Finland, Turkey, Ireland, and Portugal: 5% to 10% in Italy; and 1% to 5% in the United Kingdom, Norway, Austria, Germany, and Sweden (Reinert and Infection, 2009).

Rising resistance to antibiotics among *S. pneumoniae* isolates is a critical problem worldwide. It was reported that in China, pneumococcal infections are the leading cause of pneumonia outside the hospitals, i.e., in local communities usually in the children under the age of 5 years. In this report it also included that *S. pneumoniae* shows significant resistance toward major classes of antibiotics. The percentage of resistance in case of *S. pneumoniae* to clindamycin, erythromycin, tetracycline, and trimethoprim has been reported as 95.8%, 95.2%, 93.6%, and 66.7%, respectively. Also the resistance rate of *S. pneumoniae* to penicillin was reported as 86.9% (Wang et al., 2019).

2.5.5 Nontyphoidal salmonella-resistance to fluoroquinolones

Salmonella was first discovered in 1884 by Dr. DE Salmon. This bacterium causes invasive diseases called as invasive salmonellosis. These infections are usually treated by the fluoroquinolones. These drugs possess broadspectrum activity against various bacteria species (Karp et al., 2018). Salmonella are important foodborne pathogens that cause foodborne illness throughout the world—gastroenteritis, bacteremia, and subsequent focal infection. The US CDC identifies this pathogen as a hazard to public health (McDermott et al., 2018).

The antibacterial resistance differs among variants of nontyphoidal salmonella pathogens because some of them are highly resistant and infectious. During the period between 1990 and 2000 particular variants of MDR salmonella species appeared in some parts of the world and by the time these resistant strains emerged globally, certain salmonella variants including Typhimurium, the extra chromosomal element that confers resistance to various antimicrobials such as penicillin, sulfonamides, tetracycline, streptomycin, etc., had got extra resistance determinants and spread horizontally among other serotypes. Infections caused by non-typhoidal salmonella are common and MDR *Salmonella enterica* variant Typhimurium is thought to be highly contagious and leads to numerous invasive infections and prolonged hospitalization. This salmonella strain is highly resistant to fluoroquinolones and causes higher mortalities as compared to other variants/strains (Organization, 2014a,b). Fluoroquinolone class of antibiotics attack two basic enzymes of the bacteria —topoisomerase IV and bacterial topoisomeras II. Fluoroquinolone resistance in salmonella is attained due to the chromosomal mutations in the QRDR region (Quinolone resistance-determining region) (Karp et al., 2018). There are usually two scenarios of salmonella quinolone resistance, one scenario is the mutation in the gyrA, gyrB genes that code for bacterial topoisomerase II subunit and parC gene that codes for topoisomerase IV subunit. The second scenario of salmonella resistance involves the mutation of the genes that regulate AcrAB-TolC efflux pumps. These two mechanisms directly decrease the susceptibility of quinolones antibiotics. Approximately 45,000 cases and 400 to 600 deaths have been reported annually by the different CDCs over past few years and is still rising (Acheson and Hohmann, 2001).

2.5.6 Shigella species—resistance to fluoroquinolones

Shigella species are considered as one of the major cause of diarrhea and dysentery. Bloody diarrhea is the main diagnosis for shigella infection. This disease can be transmitted through various ways including eating raw food, intake of impure water, etc. It has been reported that this infectious species is acquiring more and more resistance day by day because of the resistance gene elements present on plasmids, integrons, integrating conjugative elements, etc. From the past few years shigella infections were usually treated by quinolones, but recently it has been seen that the resistance against quinolones among different variants of shigella is increasing at an alarming rate (Mensa et al., 2008). There are two ways of resistance in shigella species. One way is the result of mutation in gyrA, gyrB, and parC, parR genes and second type is the fluoroquinolone efflux. It has been observed that the fluoroquinolone resistance is associated with the mutation and increased drug extrusion through efflux pumps (Kim et al., 2008). A study conducted in Israel showed that the shigella species showed a significant resistance to tetracyclin, high resistance to sulfamethoxazole, and emerging resistance to fluoroquinolones (Ashkenazi et al., 2003).

2.5.7 *Neisseria gonorrhoeae*-decreased susceptibility to third-generation cephalosporins

Neisseria gonorrhoeae is a Gram-negative, diplococcus (oval shaped), nonmotile, and aerobic microbe. This bacteria causes gonorrhea, which is one of the most common sexually transmitted disease (STD). World Health Organisation (WHO) estimates gonorrhea as the major prevalent bacterial STD, and is an emerging acute infection leading to economic burden, morbidity, and mortalities (Bharara et al., 2015). The bacteria responsible for gonorrhea are also known as *gonococci*. It usually infects sexual organs and can infect other body parts also. The infection can be transmitted from mother to child leading to the blindness of new-born. According to the WHO report 2007, 6,000,000 fresh infections due to *N. gonorrhoeae* occur between the age group of 15 and 49 years globally. Antimicrobial resistance in *N. gonorrhoeae* is the emerging global threat; if left untreated, this can lead to acute reproductive organ infections, deranged pregnancies, and impotency in females. It has been reported by WHO in 2016 that there is an increase in the disease burden with 87,000,000 new cases of *N. gonorrhoeae* (Williamson et al., 2019). Each of the available antibiotic when administered against gonorrhoea follows resistance. Due to such a huge resistance, this infection is emerging in every part of the globe. In 1970s, N. gonorrhoeae-resistant strains against tetracyclin and penicillin were identified in Asia. In the middle of 1990s, highly fluoroquinone-resistant strains outraged in Asia and then emerged as a global issue. Nowadays, cephalosporins a third-line treatment is used as a first-line treatment against gonorrhoea infections. But due to resistance there is no such drug that can effectively combat such MDR infection (Organization, 2014a,b).

2.6 Economic burden of antibacterial resistance and its impacts on human health

Antibacterial resistance is considered as the primary healthcare burden of the 21st century (Woolhouse and Farrar, 2014) (Fig. 2.3). The downturn in the effectiveness of antibiotics causes significant economic and health burdens on societies. Estimates of clinical benchmarks (mostly crude mortality and morbidity) and economic indicators (direct costs, drug expenditure, and utilization of resources) are used to assess the public health burden of antibiotic resistance, which is essential to motivate policy responses. The majority of these estimates are limited to high-income nations, and they use this data from national clinical sample surveillance, prevalence or incidence surveys, and retrospective cohort groups to suit the design models (Naylor et al., 2018). The public health burden is greatly underestimated due to the large variation in surveillance data reporting and the paucity of assessments of the socioeconomic implications of antibiotic resistance (such as lower productivity due to disease). As a result, the generalizability of the results and the predicted values are limited in global estimates (de Kraker et al., 2016). An important question is whether antibiotic resistance creates an economic and health burden on hospitals (Control and Prevention, 2015). Studies on economic and health load due to antibacterial resistance are not many (Cosgrove et al., 2003; Roberts et al., 2009).

2.6.1 Health burden

Several studies have shown that bacterial infections that are sensitive or tolerant/resistant to various antibacterial agents studied have mixed results. The systematic review found a large number of publications that may be used to evaluate the effect of tolerance on a variety of health outcomes. Patients with third-generation cephalosporin-resistant *E. coli* infections

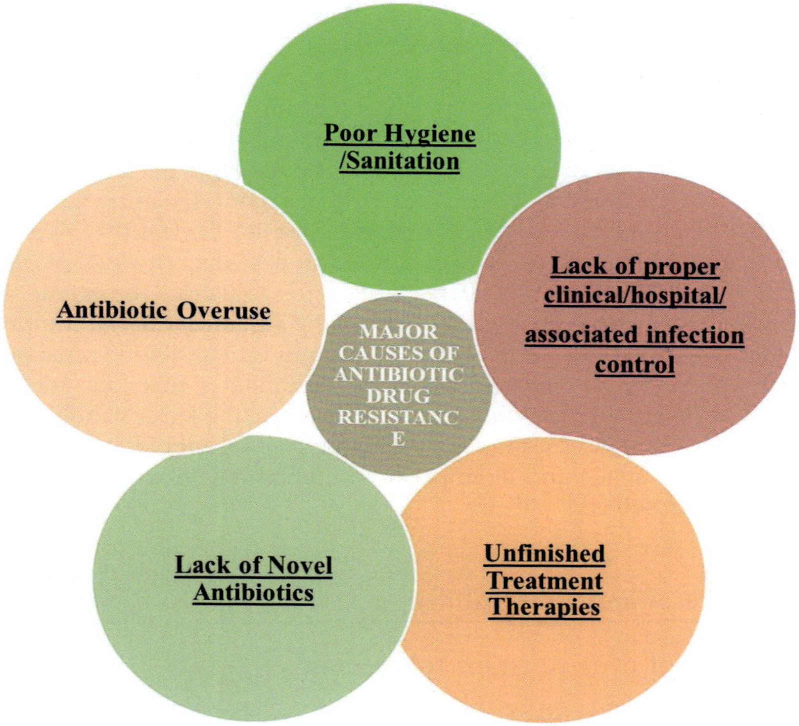

FIGURE 2.3 Diagrammatic illustration of some of the major causes of the worldwide rapidly growing antibiotic drug resistance issue.

had a significant twofold rise in all-cause mortality as well as mortality in 30 days. There was no statistically significant increase in bacterial mortality or length of stay (but the results of the two studies contributing to this finding were inconsistent). Alessandro Cassini and his associates investigated the health burden of five antibiotic-tolerant infections (noninvasive and invasive) produced by eight drug-resistant bacterial organisms having 16 patterns across the EU and the European Economic Area in the journal *Lancet* "Infectious Diseases." The results, which are expressed as disability-adjusted life years, are terrifying. In addition one study estimated that in 2015, out of the 95% of antibiotic-resistant bacterial infections, 63% were healthcare associated (Cassini et al., 2019). Every year below three million deaths were reported due to infection caused by antibiotic resistance as witnessed in the United States followed by Italy with 10,000 cases and 15,000 cases in Greece. Nearly two lakh cases of *Clostridioides difficile* infection were recorded in year 2017 in which 11,000 deaths were also recorded (Cassini et al., 2019).

2.6.2 Economic burden

Globally, microbial resistance causes seven lakh deaths each year which has been predicted to rise up to ten million with cumulative cost of hundred billion dollars by 2050 and could be extreme cause of future economic crisis (Ebi et al., 2017). Around 20 antibiotic-resistant bacteria cases were recorded on the basis of concern and threat to human health. According to a joint CDC study, figures assessed at wider level for treating infections caused by the different variants of six multidrug-resistant germs have been identified in the report and commonly found in healthcare could be considerably more than four billion dollars per year (Control and Prevention, 2015). Antibiotic-resistant bacterial infections have a high financial cost, which is difficult to estimate. The need of calculating the economic impact of disease caused by antibiotic resistance is critical. Despite the fact that resistance has been linked to negative health outcomes, present studies for assessing the load of diseases contain methodological flaws (Eber et al., 2010). Antibiotic resistance economic estimates are currently limited in scope and do not account for the significant social importance of antibiotics, potentially ignoring the full economic implications of antibiotic resistance. Advanced future research should apply macroeconomic methodologies that take into consideration broader effects associated with growing resistance, such as the lack of antibiotic effectiveness in recent therapeutics, to better assess the economic cost of antibiotic resistance. Antibiotic resistance's economic damage and its impacts on human health will remain unquantifiable until these issues are resolved.

2.7 Surveillance of antimicrobial drug resistance in disease-specific programs

2.7.1 Surveillance of drug-resistant tuberculosis

World Health Organization Global Anti-TB Drug-Resistance Project was started in 1994. Universally, 71 nations developed daily testing surveillance systems and 65 nations developed epidemiological examination surveillance systems. One of the best methods to examine the resistance status in specific drug-resistant microorganisms is the surveillance at routine-susceptibility examination of tuberculosis patients. This method should include well-organized data accumulation and examination. In addition, we have to select every area for surveillance program, otherwise TB epidemics may remain unchecked, and this may result in huge outbreaks resulting in health and wealth loss (Organization, 2009). One of the main advantages of this surveillance system is that this data can be used by the different countries that need such data. Molecular technology is currently being introduced in resistance investigations, so that it could be understandable and it can also help in reducing experimental phenomenon. It has been estimated by the WHO that there are nearly half a million multidrug-resistant tuberculosis infections annually. Globally, among the 22 high-TB load nations, India is the ranked number one possessing nearly about 3,400,000 TB infections (Ramachandran et al., 2009, Sheikh et al., 2021b, Mir). This TB-drug resistance in different nations is assisted by a security network called as Security Network SRL that currently assists 29 laboratories near six WHO sectors (Organization, 2010a).

2.7.2 Surveillance of antimalarial therapeutic efficacy and resistance

Currently, the monitoring of drug resistance to antimalarial drugs is done by one of three methods: (1) in vivo study to examine the efficacy of antimalarial drugs in different malaria patients; (2) Both inside living model system as well as outside the model studies to verify the susceptibility of the parasite to the drug; and (3) molecular tests to check for confirmed DNA nucleotide alterations and variants per cell that are directly related to antimalarial drug resistances (Nsanzabana et al., 2018). All the above three monitoring methods complete each other because they examine each and every part of the malarial drug-resistance, but there are certain outcomes of this surveillance method. One of the best qualitative methods has been developed by the WHO that estimates the efficacy of antimalarial medications, which is specifically designed for nations to discover the knowledge on the treatment efficacy of malaria drugs. Collaborative networks, like East African Malaria Treatment Surveillance Network and the Amazon Malaria Drug Resistance Surveillance Network, are collaborated specially to conduct surveillance checks on the regional basis for antimalarial resistance surveillance (Hurwitz et al., 1981).

An international surveillance network has been formed that usually collects data globally on antimalarial drug efficacy and resistance called as the Global Antimalarial Resistance Network. It also provides information about the resistance and efficacy of the said drugs. While the transfer of data between native nations is vital for observing antimalarial drug resistance, sustainable funding of surveillance activities and networks remains difficult. The technological area for molecular testing is evolving peacefully. New techniques are being developed that are cheap and simply usable. These new techniques provide a chance of deep investigation from blood smears. Novel techniques added will be beneficial in those areas where malarial cases are high and these techniques will upgrade national and international malarial drug-resistance surveillance network (Organization, 2010b).

2.7.3 Surveillance of anti-HIV drug resistance

There are two main components of the WHO Human Immunodeficiency Virus Drug Resistance Surveillance and Monitoring Strategy including

1. Check for warning signs of anti-HIV drug resistance and examining the achievement of healthcare personnel to provide ART to immunodeficient patients.
2. HIV-resistant surveillance in adults that take ART, have taken ART, untreated HIV adults, and children under the age group of 18 (Organization, 2012).

HIV drug resistance needs regular monitoring and surveillance, as it is a dreadful emerging viral infection particularly in affected areas. It also needs monitoring in delivering of ART (Organization, 2012).

2.7.4 Surveillance of anti-influenza drug resistance

The early detection of anti-influenza resistance can be obtained by testing influenza variants from the influenza affected individuals who have not taken any antiviral drug. Genotyping and phenotype testing are the two methods by which we can monitor the resistance or reduced susceptibility toward anti-influenza drugs (Control and Prevention, 2006). There are various WHO collaboration centers and several countrywise influenza testing centers, that can examine the resistance and susceptibility of different influenza variants, and they provide a baseline data seasonally. In addition, antiviral resistance can complicate clinical treatments in a number of ways, including limiting combination treatment options (Organization).

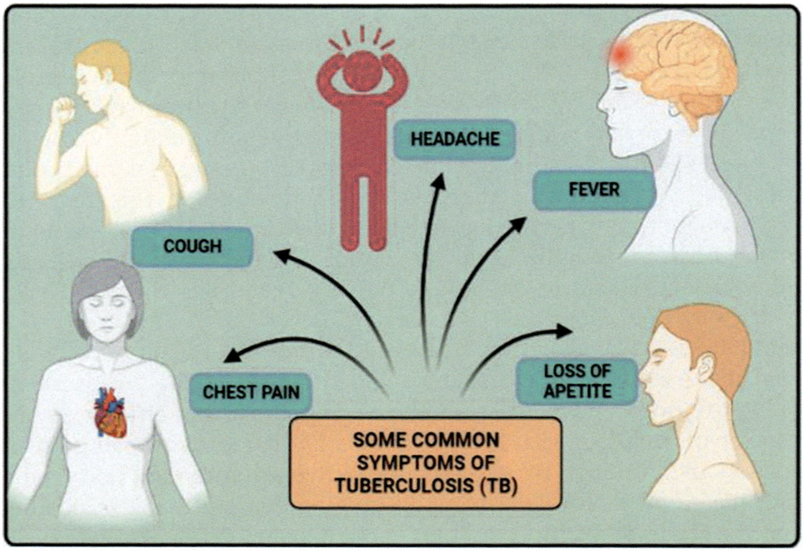

FIGURE 2.4 Diagrammatic illustration of some of the common symptoms of tuberculosis (TB). Tuberculosis represents a worldwide pandemic and is one of the most deadly infectious diseases caused by *Mycobacterium tuberculosis*, a Gram-positive, acid-fast eubacterium (Qadri et al., 2021).

2.8 Evolution of drug resistance in mycobacterium tuberclosis

A Gram-positive, acid-fast eubacterium "*Mycobacterium tuberculosis*" causes tuberculosis (TB), which is a global pandemic and a deadly infectious disease, generally spreading via cough aerosols and primarily impacting the lungs of an individual (Gygli et al., 2017; Cohen et al., 2019). Following the introduction of antibacterial drugs to treat TB, tolerance to anti-TB agents is established. A massive majority of patients treated with streptomycin developed resistance to the drug at the time of first randomized medical trial (RCT) in the 1940s. Inspite of the widespread initiation of drug combination procedures all over the world, the presence of drug tolerance was reported with growing frequency in a broad geographical region (Dormandy, 1999; Mir, 2015).

Multiple risk elements like improper utilization of numerous antibiotic agents, absence of proper treatment/improper treatment adherence, mutational defects, and lack of drug/diagnostic tool accessibility were reported to be connected with the rapid development of diverse TB resistant profiles (Georghiou et al., 2017; Manson et al., 2017). It has been reported that *M. tuberculosis* adapt multiple drug-resistant molecular processes/mechanisms which enable the pathogen to reduce the harmful impact of the majority of the antibiotic compounds (Qadri et al., 2021) (Fig. 2.4).

2.8.1 Global public health response to drug-resistant tuberclosis

As a module of the worldwide plan to bring an end to TB 2011–15, the new TB cases had been evaluated as too unsafe for MDR-TB by year 2015 (Organization, 2010a,c; Organization and Initiative, 2010). Largely, 5% of bacteriologically definite final TB cases had DST for at

TABLE 2.3 Grouping of some antituberculosis drugs by WHO (World Health Organization) (Caminero et al., 2010; WHO, 2010a,b,c).

S.No	Drug	Treatment
1.	Rifampicin, isoniazid, etc.	First-line treatment
2.	Kanamycin, levofloxacin, terizidone, etc.	Second-line treatment
3.	Delamanid, clarithromycin, etc.	Third-line treatment

least the first-line drugs rifampicin and isoniazid. Extended DST is required for all previously treated cases. In 2012, 9% of those earlier treated were diagnosed for MDR-TB. This graph has enlarged slightly in current years, but is still well below the 2012 target set by the Global Plan. In year 2012, 23% of multidrug-resistant TB cases had a DST for fluoroquinolones besides a second-line injectable drug to determine if it was multiple drug resistance. There is an imperative need for wider use of regular DSTs to upgrade the diagnosis of MDR-TB and XDR-TB. To increase DST, it is essential to increase laboratory capacity as well as quick diagnostic methods and superior diagnostic center reporting. Due to poor care and prevention, and due to a higher risk of developing drug resistance or spreading drug-resistant strains of TB bacteria XDR-TB is becoming more common around the globe. WHO reported 84,000 cases of MDR-TB in 2012, with more than half of the reported cases in South Africa, Russia, and India. The phenomenon of drug resistance was treated in all WHO regions in 2011 and 2012, except the Americas. While this is lower than the global plan target, it is still much higher than for populations receiving first-line drug treatment. From 2009 to 2012, the worldwide recruitment of MDR-TB individuals is enhanced by around 150%. In 2012, 92% of all reported cases started second-line treatment for MDR-TB. TB care requires a concerted effort on many fronts, especially in the highest burden countries, if we are to meet the targets set out in the global plan and achieve worldwide treatment accessibility. According to the WHO Report 2018 Global Tuberculosis Report, around 1.7 billion individuals worldwide are at risk of developing active TB during their lifetime and have risk of developing latent tuberculosis. WHO estimates that nearly 10 million people contracted TB in 2017, with about 1.3 million deaths. WHO calls MDR-TB a worldwide public health menace. As per the WHO Global Tuberculosis Report 2020, published on 14 October 2020, an estimated 1.4 million individuals died from TB-associated illness in the year 2019. Moreover, around 10 million estimated individuals developed TB that year, and about 14 million were treated in 2018–19 (WHO World Tuberculosis Report, 2020). Table 2.3 represents the grouping of some anti-tuberculosis drugs by WHO (World Health Organization) (Organization and Initiative, 2010; Caminero et al., 2010).

2.9 Future perspectives

We human beings are living in a bacterial world in which we are in no way capable of remaining forward of the mutation curve. So, we can say that every individual living on this Earth is under the threat of bacterial infection (Fig. 2.5). There are various classical

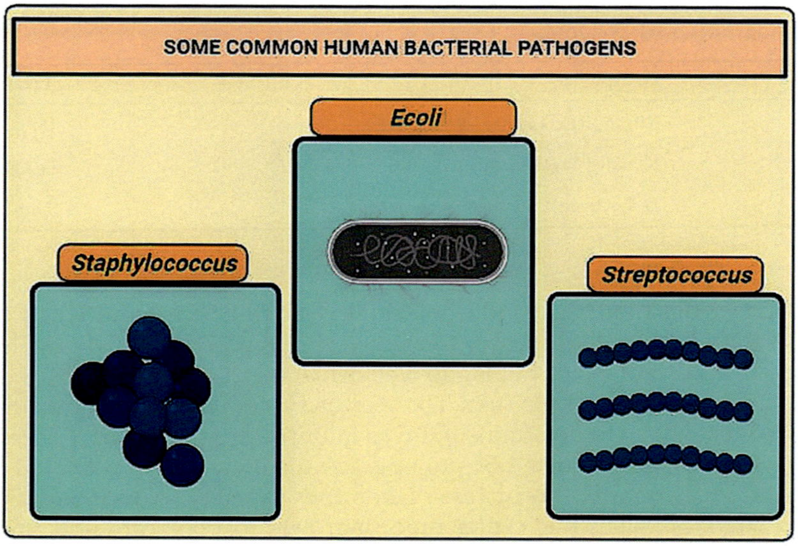

FIGURE 2.5 Diagrammatic illustration of some of the common bacterial pathogens infecting humans.

antibiotic drugs introduced more than 60 years ago that are used against such infections, but due to resistance mechanisms there is no such drug available in the market that can completely eradicate such infections. Human pathogenic bacteria evolved themselves to evade toxic effects of these antibiotics. In the current era there is no such available remedy that can break this resistance. There is urgent need of new antibacterial drugs and therapeutics that can combat such resistance that is increasing and evolving day by day, and should eliminate bacterial infections completely. We have gone through various resistant bacteria and resistance mechanisms that are responsible for emergence and increase in mortality rates worldwide. There is an urgent need of new broadspectrum agents that can kill these resistant strains and degrade these resistance mechanisms so that humanity can be saved from such dreadful infections. In the near future we will be able to withstand such resistant infections with the help of novel drugs that are under clinical trials or under studies.

References

Aarestrup, F.M., Wegener, H.C., Collignon, P.J., 2008. Resistance in bacteria of the food chain: epidemiology and control strategies. Expert Rev. Anti. Infect. Ther. 6, 733–750.

Acheson, D., Hohmann, E.L., 2001. Nontyphoidal salmonellosis. Clin. Infect. Dis. 32, 263–269.

Alam, M.M., Islam, M., Wahab, A., Billah, M.J., 2019. Antimicrobial resistance crisis and combating approaches. J. Med. 20, 38–45.

Alekshun, M.N., Levy, S.B., 2007. Molecular mechanisms of antibacterial multidrug resistance. Cell 128, 1037–1050.

Allocati, N., Masulli, M., Alexeyev, M.F., DI Ilio, C.J., 2013. *Escherichia coli* in Europe: an overview. Int. J. Environ. Res. Public Health 10, 6235–6254.

Almeida da Silva, P.E., Palomino, J.C., 2011. Molecular basis and mechanisms of drug resistance in *Mycobacterium tuberculosis*: classical and new drugs. J. Antimicrob. Chemother. 66, 1417–1430.

Arya, S.C., 2002. Global response to antimicrobial resistance. Bull. World Health Organ. 80, 420-420.

Ashkenazi, S., Levy, I., Kazaronovski, V., Samra, Z.J., 2003. Growing antimicrobial resistance of Shigella isolates. J. Antimicrob. Chemother. 51, 427–429.

Bharara, T., Bhalla, P., Rawat, D., Garg, V., Sardana, K., Chakravarti, A., 2015. Rising trend of antimicrobial resistance among *Neisseria gonorrhoeae* isolates and the emergence of *N. gonorrhoeae* isolate with decreased susceptibility to ceftriaxone. Indian J. Med. Microbiol. 33, 39–42.

Blair, J.M., Webber, M.A., Baylay, A.J., Ogbolu, D.O., Piddock, L.J., 2015. Molecular mechanisms of antibiotic resistance. Nat. Rev. Microbiol. 13, 42–51.

Blake, K.L., O'neill, A.J., 2013. Transposon library screening for identification of genetic loci participating in intrinsic susceptibility and acquired resistance to antistaphylococcal agents. J. Antimicrob. Chemother. 68, 12–16.

Caminero, J.A., Sotgiu, G., Zumla, A., Migliori, G.B., 2010. Best drug treatment for multidrug-resistant and extensively drug-resistant tuberculosis. Lancet Infect. Dis. 10, 621–629.

Cassini, A., Högberg, L.D., Plachouras, D., Quattrocchi, A., Hoxha, A., Simonsen, G.S., Colomb-Cotinat, M., Kretzschmar, M.E., Devleesschauwer, B., Cecchini, M.J., 2019. Attributable deaths and disability-adjusted life-years caused by infections with antibiotic-resistant bacteria in the EU and the European Economic Area in 2015: a population-level modelling analysis. Lancet Infect. Dis. 19, 56–66.

Chang, S., Hsieh, W.J., 1996. Current status of bacterial antibiotic resistance in Taiwan. J. Infect. Dis. 7, 83–88.

Chen, L.F., Anderson, D.J., Paterson, D.L., 2012. Overview of the epidemiology and the threat of *Klebsiella pneumoniae* carbapenemases (KPC) resistance. Infect. Drug Res. 5, 133.

Chuanchuen, R., Karkhoff-Schweizer, R.R., Schweizer, H.P., 2003. High-level triclosan resistance in *Pseudomonas aeruginosa* is solely a result of efflux. Am. J. Infect. Control. 31, 124–127.

Cohen, K.A., Manson, A.L., Desjardins, C.A., Abeel, T., Earl, A.M., 2019. Deciphering drug resistance in *Mycobacterium tuberculosis* using whole-genome sequencing: progress, promise, and challenges. Genome Med. 11, 1–18.

Centers for Disease Control and Prevention, U.S. Centers for Disease Control and Prevention, 2013. Antibiotic Resistance Threats in the United States. Centers for Disease Control and Prevention, US Department of Health and Human Services, Atlanta, GA.

Centers for Disease Control and Prevention, 2006. High levels of adamantane resistance among influenza A (H_3N_2) viruses and interim guidelines for use of antiviral agents-United States, 2005-06 influenza season. MMWR Morb. Mortal Wkly Rep. 55, 44–46.

Centers for Disease Control and Prevention, 2015. Antibiotic Resistance Threats in the United States. CDC website.

Cosgrove, S.E., Sakoulas, G., Perencevich, E.N., Schwaber, M.J., Karchmer, A.W., Carmeli, Y.J., 2003. Comparison of mortality associated with methicillin-resistant and methicillin-susceptible *Staphylococcus aureus* bacteremia: a meta-analysis. Clin. Infect. Dis. 36, 53–59.

Crofton, J., Mitchison, D.J., 1948. Streptomycin resistance in pulmonary tuberculosis. Br. Med. J. 2, 1009.

Dadgostar, P.J., 2019. Antimicrobial resistance: implications and costs. Infect Drug Resist. 12, 3903.

De Kraker, M.E., Stewardson, A.J., Harbarth, S.J., 2016. Will 10 million people die a year due to antimicrobial resistance by 2050? PLoS Med. 13, e1002184.

Dormandy, T., 1999. The White Death: A History of Tuberculosis.

Duncan, K., Barry Iii, C.E., 2004. Prospects for new antitubercular drugs. Curr. Opin. Microbiol. 7, 460–465.

Eber, M.R., Laxminarayan, R., Perencevich, E.N., Malani, A.J., 2010. Clinical and economic outcomes attributable to health care–associated sepsis and pneumonia. Arch. Intern. Med. 170, 347–353.

Ebi, K.L., Hess, J.J., Watkiss, P.J., 2017. Health risks and costs of climate variability and change. In: Disease Control Priorities, 7.

Enright, M.C., Day, N.P., Davies, C.E., Peacock, S.J., Spratt, B.G., 2000. Multilocus sequence typing for characterization of methicillin-resistant and methicillin-susceptible clones of *Staphylococcus aureus*. J. Clin. Microbiol. 38, 1008–1015.

Floyd, J.L., Smith, K.P., Kumar, S.H., Floyd, J.T., Varela, M.F., 2010. LmrS is a multidrug efflux pump of the major facilitator superfamily from *Staphylococcus aureus*. Antimicrob. Agents Chemother. 54, 5406–5412.

Georghiou, S.B., Seifert, M., Catanzaro, D.G., Garfein, R.S., Rodwell, T.C., 2017. Increased tuberculosis patient mortality associated with *Mycobacterium tuberculosis* mutations conferring resistance to second-line antituberculous drugs. J. Clin. Microbiol. 55, 1928–1937.

Guyomard-Rabenirina, S., Malespine, J., Ducat, C., Sadikalay, S., Falord, M., Harrois, D., Richard, V., Dozois, C., Breurec, S., Talarmin, A.J., 2016. Temporal trends and risks factors for antimicrobial resistant Enterobacteriaceae urinary isolates from outpatients in Guadeloupe. BMC Microbiol. 16, 1–8.

Gygli, S.M., Borrell, S., Trauner, A., Gagneux, S.J., 2017. Antimicrobial resistance in *Mycobacterium tuberculosis*: mechanistic and evolutionary perspectives. FEMS Microbiol. Rev. 41, 354–373.

Ho, M., Mcdonald, L., Lauderdale, T., Yeh, L., Chen, P., Shiau, Y.J., 1999. Surveillance of Antibiotic Resistance in Taiwan, 1998. J. Microbiol. Immunol. Infect. 32, 239–249.

Hsueh, P.-R., Liu, C.-Y., Luh, K.-T., 2002. Current status of antimicrobial resistance in Taiwan. Emerg. Infect. Dis. 8, 132.

Hu, F., Zhu, D., Wang, F., Wang, M.J., 2018. Current status and trends of antibacterial resistance in China. Clin. Infect. Dis. 67, S128–S134.

Hu, R.-M., Liao, S.-T., Huang, C.-C., Huang, Y.-W., Yang, T.-C., 2012. An inducible fusaric acid tripartite efflux pump contributes to the fusaric acid resistance in *Stenotrophomonas maltophilia*. PLoS One 7, e51053.

Hurwitz, E., Johnson, D., Campbell, C.J., 1981. Resistance of *Plasmodium falciparum* malaria to sulfadoxine-pyrimethamine ('Fansidar') in a refugee camp in Thailand. Lancet 317, 1068–1070.

Kariuki, S., Dougan, G.J., 2014. Antibacterial resistance in sub-Saharan Africa: an underestimated emergency. Ann. N. Y. Acad. Sci. 1323, 43.

Karp, B.E., Campbell, D., Chen, J.C., Folster, J.P., Friedman, C.R., 2018. Plasmid-mediated quinolone resistance in human non-typhoidal *Salmonella* infections: An emerging public health problem in the United States. Zoonoses Public Health 65, 838–849.

Kasimanickam, V., Kasimanickam, M., Kasimanickam, R.J., 2021. Antibiotics use in food animal production: escalation of antimicrobial resistance: where are we now in combating AMR? Med. Sci. 9, 14.

Kim, C., Mwangi, M., Chung, M., Milheirço, C., DE Lencastre, H., Tomasz, A.J., 2013. The mechanism of heterogeneous beta-lactam resistance in MRSA: key role of the stringent stress response. PLoS One 8, e82814.

Kim, J.-Y., Kim, S.-H., Jeon, S.-M., Park, M.-S., Rhie, H.-G., Lee, B.-K., 2008. Resistance to fluoroquinolones by the combination of target site mutations and enhanced expression of genes for efflux pumps in *Shigella flexneri* and *Shigella sonnei* strains isolated in Korea. J. Clin. Med. 14, 760–765.

Kitchel, B., Rasheed, J.K., Patel, J.B., Srinivasan, A., Navon-Venezia, S., Carmeli, Y., Brolund, A., Giske, C.G., 2009. Molecular epidemiology of KPC-producing *Klebsiella pneumoniae* isolates in the United States: clonal expansion of multilocus sequence type 258. Antimicrob. Agents Chemother. 53, 3365–3370.

Levy, S.B., Marshall, B.J., 2004. Antibacterial resistance worldwide: causes, challenges and responses. Nat. Med. 10, S122–S129.

Levy, S.J., 1982. Microbial resistance to antibiotics: an evolving and persistent problem. Lancet 320, 83–88.

Li, X., Robinson, S.M., Gupta, A., Saha, K., Jiang, Z., Moyano, D.F., Sahar, A., Riley, M.A., Rotello, V.M., 2014. Functional gold nanoparticles as potent antimicrobial agents against multi-drug-resistant bacteria. ACS Nano 8, 10682–10686.

Lupande-Mwenebitu, D., Baron, S.A., Nabti, L.Z., Lunguya-Metila, O., Lavigne, J.-P., Rolain, J.-M., Diene, S.M., 2020. Current status of resistance to antibiotics in Democratic Republic of Congo: a review. J. Glob. Antimicrob. Res.

Malinga, L.A., Stoltz, A., Van der Walt, M.J., 2016. Efflux pump mediated second-line tuberculosis drug resistance. Mycobacter. Dis. 6, 1–9.

Mandal, S.M., Roy, A., Ghosh, A.K., Hazra, T.K., Basak, A., Franco, O.L., 2014. Challenges and future prospects of antibiotic therapy: from peptides to phages utilization. Front. Pharmacol. 5, 105.

Manson, A.L., Cohen, K.A., Abeel, T., Desjardins, C.A., Armstrong, D.T., Barry, C.E., Brand, J., Chapman, S.B., Cho, S.-N., Gabrielian, A.J., 2017. Genomic analysis of globally diverse *Mycobacterium tuberculosis* strains provides insights into the emergence and spread of multidrug resistance. Nat. Genet. 49, 395–402.

Mcdermott, P.F., Zhao, S., Tate, H., 2018. Antimicrobial resistance in nontyphoidal *Salmonella*. Microb. Spectr. 6, 6.4. 16.

Mensa, L., Marco, F., Vila, J., Gascón, J., Ruiz, J., 2008. Quinolone resistance among *Shigella* spp. isolated from travellers returning from India. Clin. Microbiol. Infect. 14, 279–281.

Mir, M.A., 2013. Costimulation and Costimulatory Molecules.

Mir, M.A., 2015. Developing Costimulatory Molecules for Immunotherapy of Diseases. Academic Press.

Nathan, C., Cars, O.J., 2014. Antibiotic resistance—problems, progress, and prospects. N. Engl. J. Med. 371, 1761–1763.

Naylor, N.R., Atun, R., Zhu, N., Kulasabanathan, K., Silva, S., Chatterjee, A., Knight, G.M., Robotham, J.V., 2018. Estimating the burden of antimicrobial resistance: a systematic literature review. Antimicrob. Resist. Infect. Control 7, 1–17.

Netikul, T., Kiratisin, P.J., 2015. Genetic characterization of carbapenem-resistant enterobacteriaceae and the spread of carbapenem-resistant klebsiella pneumonia ST340 at a university hospital in Thailand. PLoS One 10, e0139116.
Nsanzabana, C., Djalle, D., Guérin, P.J., Ménard, D., González, I.J., 2018. Tools for surveillance of anti-malarial drug resistance: an assessment of the current landscape. Malar. J. 17, 1–16.
Ogawa, W., Onishi, M., Ni, R., Tsuchiya, T., Kuroda, T., 2012. Functional study of the novel multidrug efflux pump KexD from *Klebsiella pneumoniae*. Gene 498, 177–182.
Organization, World Health. Influenza A (H1N1) Virus Resistance to Oseltamivir, World Health Organization, Geneva.
Organization, W.H., 2009. Guidelines for Surveillance of Drug Resistance in Tuberculosis. World Health Organization.
Organization, W.H., 2010a. The Global Plan to Stop TB 2011-2015: Transforming the Fight towards Elimination of Tuberculosis.
Organization, W.H., 2010b. Global Report on Antimalarial Drug Efficacy and Drug Resistance: 2000-2010.
Organization, W.H., 2010c. Multidrug-resistant Tuberculosis (MDR-TB) Indicators: A Minimum Set of Indicators for the Programmatic Management of MDR-TB in National Tuberculosis Control Programmes. World Health Organization.
Organization, W.H., 2012. World Health Organization Global Strategy for the Surveillance and Monitoring of HIV Drug Resistance: An Update.
Organization, W.H., 2014a. Antimicrobial Resistance Global Report on Surveillance: 2014 Summary. World Health Organization.
Organization, W.H., 2014b. Antimicrobial Resistance: Global Report on Surveillance. World Health Organization.
Organization, W.H., 2019. Global Action Plan on Antimicrobial Resistance, 2015.
Organization, W.H., Initiative, S.T., 2010. Treatment of Tuberculosis: Guidelines. World Health Organization.
Patel, S.J., Saiman, L., 2010. Antibiotic resistance in neonatal intensive care unit pathogens: mechanisms, clinical impact, and prevention including antibiotic stewardship. Clin. Perinatol. 37, 547–563.
Paulsen, I.T., Chen, J., Nelson, K.E., Saier Jr., M.H., 2001. Comparative genomics of microbial drug efflux systems. J. Mol. Microbiol. Biotechnol. 3, 145–150.
Pérez, J., Contreras-Moreno, F.J., Marcos-Torres, F.J., Moraleda-Muñoz, A., Muñoz-Dorado, J., 2020. The antibiotic crisis: how bacterial predators can help. Comput. Struct. Biotechnol. J.
Prestinaci, F., Pezzotti, P., Pantosti, A., 2015. Antimicrobial resistance: a global multifaceted phenomenon. Pathog. Glob. Health 109, 309–318.
Qadri, H., Haseeb, A., Mir, M., 2021. Novel strategies to combat the emerging drug resistance in human pathogenic microbes. Curr. Drug Targets 22, 1–13.
Ramachandran, R., Nalini, S., Chandrasekar, V., Dave, P., Sanghvi, A., Wares, F., Paramasivan, C., Narayanan, P., Sahu, S., Parmar, M., 2009. Surveillance of drug-resistant tuberculosis in the state of Gujarat, India. Drug Resist. Surveill. 13, 1154–1160.
Randall, C.P., Mariner, K.R., Chopra, I., O'neill, A.J., 2013. The target of daptomycin is absent from *Escherichia coli* and other gram-negative pathogens. Antimicrob. Agents Chemother. 57, 637–639.
Reinert, R., 2009. The antimicrobial resistance profile of *Streptococcus pneumoniae*. J. Clin. Microbiol. Infect. 15, 7–11.
Reygaert, W.C., 2018. An overview of the antimicrobial resistance mechanisms of bacteria. AIMS Microbiol. 4, 482.
Roberts, R., Hota, B., Ahmad, I., Scott, R., 2009. Hospital and societal costs of antimicrobial-resistant infections in a Chicago teaching hospital: implications for antibiotic stewardship. J. Clin. Infect. Dis. 49, 1175–1184.
Rossi, E.D., Aínsa, J.A., Riccardi, G., 2006. Role of mycobacterial efflux transporters in drug resistance: an unresolved question. FEMS Microbiol. Rev. 30, 36–52.
Shankar, P.R., Balasubramanium, R., 2014. Antimicrobial resistance: global report on surveillance 2014. Australasian Med. J. 7, 237.
Sharma, S.K., Mohan, A.J.C., 2006. Multidrug-resistant tuberculosis: a menace that threatens to destabilize tuberculosis control. Chest 130, 261–272.
Sheikh, B.A., Bhat, B.A., Mehraj, U., Mir, W., Hamadani, S., Mir, M.A., 2021a. Development of new therapeutics to meet the current challenge of drug resistant tuberculosis. Curr. Pharmaceut. Biotechnol. 22, 480–500.
Sheikh, B.A., Bhat, B.A., Mehraj, U., Mir, W., Hamadani, S., Mir, M.A., 2021b. Development of new therapeutics to meet the current challenge of drug resistant tuberculosis. Curr. Pharm. Biotechnol. 22, 480–500.

Smith, T., Wolff, K.A., Nguyen, L.J.P.O.M.T., Organism, I.I.W.T.H., 2012. Molecular biology of drug resistance in Mycobacterium tuberculosis. Curr. Top Microbiol. Immunol. 53–80.

Stefani, S., Chung, D.R., Lindsay, J.A., Friedrich, A.W., Kearns, A.M., Westh, H., Mackenzie, F.M., 2012. Meticillin-resistant *Staphylococcus aureus* (MRSA): global epidemiology and harmonisation of typing methods. Int. J. Antimicrob. Agents 39, 273–282.

TAIT-Kamradt, A., Davies, T., Appelbaum, P., Depardieu, F., Courvalin, P., Petitpas, J., Wondrack, L., Walker, A., Jacobs, M., Sutcliffe, J., 2000. Two new mechanisms of macrolide resistance in clinical strains of *Streptococcus pneumoniae* from Eastern Europe and North America. Antimicrob. Agents Chemother. 44, 3395–3401.

Theuretzbacher, U., 2013. Global antibacterial resistance: the never-ending story. J. Glob. Antimicrob. Resist. 1, 63–69.

Tomasz, A., 1997. Antibiotic resistance in Streptococcus pneumoniae. Clin. Infect. Dis. 24, S85–S88.

Viveiros, M., Martins, M., Rodrigues, L., Machado, D., Couto, I., Ainsa, J., Amaral, L., 2012. Inhibitors of mycobacterial efflux pumps as potential boosters for anti-tubercular drugs. Expert Rev. Anti. Infect. Ther. 10, 983–998.

Walsh, C., 2003. Antibiotics: Actions, Origins, Resistance. American Society for Microbiology (ASM).

Wang, C.-Y., Chen, Y.-H., Fang, C., Zhou, M.-M., Xu, H.-M., Jing, C.-M., Deng, H.-L., Cai, H.-J., Jia, K., Han, S.-Z., 2019. Antibiotic resistance profiles and multidrug resistance patterns of *Streptococcus pneumoniae* in pediatrics: a multicenter retrospective study in mainland China. Medicine 98.

Williamson, D.A., Fairley, C.K., Howden, B.P., Chen, M.Y., Stevens, K., DE Petra, V., Denham, I., Chow, E.P., 2019. Trends and risk factors for antimicrobial-resistant Neisseria gonorrhoeae, Melbourne, Australia, 2007 to 2018. Antimicrob. Agents Chemother. 63 e01221-19.

Woolhouse, M., Farrar, J., 2014. Policy: an intergovernmental panel on antimicrobial resistance. Nature 509, 555.

Zhu, L., Lin, J., Ma, J., Cronan, J.E., Wang, H., 2010. Triclosan resistance of *Pseudomonas aeruginosa* PAO1 is due to FabV, a triclosan-resistant enoyl-acyl carrier protein reductase. Antimicrob. Agents Chemother. 54, 689–698.

CHAPTER

3

Evolution of antimicrobial drug resistance in human pathogenic fungi

Manzoor Ahmad Mir

Department of Bioresources, School of Biological Sciences, University of Kashmir, Srinagar, Jammu and Kashmir, India

3.1 Introduction

The fungal kingdom confines a wide diversity of taxa having an estimated 1.5 million species; with only around 5% of the species classified formally (Vandeputte et al., 2012). Although mostly nonpathogenic, many fungal species are pathogenic to plants, animals, and humans causing a wide range of diseases. The prevalence of fungal diseases especially in immunocompromised hosts undergoing corticosteroids/antibiotics therapy, diabetics, alcoholics, persons suffering from conditions like cancer, HIV/AIDS, etc., often accompanying higher rates of death and disease rates is significantly increasing (Sardi et al., 2013; Qadri et al., 2021a). Invasive fungal infections represent a major public health issue and often pose a great hurdle in the treatment and control procedures (Firacative, 2020). The species peculiar to *Candida*, *Cryptococcus*, and *Aspergillus* genera affect millions of individuals around the globe and have been reported to be associated with >90% fungal-related deaths (Schmiedel and Zimmerli, 2016). *Candidiasis*, *Aspergillosis*, *Cryptococcosis*, *coccidioidomycosis*, *histoplasmosis*, etc., are the most common and acute invasive human fungal infections (Vandeputte et al., 2012). Among the *Candida* species, *Candida albicans* is the principal opportunistic fungal agent, along with other pathogenic members of *Candida* viz., *Candida glabrata*, *Candida tropicalis*, etc. (MacCallum, 2012; Spampinato and Leonardi, 2013). *Candida auris* represents an evolving MDR fungal pathogen currently propagating rapidly across the world (Ademe and Girma, 2020). On the other hand, in developing countries, millions of individuals are infected by the disease cryptococcosis (caused by *Cryptococcus gattii* and *Cryptococcus neoformans*) every year with an associated death rate of around 70% (Brown et al., 2012). *C. neoformans* represent a major

Human Pathogenic Microbes
https://doi.org/10.1016/B978-0-323-96127-1.00009-7

opportunistic fungal pathogen in immunocompromised hosts, like AIDS patients, mostly in the individuals infected with human immunodeficiency virus (HIV), those undergoing immunosuppressive treatment for cancer, organ transplantation, and other related severe medical conditions (Mitchell and Perfect, 1995). Moreover, in humans especially immunocompromised individuals, Aspergillosis infection constitutes a serious health problem (Chowdhary et al., 2014). Thousands of invasive aspergillosis disease occurrences are noted yearly with the highest death rates observed if the disease is left untreated (Brown et al., 2012). Among the *Aspergillus* species, *A. fumigatus* represents an extensive member having association with the growing occurrence of invasive aspergillosis (IA) in immunocompromised hosts (Dagenais and Keller, 2009). All of these opportunistic fungal pathogens pose a great risk to life and health, and the incidence of growing drug resistance associated with all these pathogenic species is being continually documented and reported (Robbins et al., 2017). To acquire resistance and to tackle the effect of different antifungals, these fungal pathogens develop multiple resistance mechanisms and strategies (Qadri et al., 2021a). The common drug-resistance mechanisms evolved by these pathogenic microorganisms to thwart the fungicidal/fungistatic effects of different types of antifungal drugs generally include decreased intracellular drug accumulation, reduced affinity for the drug target, modifications of the drug target, and other related drug-resistance mechanisms (Vandeputte et al., 2012). The antifungals viz. polyenes, azoles, allylamines, flucytosine, and echinocandins represent some of the most commonly used drug classes (Table 3.1).

Due to insufficient availability of antifungal drugs, there are restricted treatment options available (Groll et al., 1998; Kathiravan et al., 2012). Currently, many novel antifungal drugs are reported to be undergoing preclinical/clinical studies which might help in the process of controlling the increasing emergence of antifungal drug resistance (Wiederhold et al., 2017). Many plant-based natural compounds (Table 3.2) are also being employed as effective antifungal agents. Given the association and implications of the emerging fungal drug resistance and the globally increasing infectious diseases, a complete knowledge and recognition of the different mechanistic processes governing the phenomenon of drug resistance for the proper treatment and control are required (Robbins et al., 2017; Qadri et al., 2021a).

TABLE 3.1 Some of the frequently applied antifungal classes and their respective targets (Qadri et al., 2021a,b).

S. No	Antifungal class	Example	Target
1.	Azoles	Fluconazole	Ergosterol biosynthesis/lanosterol demethylase
2.	Echinocandins	Caspofungin	Cell wall/1,3−β−glucan synthesis
3.	Polyenes	Amphotericin B	Ergosterol in the cell membrane
4.	Allylamines	Terbinafine	Ergosterol biosynthesis/squalene epoxidase
5.	Pyrimidine analogs	5-Flurocytosine	DNA/RNA synthesis

TABLE 3.2 Different plant-based bioactive agents with active antifungal properties (de Andrade et al., 2019).

S. No	Source	Bioactive agent involved	Effective against
1.	*Syzyqium cordatum*	Flavan-3-ol (flavonoid)	Different *Candida* species
2.	*Scutellaria baicalensis*	Biacalene, gallotannin (flavonoids)	Different *Candida* species
3.	*Terminalia catappa*	Punicalagin, punicalin (tannins)	Different *Candida* species
4.	*Punica granatum*	Gallagic acid, ellagic acid, punicalagin, punicalins (tannins)	*A. fumigatus, C. neoformans, C. albicans*

3.2 Molecular mechanisms governing antimicrobial drug resistance phenomenon in the human pathogenic fungi

The recurrent usage of varied antifungal classes led to the evolution of strong drug resistance in fungal pathogenic species. Different adaptive antifungal drug-resistance mechanisms like transport alteration, drug target alteration, etc., have been reported (Revie et al., 2018). Here we are presenting some of the important drug-resistance mechanisms (Figs. 3.1 and 3.2) associated with the chief category of drugs (azoles) employed as primary treatment against different fungal pathogenic species viz. *Candida, Cryptococcus,* and *Aspergillus.*

3.2.1 Transport alteration: drug efflux via overexpression/activation of membrane transporters

One of the most desirable drug-resistance mechanisms enabling the phenomenon of multidrug resistance evolved by the pathogenic fungi involves upregulation of various membrane transporters (efflux proteins/pumps), having ability to identify a wide range of chemical

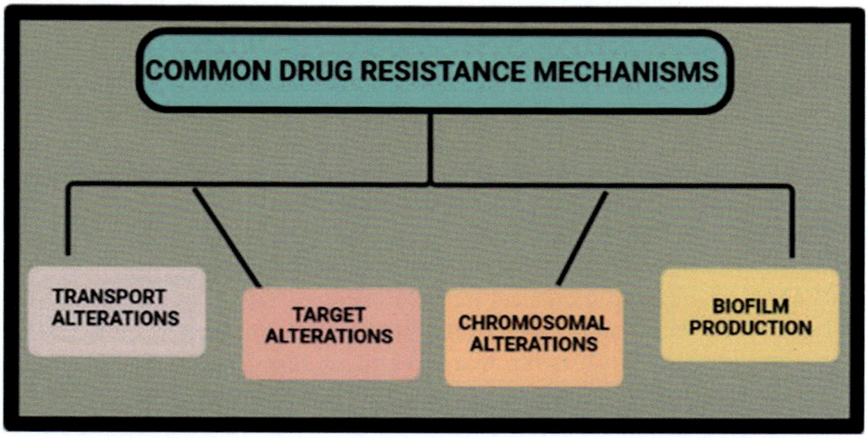

FIGURE 3.1 Common drug-resistance mechanisms adopted by different fungal pathogens to tackle the effect of different antifungals have been illustrated.

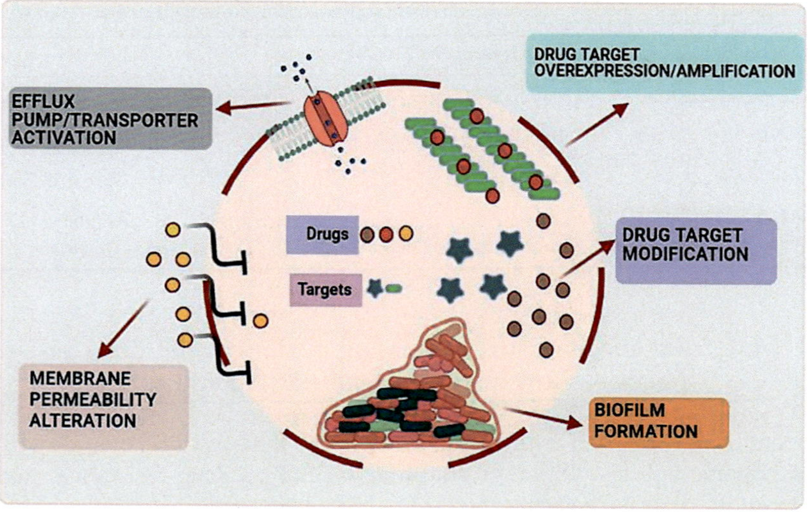

FIGURE 3.2 General drug-resistance mechanisms adopted by different fungal pathogens are represented viz. modification of the drug target, active efflux pump activation of different efflux pumps/transporters, biofilm formation, and other associated resistance mechanisms to survive the effect of antifungals.

compounds. Overall, two basic types of membrane transporters drug efflux transport systems, viz. the ATP-binding cassette (ABC) superfamily and the major facilitator superfamily (MFS) have been identified in fungi as a means of regulation of azole drug-resistance phenomenon. In general, ABC transporters represent the primary active efflux transporters mainly consisting of two transmembrane spanning domains (TMDs) and two cytoplasmic nucleotide-binding domains (NBDs) catalyzing the process of ATP hydrolysis (Cowen et al., 2015). Secondly, major facilitator transporters representing the secondary active efflux proteins use a proton gradient/electrochemical force generated across the plasma membrane as a source of energy to fuel and mediate the process of drug efflux. These transporters do not have the nucleotide-binding domains (NBDs) and consist mainly of 12 (DHA1) or 14 (DHA2) transmembrane segments (Prasad et al., 2014; Shah et al., 2015). The *C. albicans* genome sequence harbors 28 ABC transporters and also 95 MFS transporters distributed across 17 families (Gaur et al., 2008; Prasad and Goffeau, 2012). *C. glabrata* has been found to harbor around two-thirds of the total of ABC efflux proteins found in *C. albicans* (Cowen et al., 2015). Multiple ABC and MFS transporter proteins have been identified in *A. fumigatus* and *C. neoformans* (Kovalchuk and Driessen, 2010; Lamping et al., 2010).

In the pathogenic fungi, the increased expression of the ABC and MFS membrane transporters has been reported having correlation with the phenomenon of azole drug resistance. In *C. albicans*, the PDR transporters represent the most significant transporters related with azole drug resistance viz. *CDR1* (*Candida* drug resistance) and *CDR2*, besides other types of PDR transporters (*CDR3*, *CDR4*, *CDR11*, and *SNQ2*) which are not yet identified for having such antifungal-resistance association (Braun et al., 2005; Sanglard et al., 2009). The overexpression of both *CDR1* and *CDR2* genes regulates azole drug resistance via active outflow of drugs and decreased drug accumulation (Sanglard et al., 2009). The upregulation of the genes *CaCDR1*, *CaCDR2*, and *CaMDR1* has been reported in *C. albican*'s azole-resistant oral, systemic, and vaginal clinical isolates (White et al., 2002; Bhattacharya et al., 2016). Moreover, ABC transporters have been reported for their contribution toward the development of resistance against azole drugs resistance in the case of other pathogenic species including *CgCdr1*, *CgCdr2*, and *CgSnq2* in *C. glabrata*, *ABC1* in *C. krusei*, and *Afr1* in *C. neoformans* (Katiyar and Edlind, 2001; Sanguinetti et al., 2006; Coleman and Mylonakis, 2009). In the case of *A. fumigatus*, the relation between resistance against azole drugs and the phenomenon of the upregulation of drug transporters is not known. The ABC transporter *atrF* is upregulated in an azole-resistant clinical isolate; but has not strongly been assigned to have a possible role in the process of resistance (Slaven et al., 2002). In various itraconazole-resistant laboratory-derived strains of *A. fumigatus*, *AfuMDR4* is strongly upregulated (Nascimento et al., 2003). The transcriptional profiling analysis showed the expression of transporter genes induced concerning the drug voriconazole (da Silva et al., 2006), including five ABC transporter proteins/genes (*abcA−E*) and three MFS transporter proteins/genes (*mfsA−C*). It has been reported that in *A. fumigatus*, *abcA* gene (*cdr1B*) represents the single known transporter gene having an absolute part in the process of resistance against the drugs azoles (Fraczek et al., 2013).

Among the 95 MFS transporters belonging to 17 families in *C. albicans* (Gaur et al., 2008), only *MDR1* shows involvement in the resistance against azoles in case of *C. albicans* and *Candida dubliniensis*. The overexpression of *MDR1* gene has been found to contribute to the process of resistance against azoles via an active drug efflux system. In *S. cerevisiae*, heterologous expression of *MDR1* has been reported to develop resistance to the azole drug fluconazole (Sanglard et al., 1995; Sanglard et al., 1996; Lamping et al., 2007). The introduction of FLU1 (for fluconazole resistance) from *C. albicans* in *S. cerevisiae* has been found to be associated with fluconazole-specific drug efflux (Calabrese et al., 2000). In *A. fumigatus*, *AfuMDR3*, an MFS transporter has been identified to be regulated in the case of itraconazole-resistant mutant strains (Nascimento et al., 2003).

Furthermore, in all these resistant fungal pathogenic species, the overexpression of ABC/MFS transporters is regulated by distinct types of regulators. In *C. albicans*, *CDR1* and *CDR2* genes are mediated by a zinc cluster finger transcriptional regulator *TAC1*, while *MDR1* gene is known to be regulated by *MRR1* (Coste et al., 2004). Table 3.3 lists some of the significant MDR transporters/efflux pump proteins found in different fungal pathogens.

3.2.2 Target alteration: altered ergosterol biosynthesis

The process of altered ergosterol biosynthesis generally occurs via mutation and/or overexpression of different genes belonging to ergosterol biosynthetic pathway (Bhattacharya et al., 2020). The pathway is the main target of the drug azoles. Azole drugs specifically target cytochrome P450-dependent enzyme lanosterol 14α-demethylase which is also called Erg11 in yeasts. The enzyme catalyzes the oxidative elimination of lanosterol's 14α-methyl group. The azole drug binds to the ferric iron moiety of the heme-binding site rendering lanosterol—the enzyme's natural substrate—blocked, thereby damaging the entire ergosterol mechanism (Odds et al., 2003; Mir et al., 2020). In the genus *Candida*, the amino acid substitutions in the

TABLE 3.3 Important multidrug-resistant drug transporters/efflux pumps proteins present in different fungal pathogens.

S. No	Pathogenic fungal organism	Major efflux pump/transporter/protein	Type of transporter involved	Reference(s)
1.	*C. albicans*	*Cdr1p, Cdr2p*	ABC-transporter	Bhattacharya et al. (2020)
2.	*C. albicans*	*Mdr1p*	MFS-transporter	Bhattacharya et al. (2020)
3.	*C. auris*	*Cdr1p*	ABC-transporter	Bhattacharya et al. (2020)
4.	*C. glabrata*	*CgCdr1p, CgPdh1p, CgSnq2p*	ABC-transporter	Bhattacharya et al. (2020)
5.	*C. glabrata*	*CgFlr1p, CgQdr2p*	MFS-transporter	Bhattacharya et al. (2020)
6.	*C. neoformans*	*Afr1*	ABC-transporter	Coleman and Mylonakis (2009)
7.	*A. fumigatus*	*AfuMDR3*	MFS-transporter	Nascimento et al. (2003)

drug target cause inhibition of the drug-binding representing a frequent drug-resistance mechanism against azoles. Interestingly, about 140 substitutions were identified in resistant strains, mostly reported to possess a cumulative impact (Morio et al., 2010). In *C. albicans,* R467K and G464S represent the two major frequent alterations located in the proximity of the heme-binding site (Casalinuovo et al., 2004; Morio et al., 2010). A study reported that *ERG11* mutations in *C. albicans* do not display a major function than enhanced drug outflow (Perea et al., 2001). In *C. albicans,* several point mutations in the *ERG11* gene correlated with azole drug resistance have been reported in resistant clinical isolates.

Although, *ERG11* point mutations have a major effect in haploid *Candida* species like *C. glabrata* as compared in diploid *Candida* species like *C. albicans,* the system of enhanced drug efflux represents a frequent drug-resistance mechanism in *C. glabrata* like species also (Borst et al., 2005; Sanguinetti et al., 2005). In *C. neoformans,* only a few *ERG11* point mutations viz. Y145F and G484S have been described in the case of the resistant strains (Sionov et al., 2010; Bhattacharya et al., 2020). Moreover, alterations in the target site, in *A. fumigatus,* are the most frequent type of drug-resistance mechanism, having about 30 *cyp51A* reported mutations (Howard and Arendrup, 2011).

On increasing the amount/number of the drug targets, the efficacious amount of the drug is also required to be raised to saturate all the available drug target molecules, eventually resulting in enhanced drug resistance (Sanglard, 2016). *ERG11* overexpression shows correlation with the phenomenon of azole resistance in different pathogenic fungal species. Multiple azole-resistant clinical isolates of *C. albicans* display enhanced *CaERG11* expression (White et al., 2002; Flowers et al., 2012; Bhattacharya et al., 2016). Similar to other drug-resistance transcriptional regulators, GOF mutations in *UPC2* show involvement in the process of upregulation of different genes, including *ERG11* (Dunkel et al., 2008). *ERG11* gene overexpression is also reported in azole-resistant isolates of *C. glabrata, C. parapsilosis, C. tropicalis,* and *C. krusei* (Redding et al., 2002; Vandeputte et al., 2005; Pam et al., 2012; Jiang et al., 2013). Overexpression of *ERG11* showing enhanced resistance to azoles is identified in *C. auris* also (Bhattacharya et al., 2019). Moreover, in the azole-resistant *A. fumigatus* isolates, the upregulation of *Cyp51A* has also been reported, which is found to be regulated by the duplication of (34- and 42-bp) elements (*trans*-regulation) located in the *Cyp51A* promoter. The duplication is found to be connected with some specific *Cyp51A* mutations (L98H, Y121F/T289A) (Snelders et al., 2011). In the azole-resistant isolates of *A. fumigatus*, such combined mutations are found in abundance probably arising from the continuous application of azole drugs (Vermeulen et al., 2013).

Besides, in *C. albicans*, the deletion/disruption of *ERG3* gene results in enhanced resistance against azoles (Sanglard et al., 2003). In *C. glabrata* isolates, Q139A mutation in Erg3p is found correlated with azole drug resistance (Yoo et al., 2010). In *Aspergillus species*, Erg3 alterations have not yet been found related to drug resistance (Cowen et al., 2015). Moreover, in *C. albicans*, a remarkable resistance against azoles has been witnessed in the heterozygous *ERG6* gene deletion (Xu et al., 2007). In *C. glabrata ERG6* gene contributes to azole resistance because of several base pair variations causing missense mutations (Vandeputte et al., 2007).

3.2.3 Genomic alterations/genomic plasticity

Genomic alterations/variations leading to enhanced expression of drug exporters/transporters offer an alternative way to increase the efflux of drugs. Pathogenic fungal organisms possess an enormous ability to undergo the phenomenon of genomic plasticity concerning multiple environmental stresses (Revie et al., 2018). Several genetic/genomic alterations like loss of heterozygosity (LOH) and aneuploidy were reported to be correlated with resistance to the frequently employed antifungals i.e., azoles. In the clinical isolates of *C. albicans*, LOH was identified in the regions consisting of azole-resistance determinants (*CaTAC1*, *CaERG11*, and *CaMRR1*) connected with an enhanced drug-resistance phenomenon (Ford et al., 2015). It has been found that the clinical isolates of *C. albicans* may also include segmental aneuploidy, where two copies of the left arm of chromosome 5 containing *CaERG11* and *CaTAC1* form an isochromosome corresponding to resistance against azole drugs (Selmecki et al., 2006). A study reported that the sequential analysis of the isolates of *C. albicans* exhibiting resistance in patients showed that mutations in these genes often develop in the heterozygous state and become homozygous by the loss of heterozygosity (Coste et al., 2007; Selmecki et al., 2010). Of late, additional aneuploid lineages of *C. albicans* containing more copy numbers of chromosome 3 and chromosome 6 have been reported to possess decreased sensitivity to azole drugs (Hirakawa et al., 2017). Even though the molecular basis stands evasive, all these study reports point to the concept that in *C. albicans*, genomic diversity aids in the process of stress adaptation and survival (Revie et al., 2018). Duplication in *CDR1* and *ERG11* genes has also been recently reported in generationally aging *C. auris* strains associated with enhanced resilience to the commonly used azole, i.e., fluconazole (Bhattacharya et al., 2019). Chromosomal variations were described in *C. glabrata* and *C. neoformans* in association with the process of drug resistance. In *C. glabrata* resistance to azole drugs is evolved by enhancing the copy number of *ERG11* gene (Marichal et al., 1997). Furthermore, in the azole-resistant isolates of *C. glabrata*, the establishment of segmental aneuploidies and new chromosomal configurations has been described (Poláková et al., 2009). Moreover, in *C. neoformans*, resistance to azole drugs shows correlation with defined chromosomal variations, chiefly disomies of chromosome 1 and 4 and that chromosome 1 carry the azole target gene *ERG11* and ABC transporter gene

AFR1, as the two azole-resistance determinants, discovered as an adaptive system promoting azole drug resistance (Sionov et al., 2010). Therefore, the process of genomic plasticity happens to be preserved as a major adaptive system (Revie et al., 2018).

3.2.4 Development/formation of complex structures; biofilms

In the late years of the 1990s and early years of 2000s, the unique capability of the fungal organisms to evade host surfaces and develop biofilms was originally illustrated for *Candida albicans* and *Saccharomyces cerevisiae* (Hawser and Douglas, 1994; Reynolds and Fink, 2001). However, the increasing consciousness regarding the significance of biofilms in fungal organisms could be very well understood through the growing number of studies on biofilm development by other non-albican species of *Candida* (Bizerra et al., 2008; Silva et al., 2011), along with other fungal pathogenic organisms like *Cryptococcus neoformans* (Martinez and Casadevall, 2007). Moreover, the capability of biofilm formation has also been reported in various filamentous fungi, like *Aspergillus fumigatus* (Mowat et al., 2009). Biofilms show intrinsic resistance to the commonly used antifungal agents, the occupied host immune system, and additional environmental factors, rendering these biofilm-related infections an important clinical issue (Gulati and Nobile, 2016). Numerous mechanisms/processes are noted to be associated with the enhanced antifungal-resistance phenomenon of the fungal biofilms (Fig. 3.3, Vila and Rozental, 2016).

3.2.4.1 Candida *biofilms*

The species belonging to the genus *Candida* are commonly described as the causal agent of the disease Candidemia, hospital acquired pneumonia, and infections associated with urinary tract and, these infections are consistent with the application of medical instruments and the formation of biofilms on their surfaces (Vila and Rozental, 2016). Clinical *Candida* biofilms

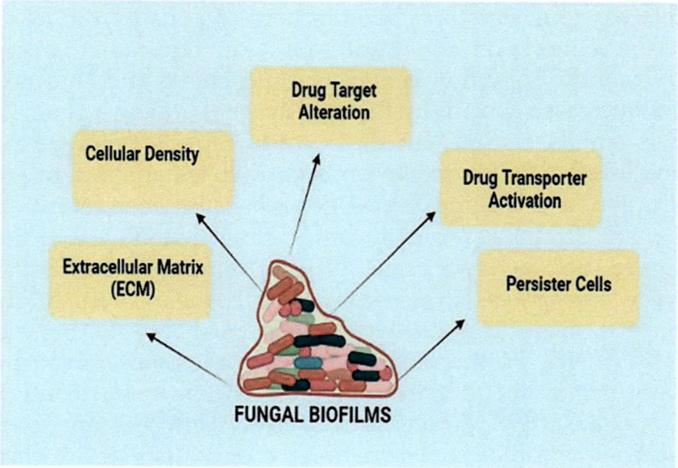

FIGURE 3.3 Flowchart representing various mechanisms associated with the enhanced antifungal resistance phenomenon of the fungal biofilms (Vila and Rozental, 2016).

develop on various medical equipment's like central venous catheters, prosthetic valves, urinary catheters, oral implements etc. (Kojic and Darouiche, 2004; Uppuluri et al., 2009). The most frequently involved medical instrument is the central venous catheter (CVC), generally employed for the dispensation of fluids, nutrients, and drugs (Seneviratne et al., 2008). Biofilm development is promoted by both vaginal/oral mucosal surfaces (Dongari-Bagtzoglou, 2008; Harriott et al., 2010). *C. albicans* represents an extensive *Candida* species involved in biofilm formation along with other non-*albicans* species developing clinically pertinent biofilms like *C. tropicalis, C. parapsilosis*, and *C. glabrata* as well (Kuhn et al., 2002; Shin et al., 2002; Kojic and Darouiche, 2004; Tumbarello et al., 2007; Bizerra et al., 2008). Furthermore, biofilm formation trait is also described for the emerging fungal pathogen *Candida auris* (Sherry et al., 2017). The important virulence trait i.e., biofilm formation in *Candida* generally includes the adhesion of the organism to the host/substrate, cell proliferation to develop a complete fungal community, and ECM (extracellular matrix) production (Chandra et al., 2001; Ramage et al., 2005; Fox and Nobile, 2012; Qadri et al., 2021b). Biofilm generation in *C. albicans* usually includes hyphae production; however, the extent of filamentation differs among different strains and niches. The non-*albicans* strains which lack the filamentation ability form biofilms made of layers of cells rooted in ECM (Silva et al., 2009). In general, the process of development of biofilms in *Candida* involves the following stages (Fig. 3.4) viz: **(1) Initial stage:** it is the adherence stage, where the fungal pathogen adheres to the surface within 1–3 h. **(2) Intermediate stage:** it is the stage concerning the formation of biofilm, commencing within 11–14 h. **(3) Maturation stage:** it is the stage wherein the ECM embeds the entire surface-adhered cells in a 3-dimensional structure within 20– 48 h. **(4) Dispersion stage:** it is the

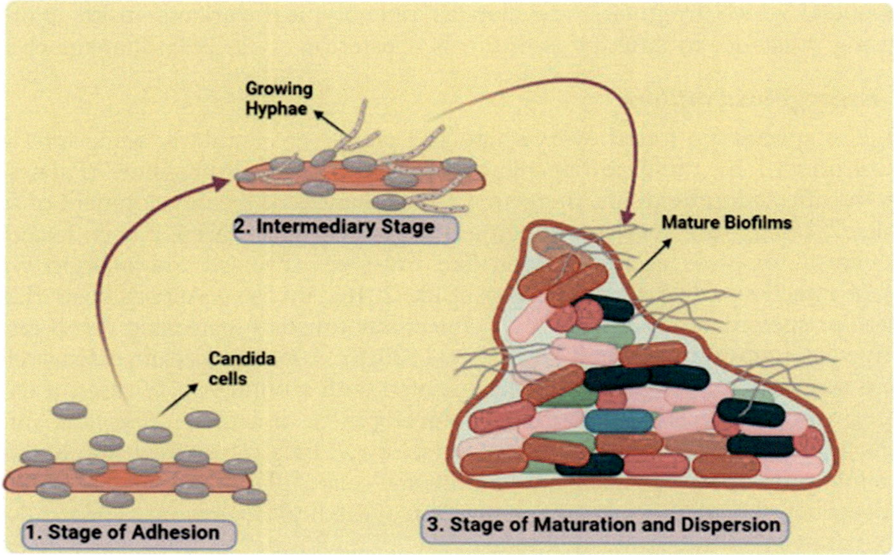

FIGURE 3.4 Different stages of the development of biofilms in *Candida* species. The three involved stages include (1) Adherence stage (2) Intermediary stage (3) Maturation and dispersion stage. The *Candida* cells undergo all these stages for the formation of biofilms as one of the important drug-resistance mechanisms.

stage wherein the majority of the superficial cells after leaving the biofilms begin to occupy the entire regions enclosing the surface, after 20 h (Chandra et al., 2001).

Multiple study reports showed that in *C. albicans*, the biofilm matrix on maturation contains a wide range of macromolecules, like protein (55%), carbohydrate (25%), lipid (15%), and nucleic acid (DNA; 5%) (Baillie and Douglas, 2000; Ramage et al., 2001; Zarnowski et al., 2014; Martinez and Casadevall, 2015). Biofilms in *Candida* have been found to possess intrinsic resistance to azole drugs, and that the involved resistance processes possess multiple factors, including activation of various efflux pumps and drug sequestration inside the large matrix system (Kumamoto, 2002; Mukherjee et al., 2003; Chandra et al., 2005; Fanning and Mitchell, 2012). It has been reported that drug sequestration inside the ECM represents major determinant of the MDR phenotype (Nett et al., 2010). In various *Candida* species, like *C. albicans*, *C. dubliniensis*, *C. glabrata*, *C. tropicalis*, and *C. parapsilosis*, the phenomenon of matrix production is strongly mediated and represents a key drug-resistance factor (Silva et al., 2009). In *C. albicans*, a thorough study concerning the vast and intricate transcriptional network governing the formation of biofilms has been illustrated (Nobile et al., 2012).

This network is believed to consist of six major transcriptional regulators viz.: *Efg1*, *Tec1*, *Bcr1*, *Ndt80*, *Brg1*, and *Rob1*, with each of them essential for the proper biofilm formation, both in vivo as well as in vitro (Nobile et al., 2012). Interestingly, all these six major master regulators are directly bound to the promoters of, and likely modulate the expression of, about 1000 target genes, and that few of them were additional transcriptional regulators, creating a system of the extensive, compound, and knotted genetic network entangled in the process of development of biofilms (Gulati and Nobile, 2016). Furthermore, a principle factor contributing toward the development of drug resistance attributes of *C. albicans* biofilms is the existence of persister cells, which are a small subgroup of cells which are metabolically inert, which irregularly develop as phenotypical variants inside biofilms, and show strong resistance to different antifungals (Costerton et al., 1995; Ramage et al., 2002).

3.2.4.2 Aspergillus *biofilms*

Aspergillus species are found everywhere and people are regularly being infected by the spores, introduced by these pathogenic organisms into the atmosphere (Patterson et al., 2016). In the affected individuals, there are greater chances of the development of severe diseases (Mir, 2015a,b). The species belonging to *Aspergillus* genera have been found to cause various clinical diseases and infections like invasive, chronic, and allergic types (Mir, 2015a,b) of infectious/diseases (Patterson et al., 2016). Invasive Aspergillosis (IA), caused by *Aspergillus* species, represents the main infectious cause of increasing death rates in persons with weak immune systems (Loussert et al., 2010). *A. fumigatus* is an extensively studied *Aspergillus* species, as it is reported to be associated with around 90% of cases of IA (Ramage et al., 2011; Sardi et al., 2014). The primary development of acute *A. fumigatus* infection includes the germination of conidia into mycelia followed by a future intrusion of the mycelial structure into pulmonary epithelial and endothelial cells (Filler and Sheppard, 2006). Several studies described the adherence and colonization of medical devices including catheters, cardiac pacemakers, prostheses, etc. (Escande et al., 2011; Jeloka et al., 2011; Ramage et al., 2012; Sardi et al., 2014). In contrast to *C. albicans*, the process of formation of biofilm is slower in *Aspergillus* (Mowat et al., 2007; Seidler et al., 2008). Many in vitro and in vivo reports have described biofilm development in *A. fumigatus*, offering resistance to a variety of antifungal agents (Seidler et al., 2008; Loussert et al., 2010; Fiori et al., 2011). *A. fumigatus* is believed to form a special extracellular matrix during biofilm development in vitro and in vivo (Beauvais et al., 2007; Loussert et al., 2010; Sheppard and Howell, 2016). Based on biochemical testing, it has been revealed that this material contains about 43% carbohydrates, 40% proteins, 14% lipids, and 3% aromatic compounds, and nucleic acids (Rajendran et al., 2013; Shopova et al., 2013; Reichhardt et al., 2015). A study reported that the polysaccharides of the ECM possess adhesion features and offer immune defense. The major extracellular matrix polysaccharides involve galactomannan and galactosaminogalactan (GAG), and among these polysaccharides, GAG has gained more attention (Loussert et al., 2010). Aspergillus biofilms have also been found to produce melanin, an immune regulator, although its proper function or contribution in the process of biofilms formation is unclear (Beauvais et al., 2007; Loussert et al., 2010; Akoumianaki et al., 2016).

3.2.4.3 Cryptococcus *biofilms*

Species belonging to *Cryptococcus* genus represent the opportunistic environmental human fungal pathogens causing severe diseases like meningoencephalitis, especially in individuals having repressed immune system, with conditions like HIV, undergoing organ transplantation, and other related conditions (Perfect et al., 2010). Infection by *Cryptococcus* species usually arise via inhalation of yeast spores in the air and is referred to as a chief pulmonary infection which then advances to diffused infectious state (Vila and Rozental, 2016). Among the *Cryptococcus* species, the major human pathogenic species include *C. neoformans* and *C. gatti*. *Cryptococcus* species can infect and generate biofilms on different prosthetic devices like peritoneal dialysis fistulas, etc. (Ramage et al., 2012). For instance, *C. neoformans* display the ability to develop biofilms on various medical devices, like vascular catheters, cerebrospinal fluid shunts, and prosthetic dialysis fistulae (Braun et al., 1994; Bach et al., 1997; Banerjee et al., 1997). A study reported that in vitro, *C. neoformans* biofilms grow fully in 24 to 48 hours and show an MDR phenotype (Martinez and Casadevall, 2005).

The *Cryptococcus* biofilms chiefly include the pathogenic yeast cells with an extensive proportion of polysaccharides composing the ECM responsible for averting its elimination by different environmental/antimicrobial agents. As the species of *Cryptococcus* represents critical environmental pathogenic fungi capable of colonizing human hosts, the process of biofilm development is an expected survival procedure in severe environmental conditions like UV radiations, antimicrobial agents, heat, dryness, etc. (Mir, 2015a,b). Although, only fewer studies on biofilm generation by different *Cryptococcus* species has been reported, the process of biofilm development in these species has been found dependent on the presence of the polysaccharide capsule, comprising chiefly of glucuronoxylomannnan (GXM) (Martinez and Casadevall, 2005; Vila and Rozental, 2016).

It is reported that *Cryptococcus neoformans* induces a protective polysaccharide capsule consisting of GXM, mannoprotein, and galactoxylomannan (Vecchiarelli, 2000; Martinez and Casadevall, 2015). At the time of biofilm development, all the capsular polysaccharides are shed into the surrounding milieu, eventually producing ECM material for proper surface adhesion and cellular cohesion processes (Martinez and Casadevall, 2005). It has been reported in a study that acapsular *C. neoformans* mutant strains have not been capable enough to generate biofilms (Martinez and Casadevall, 2005). It has been described that GXM is the chief polysaccharide found in abundance in the biofilm matrix of *C. neoformans* (Martinez and

Casadevall, 2007). GXM is reported to possess essential immunomodulatory functions and shows increased occurrence in the matrix of biofilms (Dong and Murphy, 1995; Vecchiarelli, 2000; Baena-Monroy et al., 2005; Martinez and Casadevall, 2005).

3.3 Conclusions

The incidence of fungal infections, more likely to be associated with immunocompromised individuals, is increasing rapidly. Around the globe, many individuals are often infected with different pathogenic fungal species causing invasive human fungal infections every year (Firacative, 2020). These pathogenic species mainly belong to *Candida, Aspergillus, Cryptococcus* genus. Along with various mechanisms of resistance acquired by these fungal species, biofilm formation represents another interesting mechanism acquired to survive the effect of different antifungal agents. Unfortunately, the scarcity of appropriate antifungals creates a hurdle in the proper control/treatment plans/procedures. In summary, the chapter provides a complete description of various drug-resistance mechanisms in a way providing meaningful insights about the pathogenic organisms and eventually aiding in the process of identification and development of novel targets/strategies (Sheikh et al., 2021). The proper knowledge and comprehension of the different resistance processes will aid in the production of better antifungal therapeutics and advancement of improved treatment and control procedures. Some of the novel approaches to manage the evolving phenomenon of antimicrobial resistance have been illustrated (Fig. 3.5).

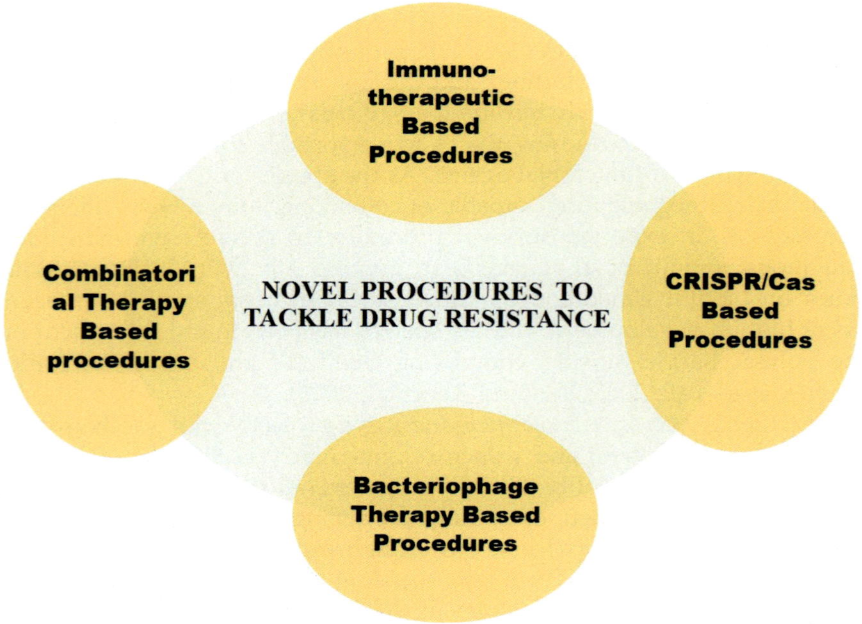

FIGURE 3.5 Diagrammatic representation of some of the novel procedures/approaches to tackle the evolving drug-resistance phenomenon (Qadri et al., 2021a,b).

References

Ademe, M., Girma, F., 2020. Candida auris: from multidrug resistance to pan-resistant strains. Infect. Drug Resist. 13, 1287.

Akoumianaki, T., Kyrmizi, I., Valsecchi, I., Gresnigt, M.S., Samonis, G., Drakos, E., Boumpas, D., Muszkieta, L., Prevost, M.-C., Kontoyiannis, D.P., 2016. *Aspergillus* cell wall melanin blocks LC3-associated phagocytosis to promote pathogenicity. Cell Host & Micr. 19 (1), 79–90.

Bach, M.C., Tally, P.W., Godofsky, E.W., 1997. Use of cerebrospinal fluid shunts in patients having acquired immunodeficiency syndrome with cryptococcal meningitis and uncontrollable intracranial hypertension. Neurosurgery 41 (6), 1280–1283.

Baena-Monroy, T., Moreno-Maldonado, V., Franco-Martínez, F., Aldape-Barrios, B., Quindós, G., Sánchez-Vargas, L.O., 2005. Candida albicans, *Staphylococcus aureus* and *Streptococcus mutans* colonization in patients wearing dental prosthesis. Med. Oral Patol. Oral Cirugía Bucal 10, E27–E39.

Baillie, G.S., Douglas, L.J., 2000. Matrix polymers of Candida biofilms and their possible role in biofilm resistance to antifungal agents. J. Antimicrob. Chemother. 46 (3), 397–403.

Banerjee, U., Gupta, K., Venugopal, P., 1997. A case of prosthetic valve endocarditis caused by *Cryptococcus neoformans* var. *neoformans*. J. Med. Vet. Mycol. 35 (2), 139–141.

Beauvais, A., Schmidt, C., Guadagnini, S., Roux, P., Perret, E., Henry, C., Paris, S., Mallet, A., Prévost, M.C., Latgé, J.P., 2007. An extracellular matrix glues together the aerial-grown hyphae of *Aspergillus fumigatus*. Cell Microbiol. 9 (6), 1588–1600.

Bhattacharya, S., Holowka, T., Orner, E.P., Fries, B.C., 2019. Gene duplication associated with increased fluconazole tolerance in *Candida auris* cells of advanced generational age. Sci. Rep. 9 (1), 1–13.

Bhattacharya, S., Sae-Tia, S., Fries, B.C., 2020. Candidiasis and mechanisms of antifungal resistance. Antibiotics 9 (6), 312.

Bhattacharya, S., Sobel, J.D., White, T.C., 2016. A combination fluorescence assay demonstrates increased efflux pump activity as a resistance mechanism in azole-resistant vaginal *Candida albicans* isolates. Antimicrob. Agents Chemother. 60 (10), 5858–5866.

Bizerra, F.C., Nakamura, C.V., de Poersch, C., Estivalet Svidzinski, T.I., Borsato Quesada, R.M., Goldenberg, S., Krieger, M.A., Yamada-Ogatta, S.F., 2008. Characteristics of biofilm formation by *Candida tropicalis* and antifungal resistance. FEMS Yeast Res. 8 (3), 442–450.

Borst, A., Raimer, M.T., Warnock, D.W., Morrison, C.J., Arthington-Skaggs, B.A., 2005. Rapid acquisition of stable azole resistance by *Candida glabrata* isolates obtained before the clinical introduction of fluconazole. Antimicrob. Agents Chemother. 49 (2), 783–787.

Braun, B.R., van Het Hoog, M., d'Enfert, C., Martchenko, M., Dungan, J., Kuo, A., Inglis, D.O., Uhl, M.A., Hogues, H., Berriman, M., 2005. A human-curated annotation of the *Candida albicans* genome. PLoS Genet. 1 (1), e1.

Braun, D.K., Janssen, D.A., Marcus, J.R., Kauffman, C.A., 1994. Cryptococcal infection of a prosthetic dialysis fistula. Am. J. Kidney Dis. 24 (5), 864–867.

Brown, G.D., Denning, D.W., Gow, N.A., Levitz, S.M., Netea, M.G., White, T.C., 2012. Hidden killers: human fungal infections. Sci. Transl. Med. 4, 165rv13.

Calabrese, D., Bille, J., Sanglard, D., 2000. A novel multidrug efflux transporter gene of the major facilitator superfamily from *Candida albicans* (FLU1) conferring resistance to fluconazole. Microbiology 146 (11), 2743–2754.

Casalinuovo, I.A., Di Francesco, P., Garaci, E., 2004. Fluconazole resistance in *Candida albicans*: a review of mechanisms. Eur. Rev. Med. Pharmacol. Sci. 8 (2), 69–77.

Chandra, J., Kuhn, D.M., Mukherjee, P.K., Hoyer, L.L., McCormick, T., Ghannoum, M.A., 2001. Biofilm formation by the fungal pathogen *Candida albicans*: development, architecture, and drug resistance. J. Bacteriol. 183 (18), 5385–5394.

Chandra, J., Zhou, G., Ghannoum, M.A., 2005. Fungal biofilms and antimycotics. Curr. Drug Targets 6 (8), 887–894.

Chowdhary, A., Sharma, C., Kathuria, S., Hagen, F., Meis, J.F., 2014. Azole-resistant *Aspergillus fumigatus* with the environmental TR46/Y121F/T289A mutation in India. J. Antimicrob. Chemother. 69 (2), 555–557.

Coleman, J.J., Mylonakis, E., 2009. Efflux in fungi: la piece de resistance. PLoS Pathog. 5 (6), e1000486.

Coste, A., Selmecki, A., Forche, A., Diogo, D., Bougnoux, M.E., d'Enfert, C., Berman, J., Sanglard, D., 2007. Genotypic evolution of azole resistance mechanisms in sequential *Candida albicans* isolates. Eukaryot. Cell 6 (10), 1889–1904.

Coste, A.T., Karababa, M., Ischer, F., Bille, J., Sanglard, D., 2004. TAC1, transcriptional activator of CDR genes, is a new transcription factor involved in the regulation of *Candida albicans* ABC transporters CDR1 and CDR2. Eukaryot. Cell 3 (6), 1639–1652.

Costerton, J.W., Lewandowski, Z., Caldwell, D.E., Korber, D.R., Lappin-Scott, H.M., 1995. Microbial biofilms. Annu. Rev. Microbiol. 49 (1), 711–745.

Cowen, L.E., Sanglard, D., Howard, S.J., Rogers, P.D., Perlin, D.S., 2015. Mechanisms of antifungal drug resistance. Cold Spr. Harb. Perspect. Med. 5 (7), a019752.

da Silva Ferreira, M.E., Malavazi, I., Savoldi, M., Brakhage, A.A., Goldman, M.H., Kim, H.S., et al., 2006. Transcriptome analysis of Aspergillus fumigatus exposed to voriconazole. Curr. Genet. 50, 32–44. https://doi.org/10.1007/s00294-006-0073-2.

de Andrade Monteiro, C., Ribeiro Alves dos Santos, J., 2019. Phytochemicals and their Antifungal Potential Against Pathogenic Yeasts. Phytochemicals in human health 2, 1–31.

Dagenais, T.R., Keller, N.P., 2009. Pathogenesis of *Aspergillus fumigatus* in invasive aspergillosis. Clin. Microbiol. Rev. 22 (3), 447–465.

Dong, Z.M., Murphy, J.W., 1995. Intravascular cryptococcal culture filtrate (CneF) and its major component, glucuronoxylomannan, are potent inhibitors of leukocyte accumulation. Infect. Immun. 63 (3), 770–778.

Dongari-Bagtzoglou, A., 2008. Mucosal biofilms: challenges and future directions. Exper. Rev. Anti-infect. Ther. 6 (2), 141–144.

Dunkel, N., Liu, T.T., Barker, K.S., Homayouni, R., Morschhaüser, J., Rogers, P.D., 2008. A gain-of-function mutation in the transcription factor Upc2p causes upregulation of ergosterol biosynthesis genes and increased fluconazole resistance in a clinical Candida albicans isolate. Eukaryot. Cell 7 (7), 1180–1190.

Escande, W., Fayad, G., Modine, T., Verbrugge, E., Koussa, M., Senneville, E., Leroy, O., 2011. Culture of a prosthetic valve excised for streptococcal endocarditis positive for *Aspergillus fumigatus* 20 years after previous A fumigatus endocarditis. Ann. Thorac. Surg. 91 (6), e92–e93.

Fanning, S., Mitchell, A.P., 2012. Fungal biofilms. PLoS Pathog. 8 (4), e1002585.

Filler, S.G., Sheppard, D.C., 2006. Fungal invasion of normally non-phagocytic host cells. PLoS Pathog. 2 (12), e129.

Fiori, B., Posteraro, B., Torelli, R., Tumbarello, M., Perlin, D.S., Fadda, G., Sanguinetti, M., 2011. In vitro activities of anidulafungin and other antifungal agents against biofilms formed by clinical isolates of different *Candida* and *Aspergillus* species. Antimicrob. Agents Chemother. 55 (6), 3031–3035.

Firacative, C., 2020. Invasive fungal disease in humans: are we aware of the real impact? Mem. Inst. Oswaldo Cruz 115.

Flowers, S.A., Barker, K.S., Berkow, E.L., Toner, G., Chadwick, S.G., Gygax, S.E., Morschhäuser, J., Rogers, P.D., 2012. Gain-of-function mutations in UPC2 are a frequent cause of ERG11 upregulation in azole-resistant clinical isolates of *Candida albicans*. Eukaryot. Cell 11 (10), 1289–1299.

Ford, C.B., Funt, J.M., Abbey, D., Issi, L., Guiducci, C., Martinez, D.A., Delorey, T., yu Li, B., White, T.C., Cuomo, C., 2015. The evolution of drug resistance in clinical isolates of *Candida albicans*. Elife 4, e00662.

Fox, E.P., Nobile, C.J., 2012. A sticky situation: untangling the transcriptional network controlling biofilm development in *Candida albicans*. Transcription 3 (6), 315–322.

Fraczek, M.G., Bromley, M., Buied, A., Moore, C.B., Rajendran, R., Rautemaa, R., Ramage, G., Denning, D.W., Bowyer, P., 2013. The cdr1B efflux transporter is associated with non-cyp51a-mediated itraconazole resistance in *Aspergillus fumigatus*. J. Antimicrob. Chemother. 68 (7), 1486–1496.

Gaur, M., Puri, N., Manoharlal, R., Rai, V., Mukhopadhayay, G., Choudhury, D., Prasad, R., 2008. MFS transportome of the human pathogenic yeast *Candida albicans*. BMC Genom. 9 (1), 1–12.

Groll, A.H., Piscitelli, S.C., Walsh, T.J., 1998. Clinical pharmacology of systemic antifungal agents: a comprehensive review of agents in clinical use, current investigational compounds, and putative targets for antifungal drug development. Adv. Pharmacol. 343–500. Academic Press Inc.

Gulati, M., Nobile, C.J., 2016. Candida albicans biofilms: development, regulation, and molecular mechanisms. Microb. Infect. 18 (5), 310–321.

Harriott, M.M., Lilly, E.A., Rodriguez, T.E., Fidel Jr., P.L., Noverr, M.C., 2010. Candida albicans forms biofilms on the vaginal mucosa. Microbiology 156 (Pt 12), 3635.

Hawser, S.P., Douglas, L.J., 1994. Biofilm formation by *Candida* species on the surface of catheter materials in vitro. Infect. Immun. 62 (3), 915–921.

Hirakawa, M.P., Chyou, D.E., Huang, D., Slan, A.R., Bennett, R.J., 2017. Parasex generates phenotypic diversity de novo and impacts drug resistance and virulence in *Candida albicans*. Genetics 207 (3), 1195–1211.

Howard, S.J., Arendrup, M.C., 2011. Acquired antifungal drug resistance in *Aspergillus fumigatus*: epidemiology and detection. Med. Mycol. 49 (Supplement_1), S90–S95.

Jeloka, T.K., Shrividya, S., Wagholikar, G., 2011. Catheter outflow obstruction due to an aspergilloma. Perit. Dial. Int. 31 (2), 211–212.

Jiang, C., Dong, D., Yu, B., Cai, G., Wang, X., Ji, Y., Peng, Y., 2013. Mechanisms of azole resistance in 52 clinical isolates of *Candida tropicalis* in China. J. Antimicrob. Chemother. 68 (4), 778–785.

Kathiravan, M.K., Salake, A.B., Chothe, A.S., Dudhe, P.B., Watode, R.P., Mukta, M.S., Gadhwe, S., 2012. The biology and chemistry of antifungal agents: a review. Bioorg. Med. Chem. 20 (19), 5678–5698.

Katiyar, S.K., Edlind, T.D., 2001. Identification and expression of multidrug resistancerelated ABC transporter genes in *Candida krusei*. Sabouraudia 39 (1), 109–116.

Kojic, E.M., Darouiche, R.O., 2004. Candida infections of medical devices. Clin. Microbiol. Rev. 17 (2), 255–267.

Kovalchuk, A., Driessen, A.J.M., 2010. Phylogenetic analysis of fungal ABC transporters. BMC Genom. 11 (1), 1–21.

Kuhn, D.M., Chandra, J., Mukherjee, P.K., Ghannoum, M.A., 2002. Comparison of biofilms formed by *Candida albicans* and *Candida parapsilosis* on bioprosthetic surfaces. Infect. Immun. 70 (2), 878–888.

Kumamoto, C.A., 2002. Candida biofilms. Curr. Opin. Microbiol. 5 (6), 608–611.

Lamping, E., Baret, P.V., Holmes, A.R., Monk, B.C., Goffeau, A., Cannon, R.D., 2010. Fungal PDR transporters: phylogeny, topology, motifs and function. Fungal Genet. Biol. 47 (2), 127–142.

Lamping, E., Monk, B.C., Niimi, K., Holmes, A.R., Tsao, S., Tanabe, K., Niimi, M., Uehara, Y., Cannon, R.D., 2007. Characterization of three classes of membrane proteins involved in fungal azole resistance by functional hyperexpression in *Saccharomyces cerevisiae*. Eukaryot. Cell 6 (7), 1150–1165.

Loussert, C., Schmitt, C., Prevost, M.C., Balloy, V., Fadel, E., Philippe, B., Kauffmann-Lacroix, C., Latgé, J.P., Beauvais, A., 2010. In vivo biofilm composition of *Aspergillus fumigatus*. Cell Microbiol. 12 (3), 405–410.

MacCallum, D.M., 2012. Hosting infection: experimental models to assay *Candida virulence*. Internat. J. Microbiol. 2012.

Marichal, P., Vanden Bossche, H., Odds, F.C., Nobels, G., Warnock, D.W., Timmerman, V., Van Broeckhoven, C., Fay, S., Mose-Larsen, P., 1997. Molecular biological characterization of an azole-resistant *Candida glabrata* isolate. Antimicrob. Agents Chemother. 41 (10), 2229–2237.

Martinez, L.R., Casadevall, A., 2005. Specific antibody can prevent fungal biofilm formation and this effect correlates with protective efficacy. Infect. Immun. 73 (10), 6350–6362.

Martinez, L.R., Casadevall, A., 2007. Cryptococcus neoformans biofilm formation depends on surface support and carbon source and reduces fungal cell susceptibility to heat, cold, and UV light. Appl. Environ. Microbiol. 73 (14), 4592–4601.

Martinez, L.R., Casadevall, A., 2015. Biofilm formation by Cryptococcus neoformans. Microbiol. Spectr. 3 (3), 3-3.

Mir, M.A., 2015a. Chapter 3 - costimulation immunotherapy in infectious diseases. In: Mir, M.A. (Ed.), Developing Costimulatory Molecules for Immunotherapy of Diseases. Academic Press, ISBN 9780128025857, pp. 83–129. https://doi.org/10.1016/B978-0-12-802585-7.00003-0.

Mir, M.A., 2015b. Chapter 4 - costimulation immunotherapy in allergies and asthma. In: Mir, M.A. (Ed.), Developing Costimulatory Molecules for Immunotherapy of Diseases. Academic Press, ISBN 9780128025857, pp. 131–184. https://doi.org/10.1016/B978-0-12-802585-7.00004-2.

Mir, M.A., Mehraj, U., Sheikh, B.A., Hamdani, S.S., 2020. Nanobodies: the "magic bullets" in therapeutics, drug delivery and diagnostics. Hum. Antibodies 28 (1), 29–51.

Mitchell, T.G., Perfect, J.R., 1995. Cryptococcosis in the era of AIDS–100 years after the discovery of *Cryptococcus neoformans*. Clin. Microbiol. Rev. 8 (4), 515–548.

Morio, F., Loge, C., Besse, B., Hennequin, C., Le Pape, P., 2010. Screening for amino acid substitutions in the *Candida albicans* Erg11 protein of azole-susceptible and azole-resistant clinical isolates: new substitutions and a review of the literature. Diagn. Microbiol. Infect. Dis. 66 (4), 373–384.

Mowat, E., Butcher, J., Lang, S., Williams, C., Ramage, G., 2007. Development of a simple model for studying the effects of antifungal agents on multicellular communities of *Aspergillus fumigatus*. J. Med. Microbiol. 56 (9), 1205–1212.

Mowat, E., Williams, C., Jones, B., McChlery, S., Ramage, G., 2009. The characteristics of *Aspergillus fumigatus* mycetoma development: is this a biofilm? Med. Mycol. 47 (Supplement_1), S120–S126.

Mukherjee, P.K., Chandra, J., Kuhn, D.M., Ghannoum, M.A., 2003. Mechanism of fluconazole resistance in *Candida albicans* biofilms: phase-specific role of efflux pumps and membrane sterols. Infect. Immun. 71 (8), 4333–4340.

Nascimento, A.M., Goldman, G.H., Park, S., Marras, S.A.E., Delmas, G., Oza, U., Lolans, K., Dudley, M.N., Mann, P.A., Perlin, D.S., 2003. Multiple resistance mechanisms among *Aspergillus fumigatus* mutants with high-level resistance to itraconazole. Antimicrob. Agents Chemother. 47 (5), 1719–1726.

Nett, J.E., Crawford, K., Marchillo, K., Andes, D.R., 2010. Role of Fks1p and matrix glucan in *Candida albicans* biofilm resistance to an echinocandin, pyrimidine, and polyene. Antimicrob. Agents Chemother. 54 (8), 3505–3508.

Nobile, C.J., Fox, E.P., Nett, J.E., Sorrells, T.R., Mitrovich, Q.M., Hernday, A.D., Tuch, B.B., Andes, D.R., Johnson, A.D., 2012. A recently evolved transcriptional network controls biofilm development in *Candida albicans*. Cell 148 (1–2), 126–138.

Odds, F.C., Brown, A.J., Gow, N.A., 2003. Antifungal agents: mechanisms of action. Trends Microbiol. 11 (6), 272–279.

Pam, V.K., Akpan, J.U., Oduyebo, O.O., Nwaokorie, F.O., Fowora, M.A., Oladele, R.O., Ogunsola, F.T., Smith, S.I., 2012. Fluconazole susceptibility and ERG11 gene expression in vaginal Candida species isolated from Lagos Nigeria. Internat. J. Mol. Epidemiol. & Genet. 3 (1), 84.

Patterson, T.F., Thompson III, G.R., Denning, D.W., Fishman, J.A., Hadley, S., Herbrecht, R., Kontoyiannis, D.P., Marr, K.A., Morrison, V.A., Nguyen, M.H., 2016. Practice guidelines for the diagnosis and management of aspergillosis: 2016 update by the Infectious Diseases Society of America. Clin. Infect. Dis. 63 (4), e1–e60.

Perea, S., López-Ribot, J.L., Kirkpatrick, W.R., McAtee, R.K., Santillán, R.A., Martınez, M., Calabrese, D., Sanglard, D., Patterson, T.F., 2001. Prevalence of molecular mechanisms of resistance to azole antifungal agents in *Candida albicans* strains displaying high-level fluconazole resistance isolated from human immunodeficiency virus-infected patients. Antimicrob. Agents Chemother. 45 (10), 2676–2684.

Perfect, J.R., Dismukes, W.E., Dromer, F., Goldman, D.L., Graybill, J.R., Hamill, R.J., Harrison, T.S., Larsen, R.A., Lortholary, O., Nguyen, M.-H., 2010. Clinical practice guidelines for the management of cryptococcal disease: 2010 update by the Infectious Diseases Society of America. Clin. Infect. Dis. 50 (3), 291–322.

Poláková, S., Blume, C., Zárate, J.Á., Mentel, M., Jørck-Ramberg, D., Stenderup, J., Piškur, J., 2009. Formation of new chromosomes as a virulence mechanism in yeast *Candida glabrata*. Proc. Natl. Acad. Sci. USA 106 (8), 2688–2693.

Prasad, R., Goffeau, A., 2012. Yeast ATP-binding cassette transporters conferring multidrug resistance. Annu. Rev. Microbiol. 66, 39–63.

Prasad, R., Shah, A.H., Dhamgaye, S., 2014. Mechanisms of drug resistance in fungi and their significance in biofilms. Antibiofilm Agen. 45–65. Springer.

Qadri, H., Haseeb, A., Mir, M., 2021a. Novel strategies to combat the emerging drug resistance in human pathogenic microbes. Curr. Drug Targets 22, 1–13.

Qadri, H., Qureshi, M.F., Mir, M.A., Shah, A.H., 2021b. Glucose-The X Factor for the survival of human fungal pathogens and disease progression in the host. Microbiol. Res. 126725.

Rajendran, R., Williams, C., Lappin, D.F., Millington, O., Martins, M., Ramage, G., 2013. Extracellular DNA release acts as an antifungal resistance mechanism in mature *Aspergillus fumigatus* biofilms. Eukaryot. Cell 12 (3), 420–429.

Ramage, G., Bachmann, S., Patterson, T.F., Wickes, B.L., López-Ribot, J.L., 2002. Investigation of multidrug efflux pumps in relation to fluconazole resistance in *Candida albicans* biofilms. J. Antimicrob. Chemother. 49 (6), 973–980.

Ramage, G., Rajendran, R., Gutierrez-Correa, M., Jones, B., Williams, C., 2011. Aspergillus biofilms: clinical and industrial significance. FEMS Microbiol. Lett. 324 (2), 89–97.

Ramage, G., Rajendran, R., Sherry, L., Williams, C., 2012. Fungal biofilm resistance. Internat. J. Microbiol. 2012.

Ramage, G., Saville, S.P., Thomas, D.P., Lopez-Ribot, J.L., 2005. Candida biofilms: an update. Eukaryot. Cell 4 (4), 633–638.

Ramage, G., Walle, K.V., Wickes, B.L., Lopez-Ribot, J.L., 2001. Characteristics of biofilm formation by *Candida albicans*. Rev. Iberoam. De. Micol. 18 (4), 163–170.

Redding, S.W., Kirkpatrick, W.R., Coco, B.J., Sadkowski, L., Fothergill, A.W., Rinaldi, M.G., Eng, T.Y., Patterson, T.F., 2002. Candida glabrata oropharyngeal candidiasis in patients receiving radiation treatment for head and neck cancer. J. Clin. Microbiol. 40 (5), 1879–1881.

Reichhardt, C., Ferreira, J.A., Joubert, L.M., Clemons, K.V., Stevens, D.A., Cegelski, L., 2015. Analysis of the *Aspergillus fumigatus* biofilm extracellular matrix by solid-state nuclear magnetic resonance spectroscopy. Eukaryot. Cell 14 (11), 1064–1072.

Revie, N.M., Iyer, K.R., Robbins, N., Cowen, L.E., 2018. Antifungal drug resistance: evolution, mechanisms and impact. Curr. Opin. Microbiol. 45, 70–76.

Reynolds, T.B., Fink, G.R., 2001. Bakers' yeast, a model for fungal biofilm formation. Science 291 (5505), 878–881.

Robbins, N., Caplan, T., Cowen, L.E., 2017. Molecular evolution of antifungal drug resistance. Annu. Rev. Microbiol. 71, 753–775.

Sanglard, D., 2016. Emerging threats in antifungal-resistant fungal pathogens. Front. Med. 3, 11.

Sanglard, D., Coste, A., Ferrari, S., 2009. Antifungal drug resistance mechanisms in fungal pathogens from the perspective of transcriptional gene regulation. FEMS Yeast Res. 9 (7), 1029–1050.

Sanglard, D., Ischer, F., Monod, M., Bille, J., 1996. Susceptibilities of *Candida albicans* multidrug transporter mutants to various antifungal agents and other metabolic inhibitors. Antimicrob. Agents Chemother. 40 (10), 2300–2305.

Sanglard, D., Ischer, F., Parkinson, T., Falconer, D., Bille, J., 2003. Candida albicans mutations in the ergosterol biosynthetic pathway and resistance to several antifungal agents. Antimicrob. Agents Chemother. 47 (8), 2404–2412.

Sanglard, D., Kuchler, K., Ischer, F., Pagani, J.L., Monod, M., Bille, J., 1995. Mechanisms of resistance to azole antifungal agents in *Candida albicans* isolates from AIDS patients involve specific multidrug transporters. Antimicrob. Agents Chemother. 39 (11), 2378–2386.

Sanguinetti, M., Posteraro, B., Fiori, B., Ranno, S., Torelli, R., Fadda, G., 2005. Mechanisms of azole resistance in clinical isolates of *Candida glabrata* collected during a hospital survey of antifungal resistance. Antimicrob. Agents Chemother. 49 (2), 668–679.

Sanguinetti, M., Posteraro, B., La Sorda, M., Torelli, R., Fiori, B., Santangelo, R., Delogu, G., Fadda, G., 2006. Role of AFR1, an ABC transporter-encoding gene, in the in vivo response to fluconazole and virulence of *Cryptococcus neoformans*. Infect. Immun. 74 (2), 1352–1359.

Sardi, J.C.O., Scorzoni, L., Bernardi, T., Fusco-Almeida, A.M., Giannini, M.J.S.M., 2013. Candida species: current epidemiology, pathogenicity, biofilm formation, natural antifungal products and new therapeutic options. J. Med. Microbiol. 62 (1), 10–24.

Sardi, J.D.C.O., Pitangui, N.D.S., Rodríguez-Arellanes, G., Taylor, M.L., Fusco-Almeida, A.M., Mendes-Giannini, M.J., 2014. Highlights in pathogenic fungal biofilms. Rev. Iberoam. De. Micol. 31 (1), 22–29.

Schmiedel, Y., Zimmerli, S., 2016. Common invasive fungal diseases: an overview of invasive candidiasis, aspergillosis, cryptococcosis, and Pneumocystis pneumonia. Swiss Med. Wkly. 146, w14281.

Seidler, M.J., Salvenmoser, S., Muller, F.-M.C., 2008. *Aspergillus fumigatus* forms biofilms with reduced antifungal drug susceptibility on bronchial epithelial cells. Antimicrob. Agents Chemother. 52 (11), 4130–4136.

Selmecki, A., Forche, A., Berman, J., 2006. Aneuploidy and isochromosome formation in drug-resistant *Candida albicans*. Science 313 (5785), 367–370.

Selmecki, A., Forche, A., Berman, J., 2010. Genomic plasticity of the human fungal pathogen *Candida albicans*. Eukaryot. Cell 9 (7), 991–1008.

Seneviratne, C.J., Jin, L., Samaranayake, L.P., 2008. Biofilm lifestyle of Candida: a mini review. Oral Dis. 14 (7), 582–590.

Shah, A.H., Rawal, M.K., Dhamgaye, S., Komath, S.S., Saxena, A.K., Prasad, R., 2015. Mutational analysis of intracellular loops identify cross talk with nucleotide binding domains of yeast ABC transporter Cdr1p. Sci. Rep. 5 (1), 1–17.

Sheikh, B.A., Bhat, B.A., Mehraj, U., Mir, W., Hamadani, S., Mir, M.A., 2021. Development of new therapeutics to meet the current challenge of drug resistant tuberculosis. Curr. Pharmaceut. Biotechnol. 22 (4), 480–500.

Sheppard, D.C., Howell, P.L., 2016. Biofilm exopolysaccharides of pathogenic fungi: lessons from bacteria. J. Biol. Chem. 291 (24), 12529–12537.

Sherry, L., Ramage, G., Kean, R., Borman, A., Johnson, E.M., Richardson, M.D., Rautemaa-Richardson, R., 2017. Biofilm-forming capability of highly virulent, multidrug-resistant *Candida auris*. Emerg. Infect. Dis. 23 (2), 328.

Shin, J.H., Kee, S.J., Shin, M.G., Kim, S.H., Shin, D.H., Lee, S.K., Suh, S.P., Ryang, D.W., 2002. Biofilm production by isolates of *Candida* species recovered from nonneutropenic patients: comparison of bloodstream isolates with isolates from other sources. J. Clin. Microbiol. 40 (4), 1244–1248.

Shopova, I., Bruns, S.M., Thywissen, A., Kniemeyer, O., Brakhage, A.A., Hillmann, F., 2013. Extrinsic extracellular DNA leads to biofilm formation and colocalizes with matrix polysaccharides in the human pathogenic fungus *Aspergillus fumigatus*. Front. Microbiol. 4, 141.

Silva, S., Henriques, M., Martins, A., Oliveira, R., Williams, D., Azeredo, J., 2009. Biofilms of non-*Candida albicans Candida* species: quantification, structure and matrix composition. Sabouraudia 47 (7), 681–689.

Silva, S., Negri, M., Henriques, M., Oliveira, R., Williams, D.W., Azeredo, J., 2011. Adherence and biofilm formation of non-*Candida albicans Candida* species. Trends Microbiol. 19 (5), 241–247.

Sionov, E., Lee, H., Chang, Y.C., Kwon-Chung, K.J., 2010. *Cryptococcus neoformans* overcomes stress of azole drugs by formation of disomy in specific multiple chromosomes. PLoS Pathog. 6 (4), e1000848.

Slaven, J.W., Anderson, M.J., Sanglard, D., Dixon, G.K., Bille, J., Roberts, I.S., Denning, D.W., 2002. Increased expression of a novel *Aspergillus fumigatus* ABC transporter gene, atrF, in the presence of itraconazole in an itraconazole resistant clinical isolate. Fungal Genet. Biol. 36 (3), 199–206.

Snelders, E., Melchers, W.J.G., Verweij, P.E., 2011. Azole resistance in *Aspergillus fumigatus*: a new challenge in the management of invasive aspergillosis? Futur. Microbiol. 6 (3), 335–347.

Spampinato, C., Leonardi, D., 2013. Candida infections, causes, targets, and resistance mechanisms: traditional and alternative antifungal agents. BioMed Res. Int. 2013.

Tumbarello, M., Posteraro, B., Trecarichi, E.M., Fiori, B., Rossi, M., Porta, R., de Gaetano Donati, K., La Sorda, M., Spanu, T., Fadda, G., 2007. Biofilm production by *Candida* species and inadequate antifungal therapy as predictors of mortality for patients with candidemia. J. Clin. Microbiol. 45 (6), 1843–1850.

Uppuluri, P., Pierce, C.G., López-Ribot, J.L., 2009. *Candida albicans* biofilm formation and its clinical consequences. Futur. Microbiol. 4 (10), 1235–1237.

Vandeputte, P., Ferrari, S., Coste, A.T., 2012. Antifungal resistance and new strategies to control fungal infections. Internat. J. Microbiol. 2012.

Vandeputte, P., Larcher, G., Berges, T., Renier, G., Chabasse, D., Bouchara, J.-P., 2005. Mechanisms of azole resistance in a clinical isolate of *Candida tropicalis*. Antimicrob. Agents Chemother. 49 (11), 4608–4615.

Vandeputte, P., Tronchin, G., Bergès, T., Hennequin, C., Chabasse, D., Bouchara, J.-P., 2007. Reduced susceptibility to polyenes associated with a missense mutation in the ERG6 gene in a clinical isolate of *Candida glabrata* with pseudohyphal growth. Antimicrob. Agents Chemother. 51 (3), 982–990.

Vecchiarelli, A., 2000. Immunoregulation by capsular components of *Cryptococcus neoformans*. Med. Mycol. 38 (6), 407–417.

Vermeulen, E., Lagrou, K., Verweij, P.E., 2013. Azole resistance in *Aspergillus fumigatus*: a growing public health concern. Curr. Opin. Infect. Dis. 26 (6), 493–500.

Vila, T.V.M., Rozental, S., 2016. Biofilm formation as a pathogenicity factor of medically important fungi. Fungal Pathogen. 1–24. InTech: Rijeka, Croatia.

White, T.C., Holleman, S., Dy, F., Mirels, L.F., Stevens, D.A., 2002. Resistance mechanisms in clinical isolates of *Candida albicans*. Antimicrob. Agents Chemother. 46 (6), 1704–1713.

Wiederhold, NP, 2017. Antifungal resistance: current trends and future strategies to combat. Infection and drug resistance 10, 249.

Xu, D., Jiang, B., Ketela, T., Lemieux, S., Veillette, K., Martel, N., Davison, J., Sillaots, S., Trosok, S., Bachewich, C., 2007. Genome-wide fitness test and mechanism-of-action studies of inhibitory compounds in *Candida albicans*. PLoS Pathog. 3 (6), e92.

Yoo, J.I., Choi, C.W., Lee, K.M., Lee, Y.S., 2010. Gene expression and identification related to fluconazole resistance of *Candida glabrata* strains. Osong Pub. Health & Res. Perspect. 1 (1), 36–41.

Zarnowski, R., Westler, W.M., Lacmbouh, G.A., Marita, J.M., Bothe, J.R., Bernhardt, J., Sahraoul, A.L.H., Fontaine, J., Sanchez, H., Hatfield, R.D., 2014. Novel entries in a fungal biofilm matrix encyclopedia. mBio. 5 e1333–e1314. PubMed PubMedCentral.

CHAPTER

4

Combating human bacterial infections: need for new antibacterial drugs and therapeutics

Manzoor Ahmad Mir[1], *Bilkees Nabi*[2], *Sushma Ahlawat*[2], *Manoj Kumawat*[3] and *Shariqa Aisha*[1]

[1]Department of Bioresources, School of Biological Sciences, University of Kashmir, Srinagar, Jammu and Kashmir, India; [2]Department of Biochemistry and Biochemical Engineering, Sam Higginbottom University of Agriculture, Technology and Sciences, Prayagraj, Uttar Pradesh, India; [3]Department of Microbiology, ICMR-NIREH, Bhopal, Madhya Pradesh, India

4.1 Introduction

Bacteria are cosmopolitan in nature; they show growth in different unfavorable as well as favorable conditions, thereby showing presence almost everywhere. They are present even inside the human gut, on skin and other surfaces also. Some bacteria are beneficial to human health, mostly those which reside in the gut, but most of the bacteria are pathogenic causing different infections or even diseases.

Bacterial infections were supposed to be the threats to the humans, but with the development of vaccines and antibiotics most of the bacterial infections are easily treated which were once considered as deadly ones, although some bacterial strains have evolved in terms of showing resistance to antibiotics which is a major concern in the present times.

4.2 Common human bacterial infections

Some commonly found bacteria which cause infections in humans are shown in Table 4.1.

Human Pathogenic Microbes
https://doi.org/10.1016/B978-0-323-96127-1.00008-5

TABLE 4.1 Some examples of bacterial infections in humans and their causative agents.

Human bacterial infections	Causative agent
Cholera	*Vibrio cholera*
Leprosy	*Mycobacterium leprae*
Tetanus	*Clostridium tetani*
Gonorrhea	*Neisseria gonorrhoeae*
Syphilis	*Treponema pallidum*
Salmonellosis	*Salmonella enteritis*
Pseudomonas infection	*Pseudomonas aeruginosa*
Pulmonary tuberculosis	*Mycobacterium tuberculosis*

The three primary human bacterial illnesses, Tuberculosis, Salmonellosis, and Cholera, will be discussed in depth in this chapter.

4.3 Salmonellosis

Salmonellosis is an infection mediated by the bacterium Salmonella. In the category of foodborne infections it is regarded as the third biggest cause of mortality. Salmonella—the causative agent of salmonellosis is ubiquitous, enteropathogenic Gram-negative facultative anaerobic in nature, which belongs to the family of enterobacteriaceae. It was Daniel E. Salmon who discovered it and was named after his name as salmonella. *Salmonella enterica* and *S. bongori* are the two species that make up the genus *Salmonella*. The *S. enterica* species is the most pathogenic which affects the humans as shown in Fig. 4.1.

Size: Salmonella has a length of 2–5 microns and a width of 0.5–1.5 microns.

Motility: shows motility by peritrichous flagella.

Genome: vary from 4460 to 4857 kb depending on type of serovar.

Based on the three morphological structures found on bacterial lipopolysaccharide (LPS), flagella, and capsular polysaccharide (Popoff et al., 2004), Salmonella genus is classified into two species:

1. *Salmonella bongori*
2. *Salmonella enterica*

S. enterica is subdivided into six subspecies: enterica, arizonae, diarizonae, salamae, houtenae, and indica, with a total of 2659 serovars. And further enterica subspecies have approximately 1547 serovars, of which 99% are infectious to humans and animals.

Salmonella species, on the other hand, are categorized into two groups based on their ability to have capacity of showing certain pathologies in humans: typhoidal and nontyphoidal salmonella (NTS).

According to the Kauffman-White scheme, there are about 2500 nontyphoidal *Salmonella* serovars in the United States that cause over one million infections each year

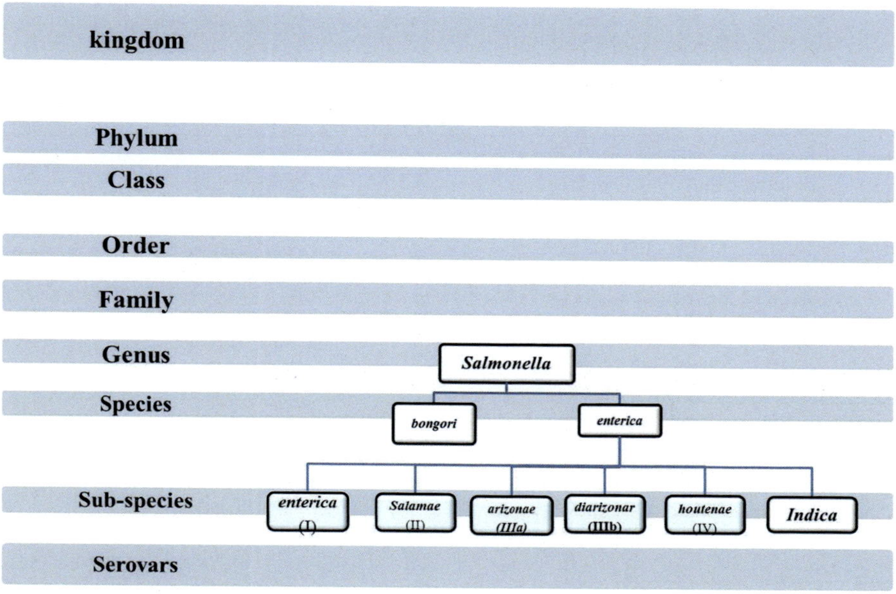

FIGURE 4.1 The taxonomic classification of salmonella it represents the taxonomic classification of salmonella, specifically salmonella enteric which is classified into six sub species and among them the most common sub-species is enteric (I) which is further divided into two types of serovars Non-typhoidal and Typhoidal.

(Roschka and Dosch, 1950; Kumawat et al., 2021). The continued outbreaks of nontyphoidal salmonellosis (including the outbreak involving contaminated peanut butter), the high incidence rates of infections worldwide, and the evolution of multidrug-resistant *Salmonella* strains (Centers for Disease Control and Prevention, 2007; Angulo and Mlbak, 2005; Hohmann, 2001) are all contributing to an increase in infection cases worldwide.

The major burden of Salmonella infections is because of non-typhoidal serotypes (Voetsch et al., 2004). Salmonella accounts for about 30% of all the deaths associated with foodborne disease in united states (Mead et al., 1999) and during diarrheal illness, salmonella has been reported as the second most prevalent bacteria isolated from the stool cultures, following *Campylobacter jejuni.* Annually about 45,000 cases are reported but it is estimated that the actual cases are about 30 folds higher. About 17.7 cases per 100,000 persons of salmonella infection were estimated in 2002, children (<5 years) being at higher risk.

Salmonella strains used to be called after the places where they were isolated, such as *Salmonella london* and *Salmonella indiana.* This technique has been superseded with a phage-typing-based nomenclature approach (Pui et al., 2011). Phage typing has turned out to be exceptionally useful in characterizing and distinguishing between *Salmonella typhimurium, virchow, enteritidis,* and *typhi* strains (Greenwood et al., 2012) Table 4.2.

TABLE 4.2 Showing the difference between typhoidal and nontyphoidal *Salmonella* (Hendriksen et al., 2011; Gal-Mor et al., 2014; Shinohara et al., 2008; Shivaprasad, 2000; Berger et al., 2010; Organization, 2015).

Parameter	Typhoidal serovars	Nontyphoidal serovars
Host range	Infect and colonize only limited narrow range of hosts.	Can infect and colonize in wide range of hosts. Some are host specific e.g., Pullorum in poultry.
Reservoir	They are highly adapted to human species. Besides only higher primates and humans serve as reservoirs for them.	Adapted to humans and animals as well as reservoirs.
Symptoms	Symptoms include a high temperature, diarrhea, vomiting, headaches, and, in the worst-case scenario, death (by typhi and sendai), enteric fever (paratyphi A, B, and C) causes diarrhea, cramps, fever, and vomiting, as well as septicemia in certain cases.	Diarrhea, abdominal cramps, vomiting
Examples	Typhi, sendai, and paratyphi A, B, and C serovars	NTS *enteritidis* and *typhimurium*
Transmission	Transmission through water, milk, raw vegetables, seafood	The major routes of transmission include animal-based foods such as beef, pork, poultry, contaminated eggs, and vegetables and fruits
Incubation period	7–14 days incubation	6–12 h incubation period
Antibiotic treatment	Fluoroquinone, chloramphenicol and amoxicillin[a] are used to treat.	Antibiotics are not advised for the treatment of systemic illness[b]
Vaccines	Vaccines available in endemic areas[c]	Not available

[a] *Depending on the pattern of antibiotic resistance in the area, the severity of the sickness, the availability of antibiotics, and the cost.*
[b] *Fluoroquinone are usually suggested if antibiotic treatment is necessary.*
[c] *Licensed vaccinations with effectiveness of 60%–80% and protection lasting up to 7 years are available.*

4.3.1 Pathogenesis and virulence

In humans, the Salmonella bacterium causes three different types of salmonellosis:

1. Noninvasive and nontyphoidal
2. Invasive and typhoidal
3. Typhoidal

The virulence of *salmonella* species resides there in both with chromosomal and plasmid genes (Chaudhary et al., 2015). The genes that determine virulence in bacteria are found in vast gene clusters called pathogenicity islands on the bacterial chromosome (SPIs). SPIs encode over 60 genes that are involved in specific interactions between bacteria and their hosts (Chaudhary et al., 2015; Lahiri et al., 2010). The infection cycle begins just after the host consumes salmonella-contaminated food or water. Salmonella enters the small intestine

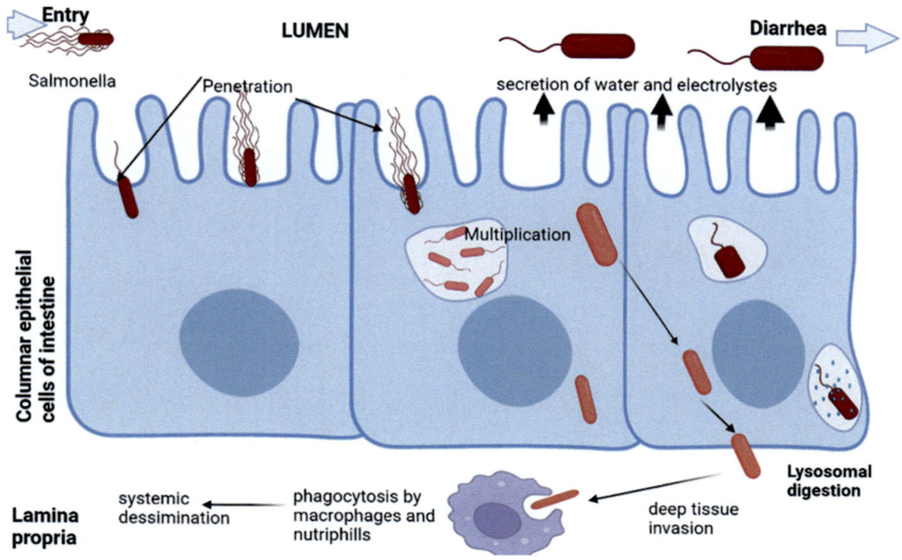

FIGURE 4.2A Showing the pathway of salmonella of reaching to epithelial cell of intestine and formation of salmonella containing sacs (SCV), and multiplication of salmonella with the cell.

after passing through the stomach and attacks the ileum's gut epithelium (Peyer's patches). They come to a stalemate and begin to proliferate until they reach a critical mass. The amount of bacteria present, their pathogenicity, and the host's immune response are likely to define the critical point (Buckle et al., 2012). *Salmonella* can be taken up by M-cells in the ileum, caught in lumen by CD18-expressing phagocytic cells that pierce the monolayer epithelial cells, or bacteria may enter into nonphagocytic enterocytes, thus forming the Salmonella-containing vacuoles (SCV). The maturing SCV moves toward Golgi apparatus where it shows selective interaction with the host endocytic pathway (Hafsa et al., 2021; Mir, 2020a,b,c,d). On reaching the perinuclear area the salmonella enclosed in SCV replicate; this stage is characterized by the development of salmonella-induced filaments (Sifs) (the tubulovesicular SCV structures). Most of the *Salmonella* infections remain localized to the intestine, where stimulation of inflammatory response results in diarrhea Fig. 4.2A and B.

The microbes enter the circulation and spread throughout the body during a 7–14 days incubation period. Some bacteria are shed in the feces, ready to infect another person (Buckle et al., 2012).

The virulence of *Salmonella* species depends on the bacterial nature, i.e., serotype and the hosts immunity. Besides, the virulence is defined by various virulence factors (Fig. 4.3).

4.3.2 Symptoms of salmonellosis

Diarrhea, stomach discomfort, low fever, chills, uneasiness, vomiting, headache, anorexia, and exhaustion are the most common symptoms of *Salmonella* infection in humans. Symptoms usually occur in 12–36 h of exposure to pathogen and usually last for 2–7 days. It is

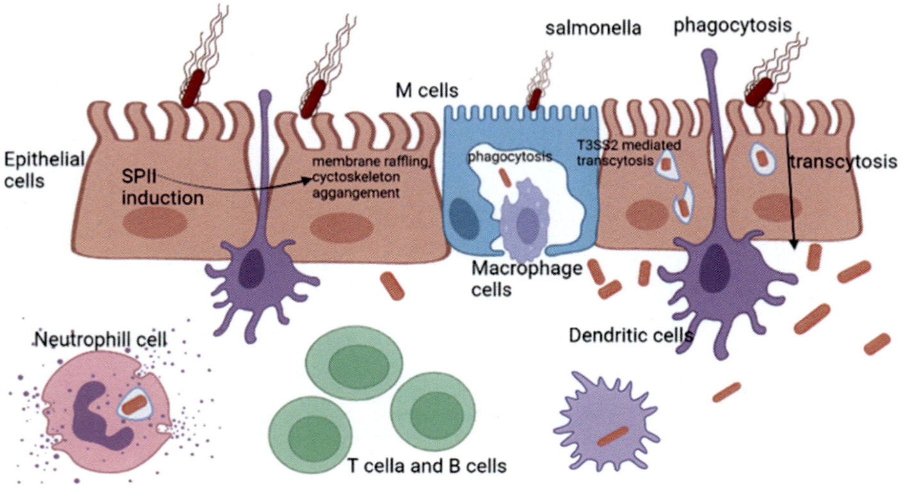

FIGURE 4.2B Sowing the adherence of salmonella to epithelial cells in luman whch then by transcytosis pass through mono-layered epithelial cells and reach payers patch, some are taken up by M cells and some salmonella bacterium face the phagocytosis by macrophages.

most severe in children and the elderly, and as a result, death is more common in these age groups. The acute stage lasts for a few weeks, but the carrier stage lasts for more than 3 months and can lead to consequences such as damage to the mucosal surfaces of the small intestine and colon, sensitivities, and severe sickness, which is especially common in malnourished people (Murphy, 1981; Mosher, 1988). Enteric fever is caused because of *S. typhi*, *S. paratyphi A*, *S. paratyphi B*, and *S. paratyphi C*.

The most frequent symptoms of enteric fever are lethargy, discomfort, high and persistent fever, body pains, abdominal cramps, uneasiness, vomiting, coughing, chills, and anorexia. It takes 7–28 days of incubation period, depending on the effective dose.

4.3.3 Source

Salmonellosis is caused mainly due to contaminated food or water; besides, it also spreads directly via fecal-oral route, for example, in patients in hospitals and institutions and among farmed animals. *Salmonella*, a naturally occurring zoonotic pathogen, has been connected to farm and wild animals including mammals, birds (poultry), reptiles, amphibians, and arthropods. *Salmonella* infections related with reptiles are more likely to result in hospitalization and involve infants than other *Salmonella* infections (Qadri et al., 2021).

They colonize in intestinal tract and other associated organs of organisms. Excreta from infected humans and animals are main source of contaminations of environment and food chain.

4.3.4 Transmission

Animal food products contaminated with animal excrement, including eggs, chickens, turkey, undercooked minced meat, dairy products, and fresh produce, are the most prevalent

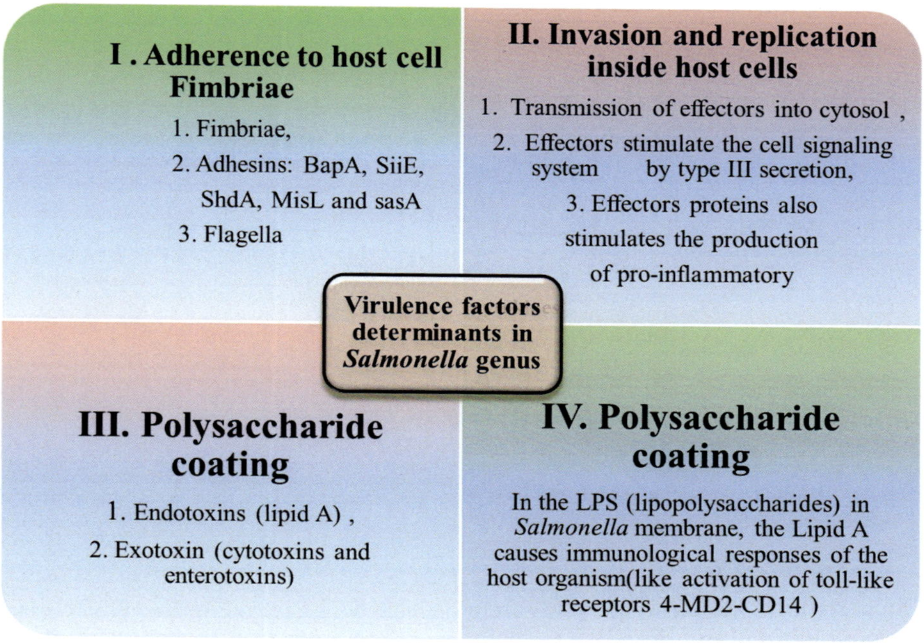

FIGURE 4.3 Represents the different virulence factors which define the pathogenicity of salmonella genus. It shows four major aspects 1. Adherence of pathogen to host cell, 2. Invasion and replication with the host cell, 3. Toxins released which help pathogen in survival and 4. Polysaccharide coating.

sources of transmission. Contamination can occur at many stages, including during slaughter, dressing, and preparation of raw meats, which are a major source of Salmonella bacterium in the human food chain. Overall, approximately 1%–3% of domestic animals are believed to carry *Salmonella* species but their prevalence can be higher in some cases. Furthermore, according to reports, roughly 50% of broilers are contaminated with Salmonella, although beef and lamb contamination is normally less than 1%, depending on the source (Ross, 1989). Vegetables are also contaminated but the prevalence and contamination level is much lower than that of meat. As far as climate is concerned, most of the foodborne infections occur during summers and the most common serotype reported are *S. typhimurium, S. enteritidis,* and *S. newport* (Control and Prevention, 2006) (Fig. 4.4).

4.3.5 Clinical manifestation and pathophysiology of salmonellosis

Out of 31 pathogens reported, *Salmonella* species has the highest capability for causing intestinal or systemic illness in humans among diarrheal and/or invasive agents, according to the World Health Organization (WHO) (viruses, protozoa, bacteria, helminths, and chemicals) (Organization, 2019; Tauxe et al., 2010). Gastroenteritis, bacteremia, endovascular infection, and localized infections are all clinical manifestations of nontyphoidal salmonellosis.

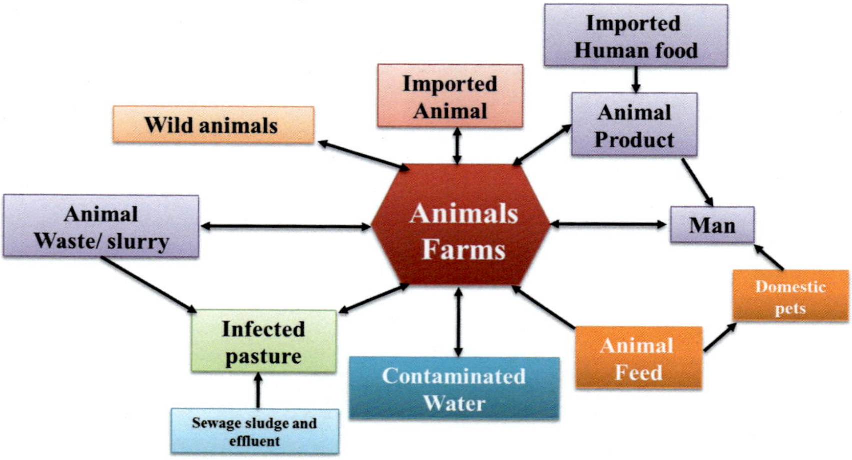

FIGURE 4.4 Shows factors contaminating the food chain and sources of transmission and infection of salmonella.

4.3.5.1 *Gastroenteritis*

The most frequent clinical manifestation of *Salmonella* is gastroenteritis, which manifests as uneasiness, vomiting, and diarrhea 6–48 h after consuming infected food or drink. Fever, chills, stomach pains, and headache are some of the other symptoms. Severe right quadrant pain can develop on occasion, simulating appendicitis (Arda et al., 2001). A clinical blood test reveals an increase in peripheral white blood cells (10,000–15,000 cells/mm^3), and a microscopic observation of feces reveals leukocytes and occasionally erythrocytes. Gastroenteritis in most cases is self-limiting and lasts for 3–7 days. Sometimes, it may require hospitalization because of severe diarrhea and dehydration. The incidence of such cases where patient needs to be admitted is 2.2 per one million persons per year. Severe symptoms are more common in older people, newborns, and severely immunocompromised individuals (e.g., transplant recipients, HIV-infected adults) (Mir, 2020a,b,c,d). After an acute infection, *Salmonella* stays in the gastrointestinal tract for around 4 weeks; neonates and children can shed bacteria for up to 7 weeks (Buchwald and Blaser, 1984).

4.3.5.2 *Bacteremia and endovascular infections*

Bacteremia is one of the prevalent clinical manifestations of *Salmonella* which affects roughly 1%–4% of immunocompromised patients (Buchwald and Blaser, 1984). Some serovars, such as *S. choleraesuis* and *S. dublin*, have a natural proclivity for causing bacteremia. Individuals (adults) with *Salmonella* gastroenteritis are more prone than toddlers to develop complicated illnesses (Shimoni et al., 1999). Infants, the geriatric, and those with impaired immune systems (such as AIDS patients, transplant recipients, and those with malignancies or autoimmune illnesses) are all at risk for bacteremia (Pegues, 2005). Salmonellosis is up to 100 times more common in HIV patients than in the regular populace (Celum et al., 1987). *Salmonellosis* cases in HIV patients have reduced, owing to the development of powerful antiretrovirals such as zidovudine, which has anti-*Salmonella* action, and the use of trimethoprim-sulfamethoxazole (TMP-SMX), which tackles both *Pneumocystis jiroveci* pneumonia and *Salmonella*. In individuals with *Salmonella* bacteremia, clinicians should look for an immunosuppressive condition or anatomic risk factor, especially if they have never had gastroenteritis or have had recurrence (e.g., kidney or biliary stone, endovascular lesion) (Hohmann, 2001).

Endovascular infection: In high-grade bacteremia, nontyphoidal salmonellosis endovascular infection is strongly suspected, especially in patients with valvular heart disease, atherosclerotic vascular disease, or aortic aneurysm. Arteritis is suspected in senior people who have a protracted fever and back, chest, or abdominal discomfort following gastroenteritis bouts. Endocarditis and arteritis are uncommon (less than 1% of cases), but they can cause life-threatening consequences such as valve rupture, endomyocardial abscess, cardiogenic shock, and aneurysm rupture. Around 10%–25% adults suffering from bacteremia face vascular complications (Pegues, 2005; Cohen et al., 1978).

4.3.5.3 *Localized infections*

Bacteremia can develop localized infections such endocarditis, pneumonia, empysema, encephalopathy, osteomyelitis, or septic arthritis in 5%–10% of patients. Cases of arthritis and arteritis are uncommon (Pegues, 2005). Patients suffering from sickle cell disease are more likely to develop *Salmonella* osteomyelitis. Furthermore, *Salmonella* can directly infect joints, causing arthritis (usually monoarticular); nevertheless, reactive polyarthritis can develop following gastroenteritis. Nontyphoidal salmonellosis (NTS) can cause hepatotoxicity, splenomegaly, necrotizing enterocolitis, cholangitis, and splenic and hepatocellular abscesses; however these are unusual gastrointestinal symptoms (Pegues, 2005; Cheng-Chi et al., 2002; Torres et al., 1994).

4.3.6 Diagnosis and treatment

Diagnosis:

- *Salmonella* bacteria can be detected in feces, tissue samples, or bodily fluids through laboratory testing.
- The test might be a culture containing bacterial isolates or a culture-independent diagnostic test (CIDT) that identifies the genetic material of bacteria.

The Centers for Disease Control and Prevention (CDC) emphasize the use of culture specimens with CIDT findings, a method known as "reflex culturing."

- Once the test results are reported to the doctor, *Salmonella* isolates are sent to state public health laboratories for serotyping and DNA fingerprinting.
- Public health laboratories then send *Salmonella* case reports to the CDC's Laboratory-based Enteric Disease Surveillance and PulseNet.
- Public health laboratories send uncommon serotypes to the Centers for Disease Control and Prevention's National Salmonella Reference Laboratory for alignment, conformation, and characterization.

Some laboratory tests used to detect *Salmonella* infections are

- Widal test method: The Widal test technique identifies the presence of *Salmonella* antigens 0 and H.
- ELISA: it detects the antibodies (IgM and IgG) against *Salmonella* surface molecules.
- Typhidot ELIZA kit: It is used to detect IgG and IgM with the sensitivity and specificity of >95%, and 75%, respectively.
- PCR: By using this method, the *Salmonella* flagellin gene (fliC), polysaccharide capsule gene, and virulence (vi) gene are focused using a molecular biology–based technique (tvi and tviB). When compared to other approaches, it is said to have higher specificity and sensitivity (Medalla et al., 2017).

Treatment: most of the patients suffering from *Salmonella* infection recover in 4–7 days and do not need antibiotics or other specific treatment. But patients need to drink plenty of fluids to compensate the fluid loss by diarrhea. In some cases patients with severe diarrhea need to be hospitalized. Antibiotics are required to treat individuals with severe illness. Antimicrobial therapy is required only for patients with severe conditions and signs of invasive disease and systemic disease (Mir, 2015a,b).

Children under 1 year should be treated with antimicrobials preferably to prevent invasion. Usually, 3–7 days of treatment is required and is reasonable (Coburn et al., 2007; Hohmann, 2001).

Fluoroquinolones, trimethoprim-sulfamethoxazole (TMP-SMZ), ampicillin, cephalosporins (e.g., ceftriaxone or cefixime), azithromycin, and aztreonam are among the antibiotics used to treat *Salmonella* infection (Kumar and Kumar, 2017). When a condition such as delirium, obtundation, amnesia, coma, or shock arises, dexamethasone, a corticosteroid medication, may be used (Kumar and Kumar, 2017; Mir and Albaradie, 2014) (Table 4.3).

Vaccines: Vaccines for NTS salmonellosis are currently unavailable, but for typhoid fever two vaccines are available: oral Vivotif and injectable typhoid vaccine. The CDC recommends typhoid fever vaccination for all Americans traveling to developing countries.

Current challenges with the use of antibiotic drugs as a therapy for *Salmonella* infections:

Bacterial infections are being treated by using antibiotics since long; however, most of the strains are developing the resistance to current antibiotics which is a major concern for the

TABLE 4.3 Some common antibiotics used with the required dose in different cases of *Salmonella* infection.

Drug class/drug name	Route	Standard dosage
Fluoroquinolone antibiotic/ceprofloxacin Ofloxacin	Oral oral	500 mg every 12 h. 200 mg every 12 h.
Macrolide/Azithromycin	Oral	250 mg twice on day 1, then 250 mg once a day for the next 2–5 days.
Cephalosporin/Ceftriaxone	Injection	1–2 g injected daily in divided doses.
Carbapenem/Ertapenem	Infusion	1 g infused daily.

Note: Dosage shown in the table is as per U.S. Food and Drug Administration (FDA).

mankind in current times (Mehraj et al., 2020). Antibiotic resistance genes that are/or have been employed in animals have been discovered not only in animal bacteria, but also in zoonotic pathogens like Salmonella and solely human pathogens like Shigella. It has been established that resistance determining factors can be transferred between unrelated bacteria, such as *Bacteroides, Salmonella,* and *E. coli* (O'Brien, 2002). Antimicrobial resistance in nontyphoidal *Salmonella* serotype is a worldwide problem. Antibiotic drug resistance can be inherited or induced when bacteria are exposed to antibiotics. Bacteria use the genetic mechanism of conjugation, transduction, and transformation to acquire antibiotic-resistant genes (Sabtu et al., 2015). *Salmonella* bacteria have become resistant to both first-line antibiotics and alternative therapeutics (Crump et al., 2015). In Malawi, 7% of *S. Typhi* infection cases were reported to be multidrug resistance in 2010, and by 2014, that percentage had jumped to 97% (Feasey et al., 2015; Wong et al., 2015). Due to their resistance to ceftriaxone and ampicillin, *S. enteridis* was discovered to be liable for 50% of ciprofloxacin-resistant infections in the United States, while *S. newport, S. typhimurium,* and *S. heidelberg* were seen to be responsible for 75% of antibiotic-resistant illnesses (Medalla et al., 2017).

Alternative treatments:

The alternative treatments used for *Salmonella* infections are probiotics (Carter et al., 2017). Probiotic bacteria display their preventative and therapeutic activities against pathogenic microorganisms through regulating inherent and adaptive immunity, acting directly on pathogens, and producing antibiotic chemicals (Oelschlaeger, 2010; Mir et al., 2021).

Probiotic bacteria present in the human gastrointestinal system include *Lactobacilli, Enterococci, Bifidobacteria, Pediococcus, E. coli, Streptococcus,* and *Leuconostoc* species (Priyodip et al., 2017; Collado et al., 2007) and are known to show efficiency in preserving the popular cultured functional foods and facilitate their distribution (Priyodip et al., 2017; Plessas et al., 2017).

The yeast also shows the therapeutic efficacy against some bacterial infections including Salmonella (typhoid, paratyphoid, and NTS). The yeast strain *S. boulardii* is currently being used as a probiotic (Kelesidis and Pothoulakis, 2012; Tomičić et al., 2016).

4.3.7 Prevention and control

Salmonella prevention and control can be achieved by taking some measures especially in agricultural practices and food handling. No single measure alone can control the salmonellosis disease. One of the most important accomplishments in the fight against zoonotic infectious illnesses was the implementation of the coordinating Salmonella control program by EU. Before 2004, almost 200,000 human salmonellosis incidences were reported in 15 EU Member States; however, due to the introduction of control measures, the number of cases was reduced to 90,000 incidents per year in all 28 EU Member States (Ehuwa et al., 2021). In 2018, the WHO announced recommendations for managing *Salmonella* across the entire food system. These recommendations were aimed at enhancing *Salmonella* surveillance efforts, addressing consumers, and training staff members on how to avoid *Salmonella* and other foodborne infections in order to improve food safety (Table 4.4). In the fight against salmonellosis, the WHO is aiming to improve the efficiency of national and regional laboratories (Ehuwa et al., 2021).

TABLE 4.4 Guidelines proposed by the WHO.

Recommendations	Objectives
Preventive methods	• From production to consumption, precautionary interventions should be applied at all levels of the food chain. • The food handler handbook's recommended *Salmonella* protection actions should be implemented. • When children interact with domestic animals such as cats, dogs, and pet reptiles, they should be monitored. • People are urged to follow national and regional foodborne illness surveillance systems to be updated about salmonellosis outbreaks and prevent them from spreading.
Guidelines for public and travellers	• Food should be adequately cooked and served hot at all times. • Use processed and boiled milk and milk products exclusively. • Food and vegetables should only be served after thorough cleaning and washing. • After contacting the animals and using restroom, the hands should be properly cleaned and sanitized.
Guidelines for food handlers	• Hygienic preparation rules should be followed. • Follow the five basic provisions, which serve as the foundation for both food handlers and customers' food safety training. • It entails maintaining cleanliness, separating raw and prepared meals, properly cooking, storing food in a proper manner and at the proper temperature, as well as the use of potable water.
Guidelines for farmers	• Maintain good personal hygiene. • Avoid fecal pollution. • Use treated and well-managed irrigation water.
Guidelines for producers of aquaculture products	• Practice good personal hygiene. • Clean pond environment. • Use hygienic harvest equipment. • Ensure healthy fishes.

4.3.7.1 Surveillance systems

The CDC has a number of surveillance methods for retrieving Salmonella-related data. They serve as a platform for gathering data on various facets of an organism's epidemiology, including the prevalence of outbreaks, antimicrobial-resistance infections, and subtypes. The surveillance systems used are mentioned below:

1. **Foodborne Disease Outbreak Surveillance System (FDOSS).** It serves to provide the information related to pathogenic agents and foods responsible for outbreaks.

   Websites:

- Foodborne Disease Outbreak Surveillance System (FDOSS)
- National Outbreak Reporting System (NORS)
- NORS Dashboard

2. **Foodborne Disease Active Surveillance Network (Food Net):** It keeps track of foodborne infection patterns and the effects of food safety policies.

   Websites:

- Food borne Disease Active Surveillance Network (Food Net)
- CDC's Emerging Infections Program

3. **Laboratory-based Enteric Disease Surveillance (LEDS):** Serotyping is included in the records of *Salmonella* isolates that cause human illnesses.

   Websites:

- Laboratory-based Enteric Disease Surveillance (LEDS)
- Enteric Diseases Laboratory Branch (EDLB)

4. **National Antimicrobial Resistance Monitoring System—Enteric Bacteria (NARMS):** Antimicrobial resistance in intestinal bacteria, such as *Salmonella* from individuals, retail meats, and animals, is recorded.

   Websites:

- National Antimicrobial Resistance Monitoring System—Enteric Bacteria (NARMS)
- US Department of Agriculture (USDA)
- Food and Drug Administration (FDA)

5. **National Molecular Subtyping Network for Foodborne Disease Surveillance (PulseNet):** it analyses and reports the illness cases nationwide to quickly identify the outbreaks.

   Websites:

- PulseNet
- Association of Public Health Laboratories (APHL)
- Pulsed-field Gel Electrophoresis (PFGE)
- PulseNet International
- Next Generation PulseNet

6. **National Notifiable Diseases Surveillance System (NNDSS)**

   Websites:

- National Notifiable Diseases Surveillance System (NNDSS)
- Data Collection and Reporting

Incidences of salmonellosis infections:

Despite the availability of several antibiotic medications, salmonellosis is the greatest cause of illness and death worldwide. *Salmonella* causes 2.8 billion cases of diarrhea each year, according to estimates and reports. *Salmonella typhi* is shown to cause 16–33 million infectious cases and 500,000 to 6,000,000 fatalities annually, whereas NTS is shown to cause 90 million infectious cases and 155,000 deaths annually, with an average frequency of 1.4 episodes per 100 people (Bula-Rudas et al., 2015). Asia had the highest number of instances, with an estimated 83.4 million illnesses, 137.7 thousand deaths, and average incidences of 4.72/100 people per year (Majowicz et al., 2010).

Some outbreaks of salmonellosis are discussed below:

A statistical analysis of data of salmonellosis outbreak in Canada from 1996 to 2005 showed salmonellosis is primarily the foodborne disease which is estimated as 15% by poultry, 15% by meat, 9% by dairy products and 66% by seafood (Ravel et al., 2009). In USA the outbreaks of salmonellosis reported during 2006–08 estimated poultry (chicken and turkey) contributing 30%, eggs 24%, pork 9% and beef 8% to be the cause (Gould et al., 2013). The medical cost of eradicating *Salmonella* in the United States was $2.17 billion in 2010 (for 1.4 million infections) (Andino and Hanning, 2015). There are 2000–7500 salmonella illness cases per 100,000 HIV-positive patients in Africa, according to reports (Feasey et al., 2012). There were 40,000 salmonellosis cases reported in Australia, with 2100 hospitalizations, 6750 sequelae, and 15 deaths (Feasey et al., 2012). Between 2000 and 2013, the National Notifiable Diseases Surveillance System (NNDSS) received 127,195 reports of *Salmonella* illness (Feasey et al., 2012). From 2000 to 2009, New Zealand faced a *Salmonella* outbreak; during study the main transmission pathway and food source were analyzed (King et al., 2011). Foodborne transmissions accounted for 63% of the 123 epidemic cases, with just one mode of transmission documented, followed by person-to-person transmission (32%), waterborne transmission (3%), and zoonotic transmission (1%). (This study supported the theory that salmonellosis is mostly a foodborne infection. In Latin America and the Caribbean, another outbreak was investigated (Pires et al., 2012). Data was collected for the period 1993 to 2010 from 20 different nations. The proportion of cases of illness caused by eggs (from 17% to 43%) and pork (4–9%) and vegetables increased from the 1990s to the 2000s, according to the findings (10%–12%). In contrast, the proportion of illnesses caused by beef products (29%–9%) and chicken (12%– 6%) has decreased over the same time period. A statistical investigation of *Salmonella* outbreaks in the United States revealed that poultry (chicken and turkey) is the most common causes of outbreaks, accounting for 30% of all cases, and eggs accounting for 24%. Other foods such as pig (9%) and beef (8%) showed lower attribution proportions (Gould et al., 2013).

4.4 Tuberculosis

Tuberculosis is an infectious disease caused by *Mycobacterium tuberculosis*. It is the prominent cause of death attributed to a single pathogen globally (Organization, 2013, MIR et al.). According to the WHO, around ten million individuals worldwide contracted tuberculosis for the first time, and 1.5 million people died because of this disease (Organization, 2013). *M. tuberculosis* infects approximately one-third of the population worldwide (Zumla et al., 2013; Mir and Al-baradie, 2013). Besides infecting one-third of the population, only 10%–20% of individuals develop active disease in their life time, rest remain asymptomatic and are referred as "latent."

M. tuberculosis is a Gram-negative, acid-fast aerobic bacillus with humans as the only reservoir. The *Mycobacterium TB* complex includes Mtb and seven closely related mycobacterial species, notably *M. bovis, M. africanum, M. microti, M. caprae, M. pinnipedii, M. canetti,* and *M. mungi.* Majority of the species, but not all, are capable of infecting humans.

Transmission: Tuberculosis is mainly spread through the air from one person to another. Microscopic droplets containing *M. tuberculosis* are discharged into the air when an infected person coughs, speaks, or sighs (Society et al., 2005). These air particles containing bacterium are called as droplet nuclei (1−5 micron in diameter) and depending on the atmosphere, it can stay suspended in the air for several hours (Wells, 1955). Inhaled *M. tuberculosis*-containing droplet nuclei migrate through the mouth or nasal passage, upper respiratory tract, and bronchi to reach the lungs, initiating infection (Fig. 4.5).

The likelihood of tuberculosis transmission is determined by the patient's characteristics as well as environmental elements. Patients with lung and throat tuberculosis, for example, are more infectious when their chest radiographs show cavitation, and they cough without covering their mouth and nose. Environmental factors that increase the possibility of tuberculosis transmission include *M. tuberculosis* exposure in a compact, enclosed room or place with recirculated air holding droplet nuclei (Jensen et al., 2005) (Fig. 4.6).

Pathogenesis: Majority of the droplet nuclei carrying *M. tuberculosis* lay in the upper respiratory tract after inhalation. When the minute droplets nuclei reach the alveoli of the lungs, they proliferate and move throughout the body via blood and lymph thereby causing infection. Tuberculosis can affect the kidneys, the brain, and the bones in addition to the lungs. Within weeks of infection, healthy persons with good immunity can stop the bacilli from multiplying, thus limiting their further spread. A person with latent tuberculosis is not considered as case of tuberculosis and is not infectious; therefore, he/she can't be the reason of transmission of tuberculosis (Control and Prevention, 2013a). But, when immune system fails to stop the multiplication of tuberculosis bacteria in the body, the latent tuberculosis progresses to develop into tuberculosis disease. According to reports, about 10% of untreated latent tuberculosis infections advance to tuberculosis illness (Horsburgh, 2004).

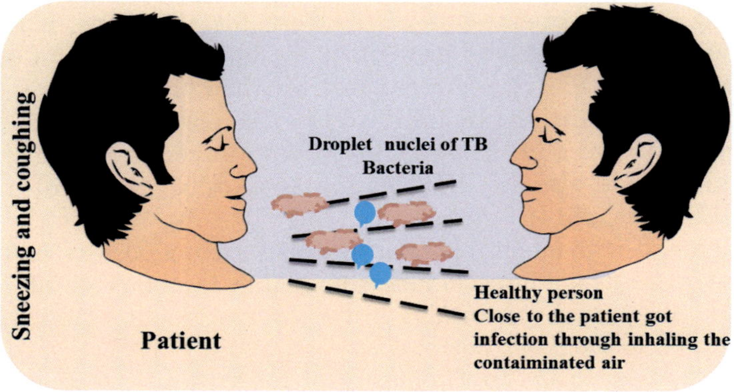

FIGURE 4.5 Represents the transmission of *M. tuberculosis* through droplet nuclei from ionfected person to healthy person.

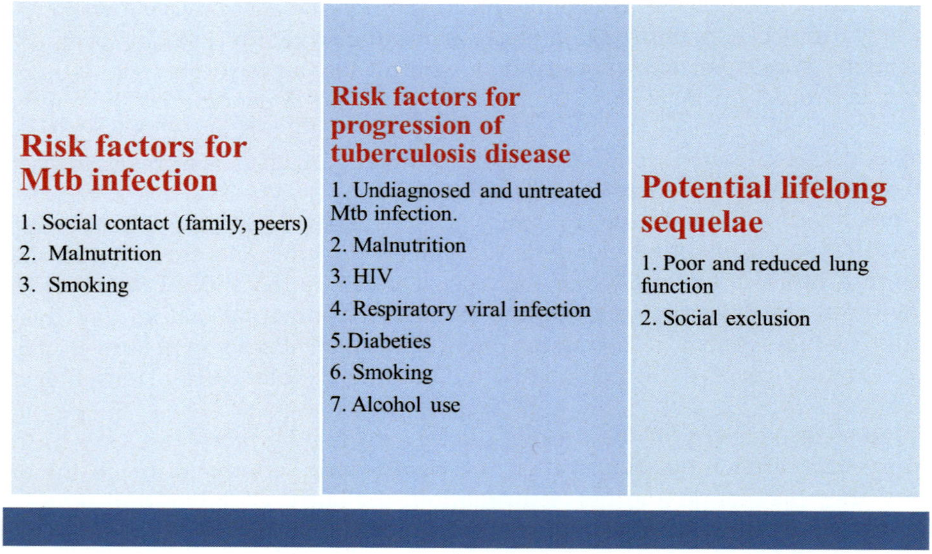

Risk factors for Mtb infection
1. Social contact (family, peers)
2. Malnutrition
3. Smoking
Risk factors for progression of tuberculosis disease
1. Undiagnosed and untreated Mtb infection.
2. Malnutrition
3. HIV
4. Respiratory viral infection
5.Diabeties
6. Smoking
7. Alcohol use
Potential lifelong sequelae
1. Poor and reduced lung function
2. Social exclusion

FIGURE 4.6 Represents different risk factors for infection, progression and potential consequences when person contracts the tuberculosis.

4.4.1 Clinical manifestation, testing and diagnosis

Cough that lasts 3 weeks or more, chest pain, exhaustion, loss of appetite, chills, fever, coughing up blood or sputum, and nocturnal sweating are all symptoms of tuberculosis.

The assays that are used to detect *M. tuberculosis* in the human body are:

1. TST test (tuberculin skin test)
2. IGRA test (interferon-gamma release assays)

These tests only reveal the presence of *M. tuberculosis* infection, not disease progression. By evaluating the immune response to tuberculosis proteins, IGRAs are used to determine whether *M. tuberculosis* is present in the body (Lewinsohn et al., 2017; Mir and Agrewala, 2007). The TST is preferred test for the children below the age of five because of difficulty in getting the blood draw (Lewinsohn et al., 2017) and IGRAs are preferred for 5 years of age and above children who have also been immunized with BCG, even though TST is also used (Pickering et al., 2012; Control and Prevention, 2010; Starke and Diseases, 2014).

A person is diagnosed with latent tuberculosis if they have a positive IGRA or TST result but their medical status does not indicate tuberculosis (Lewinsohn et al., 2017). Asking about a person's medical background, performing a medical assessment, chest radiography, and diagnostics are all used to diagnose tuberculosis (Mir, 2013).

To select the most appropriate drug regimen for tuberculosis infection treatment, a test called a drug susceptibility test (DST) is employed (Control and Prevention, 2013b).

4.4.2 Pathophysiology

TB disease targets lungs and can be at extrapulmonary sites as well. Tuberculosis most commonly affects the lungs called as pulmonary TB; sometimes TB could target almost any anatomical site including brain, kidneys bones, joints, larynx, and lymph nodes. Miliary tuberculosis develops when *M. tuberculosis* enters the bloodstream and spreads to all areas of the body, where it grows and multiplies, causing infection at many locations. It is particularly common among newborns and children under the age of five, as well as immune-compromised individuals. Tuberculosis meningitis is when TB affects the tissue around the brain or spinal cord. Headache, neck stiffness, and unconsciousness are some of the symptoms (Mir et al., 2013; Mir and Albaradie, 2013).

Treatment: The compounds having prospective antimycobacterial efficiency which are currently included in the list of WHO antituberculosis drugs such as Ethambutol, aminosalicylic acid, isoniazid, pyrazinamide, clofazimine, bedaquiline, cycloserine, rifabutin, amoxicillin-clavulanic acid, delamanid, ethionamide, streptomycin, ethionamide, moxifloxacin, meropenem, rifapentine, levofloxacin and linezolid (WHO, 2019).

Antitubercular drugs classification: Antitubercular medications are divided into four categories: A, B, C, and D (Hancock and Diamond, 2000).

- Group A: The fluoroquinolones are a subgroup of this category (high doses of levofloxacin, moxifloxacin, and gatifloxacin). These drugs are called as vital products because of bactericidal efficacy and strong safety profile.
- Group B: This category contains injectables such as (streptomycin, kanamycin, amikacin, and capreomycin). These medications have a strong bactericidal efficacy; however, they have a lower safety rating than medications in Group A.
- Group C: Ethionamide, cycloserine/terizidone, clofazimine, prothionamide, and linezolid are all members of this category. These drugs are more effective and are suggested as essential second-line treatments for tuberculosis patients who are multidrug-resistant.
- Group D: There are three subgroups within the group
  1. D1: Isoniazid, ethambutol, and pyrazinamide in massive doses
  2. D2: delamanid and bedaquiline
  3. D3: meropenem, cilastatin-imipenem, clarithromycin,clavulanate-amoxicillin, and para-aminosalicylic acid,

There are several novel drugs on the horizon that could be used as a supplement to the above-mentioned antitubercular regimen. By improving the efficacy and effectiveness against *Mycobacterium tuberculosistuberculosis*-resistant strains (such as isoniazid hydrazides and nitazoxanide) or by negating the inhibition, the inclusion of such tuberculosis medications will lead to improved outcomes (like dihydropyridomycin). In the fight against tuberculosis, many new candidate medicines are being researched to be used as monotherapy or combination therapy options. In addition, a number of pharmacological candidates with direct and/or indirect antitubercular activity, such as sildenafil, mefloquine, cilostazol, tizoxanide, metronidazole, entacapone, tolcapone, and ibuprofen, also demonstrated excellent efficacy in in vitro trials as mycobacterial monotherapy or adjunctive action (Harausz et al., 2016; Vilaplana et al., 2013; Sheikh et al., 2021) Table 4.5.

TABLE 4.5 Some promising novel chemotherapeutics.

Class of drug	Lead compound	Mechanism of action	References
Isoniazid derivatives	2-Cyano-N-(4-methylphenyl) acetamide, 2-chloro-N-(2,6-diethylphenyl) acetamide, 2-Phenylacetamide	Both mycolic acid production and cell growth are inhibited by these medicines.	Castelo-Branco et al. (2018), Loots (2014), Rajkhowa and Deka (2014)
Coumarin lead derivatives	6-((3,3-Dimethyloxiran-2-yl)-5,7-dihydroxy-8-(2-methylbutanoyl)-4-phenyl-2H-chromen-2-one (1 g), dimethyl substituted compound (1e), coumarin-oxime ether (1h), a coumarin-theophylline hybrid (3a), LSPN270, LSPN271, LSPN476, and LSPN484	Acts as a cell proliferation inhibitor, blocking cytochrome production and activating macrophages.	Reddy et al. (2018), Mangasuli et al. (2018), Adeniji et al. (2020), Pires et al. (2020), Kapp et al. (2017)
Griselimycin lead derivatives	Cyclohexyl griselimycin	It obstructs the repair and replication of DNA.	Dong et al. (2017), Kling et al. (2015), Holzgrabe (2015), Lukat et al. (2017)
Antimicrobial peptides (AMPs)	Bacteriocins (Bcn1−Bcn5), protegrin-1, nisin S, D-V13 K, cathelicidin LL37, D-LAK120-A, and D-LAK120-HP13	It controls the release of proinflammatory cytokines, calcium influx, and apoptosis.	Arranz-Trullen et al. (2017), Lee et al. (2015), Gupta et al. (2017), Rivas-Santiago et al. (2013), Jiang et al. (2011), Sosunov et al. (2007), Carroll et al. (2010)

4.4.3 Prevention and control

The resulting increasing and incidences and prevalence of TB had global implications. The global casualty count from tuberculosis had risen to 1.5 million by 2014, urging WHO to develop an action framework outlining strategies for eradicating the global TB epidemic, progressing low nations toward pre-elimination by 2035, and further advancement in the same nations toward completely eliminating TB as a public health problem by 2050 (Organization, 2020). CDC reported the national strategic plan (Cole et al., 2020; Chiswick and Miller, 2015) and in 2015, the action plan was released by U.S. government to specially addressing drug-resistant TB.

The three main fundamental approaches for tuberculosis prevention and control are listed below:

1. Identification of someone who has active tuberculosis and completion of treatment to render their ailment noninfectious.
2. Screening: Those who have come into contact with TB patients are being screened to see:
   - If they have active TB disease,
   - If they have been contaminated with *M. tuberculosis*, or
   - If they need window prophylaxis or treatment.

3. Locating those at significant risk of developing latent tuberculosis infection (LTBI) and eventually active tuberculosis disease through screening, testing, then treatment in order to identify those who potentially benefit from LTBI therapy, which is crucial for TB eradication.

Although the three primary strategies listed above are focused on individual patients, they serve larger epidemiologic purposes by reducing current and future *M. tuberculosis* transmission in the community.

4.5 Cholera

Cholera (meaning being "a gutter") is caused by the *Vibrio cholerae*. Cholera continues to be a severe hazard to humanity in many regions of the world (Legros, 2018), especially in areas where there are complicated situations such as inadequate water, sanitation, and hygiene (Spiegel et al., 2007; Shannon et al., 2019). Cholera remains a major cause of disease outbreaks worldwide (Smith et al., 2014; Ganesan et al., 2020), with enhancing rate and intensity (Cholera, 2017). An estimated 2.9 million cholera infection cases (with a 1.3–4.0 million uncertainty range) and 95,000 fatalities were reported in 69 cholera-endemic countries (21,000–143,000 uncertainty range) each year (Ali et al., 2015).

Vibrio cholerae is a Gram-negative, flagellated bacterium with a comma shape. Based on somatic O antigen, more than 206 serotypes of *V. cholerae* have already been discovered, with *V. cholerae* O1 and O139 being the epidemic serogroups (Yamai et al., 1997). Ogawa and Inaba are the two primary serotypes of *V. cholerae* O1. They can also be categorized into two biotypes, classical and El Tor, based on biochemical features and responsiveness to *V. cholerae* O1-specific bacteriophages (Palleroni et al., 1984).

4.5.1 Cause and mechanism

The bacterium *Vibrio cholerae* causes cholera, which is spread by the feces-to-mouth route. After ingestion it colonizes in the small intestine, where it secretes the 2-subunit cholera toxin (A and B subunit). The B subunit of toxin binds to epithelial cell surface and the A subunit of toxin triggers the biochemical cascade which initiates the diarrhea which in turn leads to severe dehydration (Sánchez and Holmgren, 2005). In the body, the immune system reacts to both the toxin and the bacterial cell wall (Mir, 2018). The O antigen of the cell wall lipopolysaccharide (LPS) serves as a protective antigen, particularly in vaccines (Jonson et al., 1996).

Transmission: A vulnerable person can contract cholera by ingesting germs from the environment, such as raw sewage or a fecal-infected food or water (Islam et al., 2020); this shows the occurrence of disease through the environment-to-human transmission pathway (Tien and Earn, 2010; Fung, 2014). Cholera infection can also be transmitted through infected and susceptible persons (Richterman et al., 2018; Deen et al., 2020) by consuming contaminated food (Shapiro et al., 1999; Burrowes et al., 2017) or water at point of use (POU) (Burrowes et al., 2017; Fredrick et al., 2015; Grandesso et al., 2014; Sugimoto et al., 2014) which has been contaminated by cholera infected person. This is known as human-to-human transmission pathway Fig. 4.7.

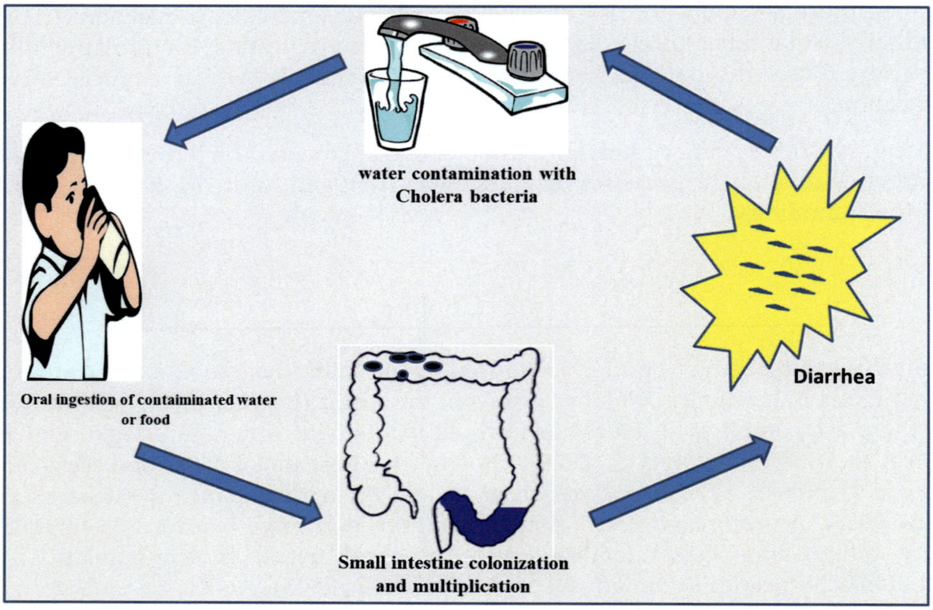

FIGURE 4.7 Showing cholera occurring through the contaminated water and its colonization site in humans and symptom.

Both the transmission pathways occur via fecal-oral routes of diarrheal disease transmission, popularly known as the F-diagram (Wagner et al., 1958). It is reported through the study on the genomics of cholera transmission that there are strong phylogenetic similarities between same household cases (Sugimoto et al., 2014; Domman et al., 2018; George et al., 2017, 2018; Rafique et al., 2016) and a study reported that about 80% of transmission occurs among people who share the same house (Domman et al., 2018).

4.5.2 Diagnosis and treatment

Diagnosis: Rapid diagnostic tests (RDTs) are used to check for the presence of the *V. cholerae* in patient stool samples and the environment and are preferred over routine culture techniques.

Treatment: Since the *Vibrios* enter through the mucosal port, so oral vaccination is preferred route of administration, as it can easily evoke the locally produced antibodies and trigger the immunological response (Holmgren and Czerkinsky, 2005). The oral vaccinations are given to the mobile population according to the prescribed dose regimen. i.e., two doses in 2–6-week intervals. During this the oral vaccine is stored and transported at 2–8°C.

The primary treatment for cholera is fluid replacement. If the patient can tolerate oral intake, ORS (oral rehydration solution) is usually chosen; however, IV fluids followed by ORS is given in cases of severe dehydration (Harris, 2012).

Antimicrobial therapy is recommended, along with fluid resuscitation, which had also been proven to lower diarrhea frequency and volume in cholera patients with extreme dehydration by approximately 50% (Nelson et al., 2011; Lindenbaum et al., 1967). Antibiotics also helped to limit *Vibrio* excretion, lowering the risk of transmission through fecal contamination when the patient returned to the community.

Vaccines: Cholera vaccines should only be used in conjunction with other preventative measures, according to the WHO. Shanchol (made by Shantha Biotech in India) and Dukoral (produced by SBL Vaccines) are the two WHO-prequalified oral cholera vaccines used against cholera (Lopez et al., 2014). Vaxchora, a single-dose live oral cholera vaccine, was licensed by the WHO in 2016 for anyone aged 18 to 64 who are traveling to a cholera endemic area (Cabrera et al., 2017). In cholera endemic countries the mass vaccination against cholera is not recommended. However, in high-risk populations such as children under the age of five, school-aged children, pregnant women, the geriatric and immunocompromised people, vaccination is required (Mir, 2015a,b).

4.5.3 Prevention and control

According to some research, considering the arena of transmission within the household and community level could aid with cholera prevention and management. WASH initiatives, Oral Cholera Vaccination (OCV), and in rare situations prophylactic medications are commonly used to prevent and control cholera (Codeco and Coelho, 2006). Community-level approaches are used to prevent cholera, with resources potentially aligned with longer-term WASH-related disease control activities (Montgomery et al., 2018). Currently global momentum is running which is supposed to tackle cholera and there is internationally agreed road map which is supposed to eliminate the disease by 2030 (Organization, 2017).

It is also documented and accepted that massive investments in water and sanitation infrastructure in the Americas and Europe resulted in the eradication of cholera and the reduction of the risk of other diarrheal diseases (George et al., 2016; Khan and Shahidullah, 1982; Najnin et al., 2017; Taylor et al., 2015; Lantagne and Yates, 2018; Patrick et al., 2013; Mahadik and Mbomena, 1983).

4.6 Development of new antimicrobial agents

The rise of antimicrobial resistance (AMR) bacteria to commonly used antibiotics has complicated and lengthened infection treatment and led to increase in the mortality rates globally (Davies and Davies, 2010; Parsonage et al., 2017; Qadri et al., 2021a,b). In view of ineffective antimicrobial agents, there is requirement of new alternative, effective, and safer antimicrobial agents against the "superbugs" like foodborne bacterial pathogens.

In 21st century, at the beginning relatively few antimicrobials drugs were developed with limited applications (Cheesman et al., 2017; Silver, 2011; Ventola, 2015). Natural sources were used to develop several antimicrobials. In developing nations, healthcare has benefited from low-cost, sometimes soil-derived medicines, but exhausting their viability necessitates discovery of alternative antibacterial compounds (Gould, 2016; Woon and Fisher, 2016;

Wohlleben et al., 2016; Clardy et al., 2009; Mir et al., 2020). However, the generation of synthetic antimicrobials from existing natural medications has been shown to have low success rates (Woon and Fisher, 2016; Payne et al., 2007). The researchers are now concentrating their efforts on finding novel ways to combat MDR. There are highly advanced technologies that have been utilized to investigate "genomics and combinatorial chemistry," but they have often been ineffective in discovering novel compounds with high antimicrobial potency (Payne et al., 2007).

Some antimicrobials introduced in recent years are listed below:

- **Tigecycline**: The medication was licensed to treat both Gram-positive and Gram-negative bacteria in 2005. It's a glycylcycline (tetracycline derivative) antibiotic given intravenously. It was developed to fight MDR pathogens; however, it has subsequently evolved antibiotic resistance (Pankey, 2005; Pankey and Steele, 2007).
- **Retapamulin**: it was introduced in 2007 and is the topical antibiotic used ion treatment of bacterial skin infections. It's the first antibiotic in a new class known as pleuromutilins to be approved for human use (Daum et al., 2007; Jacobs, 2007).
- **Fidaxomicin**: It is the first of a family of medications known as tiacumicins (macrocyclic antibiotics) to be approved by the FDA in 2011. It's a naturally occurring actinomycetes fermentation product. It is used to treat *Clostridium difficile* infections (Venugopal and Johnson, 2012; Bartsch et al., 2013).
- **Bedaquiline:** It's an oral preparation approved by the FDA in 2012 for the treatment of MDR tuberculosis (MDR-TB) in conjunction with other antitubercular medications. It is a diarylquinoline with the capacity to prevent *M. tuberculosis* from producing adenosine 5′-triphosphate (ATP) (Mahajan, 2013).
- **Televancin:** It is a vancomycin derivative that is synthesized and used to treat skin infections. It is shown to be effective in the treatment of hospital-acquired *Staphylococcus aureus* pneumonia (Saravolatz et al., 2009; Damodaran and Madhan, 2011) and was approved for licensing in 2013.
- **Plazomicin:** The FDA validated it in 2018 for the treatment of severe chronic UTI (urinary tract infection) in adults, including pyelonephritis (Dhingra et al., 2020). It is a novel aminoglycoside antibacterial medication that attaches to the 30s ribosomal subunit thus inhibiting the generation of microbial proteins. It has inhibitory effect against Enterobacteriaceae and other aerobic Gram-negative pathogens (Eljaaly et al., 2019).
- **Cephalosporin:** was developed in Japan (Zhanel et al., 2019) and approved by FDA in 2019 (Dhingra et al., 2020).Cefiderocol, a parenteral siderophore that is a beta-lactam antimicrobial, inhibits Gram-negative bacteria's cell wall formation (Bonomo, 2019).

Combinational approach of antimicrobials: Since 2015, a number of promising novel combination medicines have been developed. The FDA approved the combination of ceftazidime and avibactam in 2015 (Dhingra et al., 2020). Ceptazidime is a third-generation antibiotic with a good pharmacokinetic profile, whereas avibactam is a next-generation β-lactamase antagonist. A combination of these drugs has been approved for the treatment of hospital-acquired bacterial pneumonia, intraabdominal infections, and ventilator-associated bacterial pneumonia (Das et al., 2019; Torres et al., 2019). In 2019, the US Food and Drug Administration (FDA) approved the three-drug combination imipenem-cilastatin/relebactam for the treatment of hospital-acquired bacterial infections in adults

(Pourali, 2020). Since 1985, the imipenem-cilastatin combination is being used in therapeutic purposes (Food, 2020), the addition of relebactam which is novel beta-lactamase inhibitor in diazabicyclooctane class was new to this combination (Pourali, 2020; Sakoulas, 2019). This three-drug combination of imipenem, cilastatin, and relebactam provides new hope for long-term control of MRD pathogenic microorganisms (Smith et al., 2020; Zhanel et al., 2018; Kaye et al., 2020).

4.7 Future perspectives

Different disciplines such as molecular genetics, immunology, microbiology, and cell biology have made significant contributions to our understanding of pathogen-host interactions in recent years. Researchers discovered several intricate interactions between microbial diseases and higher organisms. Future study, perhaps, will give a more comprehensive picture of the interplay between immunity and bacteria in various infected organs. Our knowledge of how pathogens like *Salmonella* interact with innate and adaptive immunity in humans, bypassing host defense systems, is currently lacking. Therefore, the ongoing research focusing on bacterial pathogen and host immunity relationship has desired potential to uncover complex questions related to the improvement of prevention and treatment strategies aimed at combating the bacterial infectious diseases in near future. Additionally, both the public and private sectors should be committed in efforts to control salmonellosis, TB, cholera, and other foodborne bacterial infectious illnesses. The food production and processing controls should be more focused than testing finished foods and water as well. People must be trained about fundamental food safety procedures both formally and informally. Platforms that can serve as effective means of disseminating scientific information and boosting public awareness about bacterial infection diseases are needed.

References

Adeniji, A.A., Knoll, K.E., Loots, D.T., 2020. Potential anti-TB investigational compounds and drugs with repurposing potential in TB therapy: a conspectus. Appl. Microbiol. Biotechnol. 104, 5633–5662.

Ali, M., Nelson, A.R., Lopez, A.L., Sack, D.A., 2015. Updated global burden of cholera in endemic countries. PLoS Negl. Trop. Dis. 9, e0003832.

Andino, A., Hanning, I., 2015. Salmonella enterica: survival, colonization, and virulence differences among serovars. *The Scientific World Journal*. 2015, 520179.

Angulo, F.J., Mølbak, K., 2005. Human health consequences of antimicrobial drug—resistant Salmonella and other foodborne pathogens. Clin. Infect. Dis. 41, 1613–1620.

Arda, I.S., Ergin, F., Varan, B., Demirhan, B., Aslan, H., Ozyaylali, I., 2001. Acute abdomen caused by *Salmonella typhimurium* infection in children. J. Pediatr. Surg. 36, 1849–1852.

Arranz-Trullen, J., Lu, L., Pulido, D., Bhakta, S., Boix, E., 2017. Host antimicrobial peptides: the promise of new treatment strategies against tuberculosis. Front. Immunol. 8, 1499.

Bartsch, S.M., Umscheid, C.A., Fishman, N., Lee, B.Y., 2013. Is fidaxomicin worth the cost? An economic analysis. Clin. Infect. Dis. 57, 555–561.

Berger, C.N., Sodha, S.V., Shaw, R.K., Griffin, P.M., Pink, D., Hand, P., Frankel, G., 2010. Fresh fruit and vegetables as vehicles for the transmission of human pathogens. Environ. Microbiol. 12, 2385–2397.

Bonomo, R.A., 2019. Cefiderocol: a novel siderophore cephalosporin defeating carbapenem-resistant pathogens. Clin. Infect. Dis. 69, S519–S520.

Buchwald, D.S., Blaser, M.J., 1984. A review of human salmonellosis: II. Duration of excretion following infection with nontyphi *Salmonella*. Rev. Infect. Dis. 6, 345–356.

Buckle, G.C., Walker, C.L., Black, R.E., 2012. Typhoid fever and paratyphoid fever: systematic review to estimate global morbidity and mortality for 2010. J Glob Health 2, 010401.

Bula-Rudas, F.J., Rathore, M.H., Maraqa, N.F., 2015. Salmonella infections in childhood. Adv. Pediatr. 62, 29–58.

Burrowes, V., Perin, J., Monira, S., Sack, D.A., Rashid, M.U., Mahamud, T., Rahman, Z., Mustafiz, M., Bhuyian, S.I., Begum, F., Zohura, F., Biswas, S., Parvin, T., Hasan, T., Zhang, X., Sack, B.R., Saif-Ur-Rahman, K.M., Alam, M., George, C.M., 2017. Risk factors for household transmission of *Vibrio cholerae* in dhaka, Bangladesh (CHoBI7 trial). Am. J. Trop. Med. Hyg. 96, 1382–1387.

Cabrera, A., Lepage, J.E., Sullivan, K.M., Seed, S.M., 2017. Vaxchora: a single-dose oral cholera vaccine. Ann. Pharmacother. 51, 584–589.

Carroll, J., Field, D., O'connor, P.M., Cotter, P.D., Coffey, A., Hill, C., Ross, R.P., O'mahony, J., 2010. Gene encoded antimicrobial peptides, a template for the design of novel anti-mycobacterial drugs. Bioeng. Bugs 1, 408–412.

Castelo-Branco, F.S., DE Lima, E.C., Domingos, J.L.O., Pinto, A.C., Lourenco, M.C.S., Gomes, K.M., Costa-Lima, M.M., Araujo-Lima, C.F., Aiub, C.A.F., Felzenszwalb, I., Costa, T., Penido, C., Henriques, M.G., Boechat, N., 2018. New hydrazides derivatives of isoniazid against *Mycobacterium tuberculosis*: higher potency and lower hepatocytotoxicity. Eur. J. Med. Chem. 146, 529–540.

Celum, C.L., Chaisson, R.E., Rutherford, G.W., Barnhart, J.L., Echenberg, D.F., 1987. Incidence of salmonellosis in patients with AIDS. J. Infect. Dis. 156, 998–1002.

Centers for Disease Control and Prevention, 2007. Multistate outbreak of Salmonella serotype Tennessee infections associated with peanut butter—United States, 2006–2007. MMWR Morb. Mortal. Wkly. Rep. 56, 521–524.

Chaudhary, J., Nayak, J., Brahmbhatt, M., Makwana, P., 2015. Virulence genes detection of Salmonella serovars isolated from pork and slaughterhouse environment in Ahmedabad, Gujarat. Vet. World 8, 121.

Cheesman, M.J., Ilanko, A., Blonk, B., Cock, I.E., 2017. Developing new antimicrobial therapies: are synergistic combinations of plant extracts/compounds with conventional antibiotics the solution? Phcog. Rev. 11, 57–72.

Cheng-Chi, L., Poon, S.-K., Chen, G.-H., 2002. Spontaneous gas-forming liver abscess caused by Salmonella within hepatocellular carcinoma: a case report and review of the literature. Dig. Dis. Sci. 47, 586.

Chiswick, B.R., Miller, P.W., 2015. International Migration and the Economics of Language. Handbook of the Economics of International Migration. Elsevier.

Cholera, W., 2017. Weekly Epidemiological Record, p. 2018. View Article.

Clardy, J., Fischbach, M.A., Currie, C.R., 2009. The natural history of antibiotics. Curr. Biol. 19, R437–R441.

Coburn, B., Grassl, G.A., Finlay, B.B., 2007. Salmonella, the host and disease: a brief review. Immunol. Cell Biol. 85, 112–118.

Codeco, C.T., Coelho, F.C., 2006. Trends in cholera epidemiology. PLoS Med. 3, e42.

Cohen, P.S., O'brien, T.F., Schoenbaum, S.C., Medeiros, A.A., 1978. The risk of endothelial infection in adults with salmonella bacteremia. Ann. Intern. Med. 89, 931–932.

Cole, B., Nilsen, D.M., Will, L., Etkind, S.C., Burgos, M., Chorba, T., 2020. Essential components of a public health tuberculosis prevention, control, and elimination program: recommendations of the advisory council for the elimination of tuberculosis and the national tuberculosis controllers association. MMWR Recomm. Rep. 69, 1.

Centers for Disease Control and Prevention, 2006. Preliminary FoodNet data on the incidence of infection with pathogens transmitted commonly through food–10 States, United States, 2005. MMWR Morb. Mortal. Wkly. Rep. 55, 392–395.

Centers for Disease Control and Prevention, 2010. Updated guidelines for using interferon gamma release assays to detect *Mycobacterium tuberculosis* infection-United States, 2010. MMWR Recomm. Rep. 25, 1–25.

Centers for Disease Control and Prevention, 2013a. Core Curriculum on Tuberculosis: What the Clinician Should Know. National Center for HIV/Aids, Viral Hepatitis, STD, and TB Prevention Division of Tuberculosis Elimination, USA.

Centers for Disease Control and Prevention, 2013b. Latent Tuberculosis Infection: A Guide for Primary Health Care Providers. Centers for Disease Control and Prevention, Atlanta, pp. 1–40.

Crump, J.A., Sjolund-Karlsson, M., Gordon, M.A., Parry, C.M., 2015. Epidemiology, clinical presentation, laboratory diagnosis, antimicrobial resistance, and antimicrobial management of invasive Salmonella infections. Clin. Microbiol. Rev. 28, 901–937.

Damodaran, S.E., Madhan, S., 2011. Telavancin: a novel lipoglycopeptide antibiotic. J. Pharmacol. Pharmacother. 2, 135–137.

Das, S., Li, J., Riccobene, T., Carrothers, T.J., Newell, P., Melnick, D., Critchley, I.A., Stone, G.G., Nichols, W.W., 2019. Dose selection and validation for ceftazidime-avibactam in adults with complicated intra-abdominal infections, complicated urinary tract infections, and nosocomial pneumonia. Antimicrob. Agents Chemother. 63.

Daum, R.S., Kar, S., Kirkpatrick, P., 2007. Fresh from the pipeline. Nat. Rev. Drug Discov. 6, 865.

Davies, J., Davies, D., 2010. Origins and evolution of antibiotic resistance. Microbiol. Mol. Biol. Rev. 74, 417–433.

Deen, J., Mengel, M.A., Clemens, J.D., 2020. Epidemiology of cholera. Vaccine 38, A31–A40.

Dhingra, S., Rahman, N.A.A., Peile, E., Rahman, M., Sartelli, M., Hassali, M.A., Islam, T., Islam, S., Haque, M., 2020. Microbial resistance movements: an overview of global public health threats posed by antimicrobial resistance, and how best to counter. Front. Public Health 8, 531.

Domman, D., Chowdhury, F., Khan, A.I., Dorman, M.J., Mutreja, A., Uddin, M.I., Paul, A., Begum, Y.A., Charles, R.C., Calderwood, S.B., 2018. Defining endemic cholera at three levels of spatiotemporal resolution within Bangladesh. Nat. Genet. 50, 951–955.

Dong, M., Pfeiffer, B., Altmann, K.H., 2017. Recent developments in natural product-based drug discovery for tuberculosis. Drug Discov. Today 22, 585–591.

Ehuwa, O., Jaiswal, A.K., Jaiswal, S., 2021. Salmonella, food safety and food handling practices. Foods 10, 907.

Eljaaly, K., Alharbi, A., Alshehri, S., Ortwine, J.K., Pogue, J.M., 2019. Plazomicin: a novel aminoglycoside for the treatment of resistant gram-negative bacterial infections. Drugs 79, 243–269.

Feasey, N.A., Dougan, G., Kingsley, R.A., Heyderman, R.S., Gordon, M.A., 2012. Invasive non-typhoidal salmonella disease: an emerging and neglected tropical disease in Africa. Lancet 379, 2489–2499.

Feasey, N.A., Gaskell, K., Wong, V., Msefula, C., Selemani, G., Kumwenda, S., Allain, T.J., Mallewa, J., Kennedy, N., Bennett, A., Nyirongo, J.O., Nyondo, P.A., Zulu, M.D., Parkhill, J., Dougan, G., Gordon, M.A., Heyderman, R.S., 2015. Rapid emergence of multidrug resistant, H58-lineage *Salmonella typhi* in Blantyre, Malawi. PLoS Neglected Trop. Dis. 9, e0003748.

Food, U., 2020. Drug administration. Drugs@ FDA: FDA-approved drugs. New Drug Application (NDA) 50587.

Fredrick, T., Ponnaiah, M., Murhekar, M.V., Jayaraman, Y., David, J.K., Vadivoo, S., Joshua, V., 2015. Cholera outbreak linked with lack of safe water supply following a tropical cyclone in Pondicherry, India, 2012. J. Health Popul. Nutr. 33, 31.

Fung, I.C.-H., 2014. Cholera transmission dynamic models for public health practitioners. Emerg. Themes Epidemiol. 11, 1–11.

Gal-Mor, O., Boyle, E.C., Grassl, G.A., 2014. Same species, different diseases: how and why typhoidal and non-typhoidal *Salmonella enterica* serovars differ. Front. Microbiol. 5, 391.

Ganesan, D., Gupta, S.S., Legros, D., 2020. Cholera surveillance and estimation of burden of cholera. Vaccine 38, A13–A17.

George, C.M., Hasan, K., Monira, S., Rahman, Z., Saif-Ur-Rahman, K., Rashid, M.-U., Zohura, F., Parvin, T., Islam Bhuyian, M.S., Mahmud, M.T., 2018. A prospective cohort study comparing household contact and water *Vibrio cholerae* isolates in households of cholera patients in rural Bangladesh. PLoS Negl. Trop. Dis. 12, e0006641.

George, C.M., Monira, S., Sack, D.A., Rashid, M.-U., Saif-Ur-Rahman, K., Mahmud, T., Rahman, Z., Mustafiz, M., Bhuyian, S.I., Winch, P.J., 2016. Randomized controlled trial of hospital-based hygiene and water treatment intervention (CHoBI7) to reduce cholera. Emerg. Infect. Dis. 22, 233.

George, C.M., Rashid, M., Almeida, M., Saif-Ur-Rahman, K., Monira, S., Bhuyian, M.S.I., Hasan, K., Mahmud, T.T., Li, S., Brubaker, J., 2017. Genetic relatedness of *Vibrio cholerae* isolates within and between households during outbreaks in Dhaka, Bangladesh. BMC Genom. 18, 1–10.

Gould, K., 2016. Antibiotics: from prehistory to the present day. J. Antimicrob. Chemother. 71, 572–575.

Gould, L.H., Walsh, K.A., Vieira, A.R., Herman, K., Williams, I.T., Hall, A.J., Cole, D., 2013. Surveillance for foodborne disease outbreaks—United States, 1998–2008. Morb. Mortal. Wkly. Rep. Surveill. Summ. 62, 1–34.

Grandesso, F., Allan, M., Jean-Simon, P., Boncy, J., Blake, A., Pierre, R., Alberti, K., Munger, A., Elder, G., Olson, D., 2014. Risk factors for cholera transmission in Haiti during inter-peak periods: insights to improve current control strategies from two case-control studies. Epidemiol. Infect. 142, 1625–1635.

Greenwood, D., Slack, R.C., Barer, M.R., Irving, W.L., 2012. Medical Microbiology E-Book: A Guide to Microbial Infections: Pathogenesis, Immunity, Laboratory Diagnosis and Control. With STUDENT CONSULT Online Access. Elsevier Health Sciences.

Gupta, S., Winglee, K., Gallo, R., Bishai, W.R., 2017. Bacterial subversion of cAMP signalling inhibits cathelicidin expression, which is required for innate resistance to *Mycobacterium tuberculosis*. J. Pathol. 242, 52–61.

Hancock, R.E., Diamond, G., 2000. The role of cationic antimicrobial peptides in innate host defences. Trends Microbiol. 8, 402–410.

Hafsa, Q., Haseeb, A., Mir, M.A., 2021. Chapter, Role of immunogenetics polymorphisms in infectious diseases. Book, A Molecular Approach to Immunogenetics. Elsevier Publishers, USA. https://doi.org/10.1016/B978-pl0-323-90053-9.00006-3.

Harausz, E.P., Chervenak, K.A., Good, C.E., Jacobs, M.R., Wallis, R.S., Sanchez-Felix, M., Boom, W.H., 2016. Activity of nitazoxanide and tizoxanide against *Mycobacterium tuberculosis* in vitro and in whole blood culture. Tuberculosis 98, 92–96.

Harris, J., 2012. F, Ryan, ET, and Calderwood, SB. Cholera. Lancet 379, 2466–2476.

Hendriksen, R.S., Vieira, A.R., Karlsmose, S., Lo Fo Wong, D.M., Jensen, A.B., Wegener, H.C., Aarestrup, F.M., 2011. Global monitoring of *Salmonella* serovar distribution from the world health organization global foodborne infections Network country data bank: results of quality assured laboratories from 2001 to 2007. Foodborne Pathog. Dis. 8, 887–900.

Hohmann, E.L., 2001. Nontyphoidal salmonellosis. Clin. Infect. Dis. 32, 263–269.

Holmgren, J., Czerkinsky, C., 2005. Mucosal immunity and vaccines. Nat. Med. 11, S45–S53.

Holzgrabe, U., 2015. New griselimycins for treatment of tuberculosis. Chem. Biol. 22, 981–982.

Horsburgh Jr., C.R., 2004. Priorities for the treatment of latent tuberculosis infection in the United States. N. Engl. J. Med. 350, 2060–2067.

Islam, M.S., Zaman, M., Islam, M.S., Ahmed, N., Clemens, J., 2020. Environmental reservoirs of *Vibrio cholerae*. Vaccine 38, A52–A62.

Jacobs, M.R., 2007. Retapamulin: a semisynthetic pleuromutilin compound for topical treatment of skin infections in adults and children. Future Microbiol. 2 (6), 591–600.

Jensen, P.A., Lambert, L.A., Iademarco, M.F., Ridzon, R., 2005. Guidelines for preventing the transmission of *Mycobacterium tuberculosis* in health-care settings. MMWR Recomm. Rep. 54, 1–141, 2005.

Jiang, Z., P Higgins, M., Whitehurst, J., O Kisich, K., I Voskuil, M., S Hodges, R., 2011. Anti-tuberculosis activity of α-helical antimicrobial peptides: de novo designed L-and D-enantiomers versus L-and D-LL37. Protein Pept. Lett. 18, 241–252.

Jonson, G., Osek, J., Svennerholm, A.-M., Holmgren, J., 1996. Immune mechanisms and protective antigens of *Vibrio cholerae* serogroup O139 as a basis for vaccine development. Infect. Immun. 64, 3778–3785.

Kapp, E., Visser, H., Sampson, S.L., Malan, S.F., Streicher, E.M., Foka, G.B., Warner, D.F., Omoruyi, S.I., Enogieru, A.B., Ekpo, O.E., 2017. Versatility of 7-substituted coumarin molecules as antimycobacterial agents, neuronal enzyme inhibitors and neuroprotective agents. Molecules 22, 1644.

Kaye, K.S., Boucher, H.W., Brown, M.L., Aggrey, A., Khan, I., Joeng, H.-K., Tipping, R.W., Du, J., Young, K., Butterton, J.R., 2020. Comparison of treatment outcomes between analysis populations in the RESTORE-IMI 1 phase 3 trial of imipenem-cilastatin-relebactam versus colistin plus imipenem-cilastatin in patients with imipenem-nonsusceptible bacterial infections. Antimicrob. Agents Chemother. 64, e02203–e02219.

Kelesidis, T., Pothoulakis, C., 2012. Efficacy and safety of the probiotic Saccharomyces boulardii for the prevention and therapy of gastrointestinal disorders. Therap. Adv. Gastroenterol. 5, 111–125.

Khan, M.U., Shahidullah, M., 1982. Role of water and sanitation in the incidence of cholera in refugee camps. Trans. R. Soc. Trop. Med. Hyg. 76, 373–377.

King, N., Lake, R., Campbell, D., 2011. Source attribution of nontyphoid salmonellosis in New Zealand using outbreak surveillance data. J. Food Protect. 74, 438–445.

Kling, A., Lukat, P., Almeida, D.V., Bauer, A., Fontaine, E., Sordello, S., Zaburannyi, N., Herrmann, J., Wenzel, S.C., König, C., 2015. Targeting DnaN for tuberculosis therapy using novel griselimycins. Science 348, 1106–1112.

Kumar, P., Kumar, R., 2017. Enteric fever. Indian J. Pediatr. 84, 227–230.

Kumawat, M., Chaudhary, D., Nabi, B., Kumar, M., Sarma, D.K., Shubham, S., Karuna, I., Ahlawat, N., Ahlawat, S., 2021. Purification and characterization of Cyclophilin: a protein associated with protein folding in Salmonella Typhimurium. Arch. Microbiol. 203, 5509–5517.

Lahiri, A., Lahiri, A., Iyer, N., Das, P., Chakravortty, D., 2010. Visiting the cell biology of Salmonella infection. Microb. Infect. 12, 809–818.

Lantagne, D., Yates, T., 2018. Household water treatment and cholera control. J. Infect. Dis. 218, S147–S153.

Lee, I.G., Lee, S.J., Chae, S., Lee, K.Y., Kim, J.H., Lee, B.J., 2015. Structural and functional studies of the *Mycobacterium tuberculosis* VapBC30 toxin-antitoxin system: implications for the design of novel antimicrobial peptides. Nucleic Acids Res. 43, 7624–7637.

Legros, D. Partners of the Global Task Force on Cholera Control. Global Cholera Epidemiology: Opportunities to Reduce the Burden of Cholera, 2030.

Lewinsohn, D.M., Leonard, M.K., Lobue, P.A., Cohn, D.L., Daley, C.L., Desmond, E., Keane, J., Lewinsohn, D.A., Loeffler, A.M., Mazurek, G.H., 2017. Official American thoracic society/infectious diseases society of America/centers for disease control and prevention clinical practice guidelines: diagnosis of tuberculosis in adults and children. Clin. Infect. Dis. 64, e1–e33.

Lindenbaum, J., Greenough, W.B., Islam, M., 1967. Antibiotic therapy of cholera. Bull. World Health Organ. 36, 871.

Loots, D.T., 2014. An altered *Mycobacterium tuberculosis* metabolome induced by katG mutations resulting in isoniazid resistance. Antimicrob. Agents Chemother. 58, 2144–2149.

Lopez, A.L., Gonzales, M.L.A., Aldaba, J.G., Nair, G.B., 2014. Killed oral cholera vaccines: history, development and implementation challenges. Ther. Adv. Vaccines 2, 123–136.

Lukat, P., Katsuyama, Y., Wenzel, S., Binz, T., Konig, C., Blankenfeldt, W., Bronstrup, M., Muller, R., 2017. Biosynthesis of methyl-proline containing griselimycins, natural products with anti-tuberculosis activity. Chem. Sci. 8, 7521–7527.

Mahadik, V.J., Mbomena, J., 1983. Impact of health education programme on knowledge, attitude and practice (KAP) of people in cholera affected areas of Luapula Province–Zambia. Med. J. Zambia 17, 32–38.

Mahajan, R., 2013. Bedaquiline: first FDA-approved tuberculosis drug in 40 years. Int. J. Appl. Basic Med. Res. 3, 1–2.

Majowicz, S.E., Musto, J., Scallan, E., Angulo, F.J., Kirk, M., O'brien, S.J., Jones, T.F., Fazil, A., Hoekstra, R.M., International Collaboration on Enteric Disease 'Burden of Illness, Studies, 2010. The global burden of nontyphoidal *Salmonella* gastroenteritis. Clin. Infect. Dis. 50, 882–889.

Mangasuli, S.N., Hosamani, K.M., Devarajegowda, H.C., Kurjogi, M.M., Joshi, S.D., 2018. Synthesis of coumarin-theophylline hybrids as a new class of anti-tubercular and anti-microbial agents. Eur. J. Med. Chem. 146, 747–756.

Mead, P.S., Slutsker, L., Dietz, V., Mccaig, L.F., Bresee, J.S., Shapiro, C., Griffin, P.M., Tauxe, R.V., 1999. Food-related illness and death in the United States. Emerg. Infect. Dis. 5, 607–625.

Medalla, F., Gu, W., Mahon, B.E., Judd, M., Folster, J., Griffin, P.M., Hoekstra, R.M., 2017. Estimated incidence of antimicrobial drug–resistant nontyphoidal Salmonella infections, United States, 2004–2012. Emerg. Infect. Dis. 23, 29.

Mir, M., 2015a. Chapter 1-introduction to costimulation and costimulatory molecules. In: Manzoor, A.M. (Ed.), Developing Costimulatory Molecules for Immunotherapy of Diseases. Academic Press, ISBN 9780128025857, pp. 1–43. https://doi.org/10.1016/B978-0-12-802585-7.00001-7.

Mir, M., 2015b. Chapter 2 - concept of reverse costimulation and its role in diseases. In: Manzoor, A.M. (Ed.), Developing Costimulatory Molecules for Immunotherapy of Diseases. Academic Press, ISBN 9780128025857, pp. 45–81. https://doi.org/10.1016/B978-0-12-802585-7.00002-9.

Mir, M.A., 2020a. Chapter-9, Vaccines and Their Properties. Book; Basics and Fundamentals of Immunology. Nova Biomedical Publishers, New York, USA, 978-1536166392.

Mir, M.A., 2020b. Cytokines and Their Therapeutic Potential. Nova Biomedical Publishers, New York, USA, ISBN 978-1536170177.

Mir, M.A., 2020c. The Fundamentals of Hypersensitivities and Allergies. Nova Biomedical Publishers, New York, USA, ISBN 978-1536170122.

Mir, M.A., 2020d. Chapter-1, Introduction to Immunology Book; Basics and Fundamentals of Immunology. Nova Biomedical Publishers, New York, USA, 978-1536166392.

Mir, M.A., Agrewala, J.N., 2007. Influence of CD80 and CD86 co-stimulation in the modulation of the activation of antigen presenting cells. Curr. Immunol. Rev. 3, 160–169.

Mir, M.A., AL-Baradie, R., 2013. Tuberculosis time bomb-A global emergency: need for alternative vaccines. Majmaah J. Heal. Sci. 1, 77–82.

Mir, M.A., Albaradie, R.S., 2014. Inflammatory mechanisms as potential therapeutic targets in stroke. Adv. Neuroimmune Biol. 5, 199–216.

Mir, M.A., Bhat, B.A., Sheikh, B.A., Rather, G.A., Mehraj, S., Mir, W.R., 2021. Nanomedicine in human health therapeutics and drug delivery: nanobiotechnology and nanobiomedicine. Applications of Nanomaterials in Agriculture, Food Science, and Medicine (IGI Global).

Mir, M.A., Mehraj, U., Sheikh, B.A., Hamdani, S.S., 2020. Nanobodies: the "magic bullets" in therapeutics, drug delivery and diagnostics. Hum. Antibodies 28, 29–51.

Montgomery, M., Jones, M.W., Kabole, I., Johnston, R., Gordon, B., 2018. Better health for everyone. Bull. World Health Organ. 96, 371–371A.

Mosher, W.D., 1988. Fertility and family planning in the United States: insights from the national survey of family growth. Fam. Plann. Perspect. 207–217.

Murphy, F.X., 1981. Catholic perspectives on populations issues II. Popul. Bull. 35, 1–44.

Najnin, N., Leder, K., Qadri, F., Forbes, A., Unicomb, L., Winch, P.J., Ram, P.K., Leontsini, E., Nizame, F.A., Arman, S., 2017. Impact of adding hand-washing and water disinfection promotion to oral cholera vaccination on diarrhoea-associated hospitalization in Dhaka, Bangladesh: evidence from a cluster randomized control trial. Int. J. Epidemiol. 46, 2056–2066.

Nelson, E.J., Nelson, D.S., Salam, M.A., Sack, D.A., 2011. Antibiotics for both moderate and severe cholera. N. Engl. J. Med. 364, 5–7.

O'brien, T.F., 2002. Emergence, spread, and environmental effect of antimicrobial resistance: how use of an antimicrobial anywhere can increase resistance to any antimicrobial anywhere else. Clin. Infect. Dis. 34, S78–S84.

Oelschlaeger, T.A., 2010. Mechanisms of probiotic actions—a review. Int. J. Med. Microbiol. 300, 57–62.

Organization, W.H., 2013. Global Tuberculosis Report 2013. World Health Organization.

Organization, W.H., 2015. WHO Estimates of the Global Burden of Foodborne Diseases: Foodborne Disease Burden Epidemiology Reference Group 2007–2015. World Health Organization.

Organization, W.H., 2017. Ending Cholera a Global Roadmap to 2030. Ending Cholera a Global Roadmap to 2030.

Organization, W.H., 2019. Foodborne Disease Burden Epidemiology Reference Group. 2015. WHO Estimates of the Global Burden of Foodborne Diseases. World Health Organization, Geneva, Switzerland.

Organization, W.H., 2020. Tuberculosis Surveillance and Monitoring in Europe 2020: 2018 Data.

Palleroni, N., Krieg, N., Holt, J., 1984. Bergey's Manual of Systematic Bacteriology. The Willian and Wilkins Co, Baltimore.

Pankey, G.A., 2005. Tigecycline. J. Antimicrob. Chemother. 56, 470–480.

Pankey, G.A., Steele, R.W., 2007. Tigecycline: a single antibiotic for polymicrobial infections. Pediatr. Infect. Dis. J. 26, 77–78.

Parsonage, B., Hagglund, P.K., Keogh, L., Wheelhouse, N., Brown, R.E., Dancer, S.J., 2017. Control of antimicrobial resistance requires an ethical approach. Front. Microbiol. 8, 2124.

Patrick, M., Berendes, D., Murphy, J., Bertrand, F., Husain, F., Handzel, T., 2013. Access to safe water in rural Artibonite, Haiti 16 months after the onset of the cholera epidemic. Am. J. Trop. Med. Hyg. 89, 647–653.

Payne, D.J., Gwynn, M.N., Holmes, D.J., Pompliano, D.L., 2007. Drugs for bad bugs: confronting the challenges of antibacterial discovery. Nat. Rev. Drug Discov. 6, 29–40.

Pegues, D., 2005. Salmonella species, including *Salmonella typhi*. Princ. Pract. Infect. Dis. 2636–2654.

Pickering, L.K., Baker, C.J., Kimberlin, D.W., 2012. Red book (2012): Report of the Committee on Infectious Diseases. American Academy of Pediatrics, USA, 29.

Pires, C.T., Scodro, R.B., Cortez, D.A., Brenzan, M.A., Siqueira, V.L., CALEFFI-Ferracioli, K.R., Vieira, L.C., Monteiro, J.L., Correa, A.G., Cardoso, R.F., 2020. Structure–activity relationship of natural and synthetic coumarin derivatives against *Mycobacterium tuberculosis*. Future Med. Chem. 12, 1533–1546.

Pires, S.M., Vieira, A.R., Perez, E., Wong, D.L.F., Hald, T., 2012. Attributing human foodborne illness to food sources and water in Latin America and the Caribbean using data from outbreak investigations. Int. J. Food Microbiol. 152, 129–138.

Plessas, S., Nouska, C., Mantzourani, I., Kourkoutas, Y., Alexopoulos, A., Bezirtzoglou, E., 2017. Microbiological exploration of different types of kefir grains. Fermentatio 3, 1.

Popoff, M.Y., Bockemühl, J., Gheesling, L.L., 2004. Supplement 2002 (no. 46) to the kauffmann–white scheme. Res. Microbiol. 155, 568–570.

Pourali, S., 2020. Imipenem, cilastatin, relebactam (recarbrio). Infect. Dis. Alert 39.

Priyodip, P., Prakash, P.Y., Balaji, S., 2017. Phytases of probiotic bacteria: characteristics and beneficial aspects. Indian J. Microbiol. 57, 148–154.

Pui, C., Wong, W., Chai, L., Tunung, R., Jeyaletchumi, P., Hidayah, N., Ubong, A., Farinazleen, M., Cheah, Y., Son, R., 2011. Salmonella: a foodborne pathogen. Int. Food Res. J. 18.

Qadri, H., Haseeb, A., Mir, M., 2021a. Novel strategies to combat the emerging drug resistance in human pathogenic microbes. Curr. Drug Targets 22, 1–13.

Qadri, H., Qureshi, M.F., Mir, M.A., Shah, A.H., 2021b. Glucose-the X factor for the survival of human fungal pathogens and disease progression in the host. Microbiol. Res. 247, 126725. https://doi.org/10.1016/j.micres.2021.126725.

Rafique, R., Rashid, M.-U., Monira, S., Rahman, Z., Mahmud, M., Mustafiz, M., SAIF-UR-Rahman, K., Johura, F.-T., Islam, S., Parvin, T., 2016. Transmission of infectious *Vibrio cholerae* through drinking water among the household contacts of cholera patients (CHoBI7 trial). Front. Microbiol. 7, 1635.

Rajkhowa, S., Deka, R.C., 2014. DFT based QSAR/QSPR models in the development of novel anti-tuberculosis drugs targeting *Mycobacterium tuberculosis*. Curr. Pharmaceut. Des. 20, 4455–4473.

Ravel, A., Greig, J., Tinga, C., Todd, E., Campbell, G., Cassidy, M., Marshall, B., Pollari, F., 2009. Exploring historical Canadian foodborne outbreak data sets for human illness attribution. J. Food Protect. 72, 1963–1976.

Reddy, D.S., Kongot, M., Netalkar, S.P., Kurjogi, M.M., Kumar, R., Avecilla, F., Kumar, A., 2018. Synthesis and evaluation of novel coumarin-oxime ethers as potential anti-tubercular agents: their DNA cleavage ability and BSA interaction study. Eur. J. Med. Chem. 150, 864–875.

Richterman, A., Sainvilien, D.R., Eberly, L., Ivers, L.C., 2018. Individual and household risk factors for symptomatic cholera infection: a systematic review and meta-analysis. J. Infect. Dis. 218, S154–S164.

Rivas-Santiago, B., Rivas Santiago, C.E., Castaneda-Delgado, J.E., Leon-Contreras, J.C., Hancock, R.E., Hernandez-Pando, R., 2013. Activity of LL-37, CRAMP and antimicrobial peptide-derived compounds E2, E6 and CP26 against *Mycobacterium tuberculosis*. Int. J. Antimicrob. Agents 41, 143–148.

Roschka, R., Dosch, F., 1950. [Occurrence of unusual types of Salmonella in Austria, with a note on Kauffmann and Edward's simplified, rapid blood test for the diagnosis of Salmonella infection]. Wien Klin. Wochenschr. 62, 720–724.

Ross, J.A., 1989. Contraception: short-term vs. long-term failure rates. Fam. Plann. Perspect. 21, 275–277.

Sabtu, N., Enoch, D., Brown, N., 2015. Antibiotic resistance: what, why, where, when and how? Br. Med. Bull. 116.

Sakoulas, G., 2019. FDA approves imipenem-cilastatin/relebactam (recarbrio) for complicated intra-abdominal or urinary tract infections. In: *NEJM Journal Watch*. Available online at: https://www.jwatch.org/na49571/2019/08/02/fda-approvesimipenem-cilastatin-relebactam-recarbrio.

Sánchez, J., Holmgren, J., 2005. Virulence factors, pathogenesis and vaccine protection in cholera and ETEC diarrhea. Curr. Opin. Immunol. 17, 388–398.

Saravolatz, L.D., Stein, G.E., Johnson, L.B., 2009. Telavancin: a novel lipoglycopeptide. Clin. Infect. Dis. 49, 1908–1914.

Shannon, K., Hast, M., Azman, A.S., Legros, D., Mckay, H., Lessler, J., 2019. Cholera prevention and control in refugee settings: successes and continued challenges. PLoS Negl. Trop. Dis. 13, e0007347.

Shapiro, R.L., Otieno, M.R., Adcock, P.M., PHILLIPS-Howard, P.A., Hawley, W.A., Kumar, L., Waiyaki, P., Nahlen, B.L., Slutsker, L., 1999. Transmission of epidemic *Vibrio cholerae* O1 in rural western Kenya associated with drinking water from Lake Victoria: an environmental reservoir for cholera? Am. J. Trop. Med. Hyg. 60, 271–276.

Sheikh, B.A., Bhat, B.A., Mehraj, U., Mir, W., Hamadani, S., Mir, M.A., 2021. Development of new therapeutics to meet the current challenge of drug resistant tuberculosis. Curr. Pharmaceut. Biotechnol. 22, 480–500.

Shimoni, Z., Pitlik, S., Leibovici, L., Samra, Z., Konigsberger, H., Drucker, M., Agmon, V., Ashkenazi, S., Weinberger, M., 1999. Nontyphoid Salmonella bacteremia: age-related differences in clinical presentation, bacteriology, and outcome. Clin. Infect. Dis. 28, 822–827.

Shinohara, N.K.S., Barros, V.B.D., Jimenez, S.M.C., Machado, E.D.C.L., Dutra, R.A.F., Lima Filho, J.L.D., 2008. Salmonella spp., importante agente patogênico veiculado em alimentos. Ciência Saúde Coletiva 13, 1675–1683.

Shivaprasad, H.L., 2000. Fowl typhoid and pullorum disease. Rev. Sci. Tech. 19, 405–424.

Silver, L.L., 2011. Challenges of antibacterial discovery. Clin. Microbiol. Rev. 24, 71–109.

Smith, J.R., Rybak, J.M., Claeys, K.C., 2020. Imipenem-cilastatin-relebactam: a novel beta-Lactam-beta-Lactamase inhibitor combination for the treatment of multidrug-resistant gram-negative infections. Pharmacotherapy 40, 343–356.

Smith, K.F., Goldberg, M., Rosenthal, S., Carlson, L., Chen, J., Chen, C., Ramachandran, S., 2014. Global rise in human infectious disease outbreaks. J. R. Soc. Interface 11, 20140950.

Society, A.T., Control, C.F.D., 2005. American thoracic society/centers for disease control and prevention/infectious diseases society of America: controlling tuberculosis in the United States. Am. J. Respir. Crit. Care Med. 172, 1169–1227.

Sosunov, V., Mischenko, V., Eruslanov, B., Svetoch, E., Shakina, Y., Stern, N., Majorov, K., Sorokoumova, G., Selishcheva, A., Apt, A., 2007. Antimycobacterial activity of bacteriocins and their complexes with liposomes. J. Antimicrob. Chemother. 59, 919–925.

Spiegel, P.B., Le, P., Ververs, M.-T., Salama, P., 2007. Occurrence and overlap of natural disasters, complex emergencies and epidemics during the past decade (1995–2004). Conflict Health 1, 1–9.

Starke, J.R., Diseases, C.O.I., 2014. Interferon-γ release assays for diagnosis of tuberculosis infection and disease in children. Pediatrics 134, e1763–e1773.

Sugimoto, J.D., Koepke, A.A., Kenah, E.E., Halloran, M.E., Chowdhury, F., Khan, A.I., Larocque, R.C., Yang, Y., Ryan, E.T., Qadri, F., 2014. Household transmission of *Vibrio cholerae* in Bangladesh. PLoS Negl. Trop. Dis. 8, e3314.

Tauxe, R.V., Doyle, M.P., Kuchenmuller, T., Schlundt, J., Stein, C.E., 2010. Evolving public health approaches to the global challenge of foodborne infections. Int. J. Food Microbiol. 139 (Suppl. 1), S16–S28.

Taylor, D.L., Kahawita, T.M., Cairncross, S., Ensink, J.H., 2015. The impact of water, sanitation and hygiene interventions to control cholera: a systematic review. PLoS One 10, e0135676.

Tien, J.H., Earn, D.J., 2010. Multiple transmission pathways and disease dynamics in a waterborne pathogen model. Bull. Math. Biol. 72, 1506–1533.

Tomičić, Z.M., Čolović, R.R., Čabarkapa, I.S., Vukmirović, Đ.M., Đuragić, O.M., Tomičić, R.M., 2016. Beneficial properties of probiotic yeast Saccharomyces boulardii. Food Feed Res. 43, 103–110.

Torres, A., Rank, D., Melnick, D., Rekeda, L., Chen, X., Riccobene, T., Critchley, I.A., Lakkis, H.D., Taylor, D., Talley, A.K., 2019. Randomized trial of ceftazidime-avibactam vs meropenem for treatment of hospital-acquired and ventilator-associated bacterial pneumonia (REPROVE): analyses per US FDA-specified end points. Open Forum Infect. Dis. 6, ofz149.

Torres, J.R., Gotuzzo, E., Isturiz, R., Elster, C., Wolff, M., Northland, R., Christenson, B., Clara, L., 1994. Salmonellal splenic abscess in the antibiotic era: a Latin American perspective. Clin. Infect. Dis. 19, 871–875.

Ventola, C.L., 2015. The antibiotic resistance crisis: part 1: causes and threats. Pharm. Ther. 40, 277.

Venugopal, A.A., Johnson, S., 2012. Fidaxomicin: a novel macrocyclic antibiotic approved for treatment of *Clostridium difficile* infection. Clin. Infect. Dis. 54, 568–574.

Vilaplana, C., Marzo, E., Tapia, G., Diaz, J., Garcia, V., Cardona, P.-J., 2013. Ibuprofen therapy resulted in significantly decreased tissue bacillary loads and increased survival in a new murine experimental model of active tuberculosis. J. Infect. Dis. 208, 199–202.

Voetsch, A.C., Van Gilder, T.J., Angulo, F.J., Farley, M.M., Shallow, S., Marcus, R., Cieslak, P.R., Deneen, V.C., Tauxe, R.V., Group, E.I.P.F.W., 2004. FoodNet estimate of the burden of illness caused by nontyphoidal Salmonella infections in the United States. Clin. Infect. Dis. 38, S127–S134.

Wagner, E.G., Lanoix, J.N., Organization, W.H., 1958. Excreta Disposal for Rural Areas and Small Communities. World Health Organization.

Wells, W.F., 1955. Airborne Contagion and Air Hygiene. An Ecological Study of Droplet Infections. *Airborne Contagion and Air Hygiene*. An Ecological Study of Droplet Infections.

Wohlleben, W., Mast, Y., Stegmann, E., Ziemert, N., 2016. Antibiotic drug discovery. Microb. Biotechnol. 9, 541–548.

Wong, V.K., Baker, S., Pickard, D.J., Parkhill, J., Page, A.J., Feasey, N.A., Kingsley, R.A., Thomson, N.R., Keane, J.A., Weill, F.X., Edwards, D.J., Hawkey, J., Harris, S.R., Mather, A.E., Cain, A.K., Hadfield, J., Hart, P.J., Thieu, N.T., Klemm, E.J., Glinos, D.A., Breiman, R.F., Watson, C.H., Kariuki, S., Gordon, M.A., Heyderman, R.S., Okoro, C., Jacobs, J., Lunguya, O., Edmunds, W.J., Msefula, C., Chabalgoity, J.A., Kama, M., Jenkins, K., Dutta, S., Marks, F., Campos, J., Thompson, C., Obaro, S., Maclennan, C.A., Dolecek, C., Keddy, K.H., Smith, A.M., Parry, C.M., Karkey, A., Mulholland, E.K., Campbell, J.I., Dongol, S., Basnyat, B., Dufour, M., Bandaranayake, D., Naseri, T.T., Singh, S.P., Hatta, M., Newton, P., Onsare, R.S., Isaia, L., Dance, D., Davong, V., Thwaites, G., Wijedoru, L., Crump, J.A., DE Pinna, E., Nair, S., Nilles, E.J., Thanh, D.P., Turner, P., Soeng, S., Valcanis, M., Powling, J., Dimovski, K., Hogg, G., Farrar, J., Holt, K.E., Dougan, G., 2015. Phylogeographical analysis of the dominant multidrug-resistant H58 clade of Salmonella Typhi identifies inter- and intracontinental transmission events. Nat. Genet. 47, 632–639.

Woon, S.A., Fisher, D., 2016. Antimicrobial agents - optimising the ecological balance. BMC Med. 14, 114.

Yamai, S., Okitsu, T., Shimada, T., Katsube, Y., 1997. Distribution of serogroups of *Vibrio cholerae* non-O1 non-O139 with specific reference to their ability to produce cholera toxin, and addition of novel serogroups. Kansenshogaku Zasshi 71, 1037–1045. The Journal of the Japanese Association for Infectious Diseases.

Zhanel, G.G., Golden, A.R., Zelenitsky, S., Wiebe, K., Lawrence, C.K., Adam, H.J., Idowu, T., Domalaon, R., Schweizer, F., Zhanel, M.A., Lagace-Wiens, P.R.S., Walkty, A.J., Noreddin, A., Lynch III, J.P., Karlowsky, J.A., 2019. Cefiderocol: a siderophore cephalosporin with activity against carbapenem-resistant and multidrug-resistant gram-negative bacilli. Drugs 79, 271–289.

Zhanel, G.G., Lawrence, C.K., Adam, H., Schweizer, F., Zelenitsky, S., Zhanel, M., Lagacé-Wiens, P.R., Walkty, A., Denisuik, A., Golden, A., 2018. Imipenem–relebactam and meropenem–vaborbactam: two novel carbapenem-β-lactamase inhibitor combinations. Drugs 78, 65–98.

Zumla, A., Raviglione, M., Hafner, R., Von Reyn, C.F., 2013. Current concepts. N. Engl. J. Med. 368, 745–755.

Further reading

Altenhoefer, A., Oswald, S., Sonnenborn, U., Enders, C., Schulze, J., Hacker, J., Oelschlaeger, T.A., 2004. The probiotic *Escherichia coli* strain Nissle 1917 interferes with invasion of human intestinal epithelial cells by different enteroinvasive bacterial pathogens. FEMS Immunol. Med. Microbiol. 40, 223–229.

Carter, A., Adams, M., LA Ragione, R.M., Woodward, M.J., 2017. Colonisation of poultry by Salmonella Enteritidis S1400 is reduced by combined administration of Lactobacillus salivarius 59 and Enterococcus faecium PXN-33. Vet. Microbiol. 199, 100–107.

Collado, M., Meriluoto, J., Salminen, S., 2007. Role of commercial probiotic strains against human pathogen adhesion to intestinal mucus. Lett. Appl. Microbiol. 45, 454–460.

Forkus, B., Ritter, S., Vlysidis, M., Geldart, K., Kaznessis, Y.N., 2017. Antimicrobial probiotics reduce *Salmonella enterica* in Turkey gastrointestinal tracts. Sci. Rep. 7, 1–9.

Higgins, J.P., Higgins, S.E., Vicente, J.L., Wolfenden, A.D., Tellez, G., Hargis, B.M., 2007. Temporal effects of lactic acid bacteria probiotic culture on Salmonella in neonatal broilers. Poultry Sci. 86, 1662–1666.

Higgins, J.P., Higgins, S.E., Wolfenden, A.D., Henderson, S.N., TORRES-Rodriguez, A., Vicente, J.L., Hargis, B.M., Tellez, G., 2010. Effect of lactic acid bacteria probiotic culture treatment timing on Salmonella Enteritidis in neonatal broilers. Poultry Sci. 89, 243–247.

Huang, I.F., Lin, I.C., Liu, P.F., Cheng, M.F., Liu, Y.C., Hsieh, Y.D., Chen, J.J., Chen, C.L., Chang, H.W., Shu, C.W., 2015. Lactobacillus acidophilus attenuates Salmonella-induced intestinal inflammation via TGF-beta signaling. BMC Microbiol. 15, 203.

Jiang, Y., Kong, Q., Roland, K.L., Wolf, A., Curtiss Iii, R., 2014. Multiple effects of *Escherichia coli* Nissle 1917 on growth, biofilm formation, and inflammation cytokines profile of *Clostridium perfringens* type A strain CP4. Pathog. Dis. 70, 390–400.

Kamada, N., Maeda, K., Inoue, N., Hisamatsu, T., Okamoto, S., Hong, K.S., Yamada, T., Watanabe, N., Tsuchimoto, K., Ogata, H., Hibi, T., 2008. Nonpathogenic *Escherichia coli* strain Nissle 1917 inhibits signal transduction in intestinal epithelial cells. Infect. Immun. 76, 214–220.

Manzoor, A.M., 2015. Developing Costimulatory Molecules for Immunotherapy of Diseases. Academic Press, ISBN 9780128025857. https://doi.org/10.1016/B978-0-12-802585-7.00011-X. Pages xix-xx.

Manzoor, A.M., 2013a. Costimulation and Costimulatory Molecules in Cancer and Tuberculosis. Publisher: NOVA Science Publisher, New York, USA, pp. 245–272. No. 978-3-659-39067-8; Book Title: Cancer and Infectious Diseases; Chapter-3, Page No.s 141-190; LAP Publishers GermanyManzoor A Mir and Javed N Agrewala (2008) Dietary Polyphenols in Modulation of the Immune System, ISBN No. 978-1604563498, DOI: 10.13140/RG.2.1.4980.9449, Book Title : Polyphenols and Health: New and Recent Advances, Edition: First, Chapter: 10.

Manzoor, A.M., 2013b. Cancer and Infectious Diseases Chapter-1. Book Title: Cancer and Infectious Diseases. LAP Publishers Germany, pp. 1–43. ISBN No. 978-3-659-39067-8.

Manzoor, A.M., 2018. Impact of co-signaling on the survival of intracellular pathogens in antigen presenting cells. Int. J. Adv. Res. Sci. Eng. 7 (4), 3166–3183, 2018: 360.

Menconi, A., Wolfenden, A.D., Shivaramaiah, S., Terraes, J.C., Urbano, T., Kuttel, J., Kremer, C., Hargis, B.M., Tellez, G., 2011. Effect of lactic acid bacteria probiotic culture for the treatment of *Salmonella enterica* serovar Heidelberg in neonatal broiler chickens and Turkey poults. Poultry Sci. 90, 561–565.

Mir, M., 2015c. Chapter 6 - T-cell costimulation and its applications in diseases. In: Manzoor, A.M. (Ed.), Developing Costimulatory Molecules for Immunotherapy of Diseases. Academic Press, ISBN 9780128025857, pp. 255–292. https://doi.org/10.1016/B978-0-12-802585-7.00006-6.

Mir, M., 2015d. Chapter 5 - costimulation in lymphomas and cancers. In: Manzoor, A.M. (Ed.), Developing Costimulatory Molecules for Immunotherapy of Diseases. Academic Press, ISBN 9780128025857, pp. 185–254. https://doi.org/10.1016/B978-0-12-802585-7.00005-4.

Mir, M., 2015e. Chapter 4 - costimulation immunotherapy in allergies and asthma. In: Manzoor, A.M. (Ed.), Developing Costimulatory Molecules for Immunotherapy of Diseases. Academic Press, ISBN 9780128025857, pp. 131–184. https://doi.org/10.1016/B978-0-12-802585-7.00004-2.

Mir, M., 2015f. Chapter 3 - costimulation immunotherapy in infectious diseases. In: Manzoor, A.M. (Ed.), Developing Costimulatory Molecules for Immunotherapy of Diseases. Academic Press, ISBN 9780128025857, pp. 83–129. https://doi.org/10.1016/B978-0-12-802585-7.00003-0.

Mir, M., Albaradeh, R. & Agrewala, J. INNATE–EFFECTOR IMMUNE RESPONSE ELICITATION AGAINST TUBERCULOSIS THROUGH ANTI-B7-1 (CD80) AND ANTI-B7-2 (CD86) SIGNALING IN MACROPHAGES.

Okuneye, O., Oloso, N., Adekunle, O., Ogunfolabo, L., Fasanmi, O., 2016. Protective properties of probiotics on commercial broilers experimentally infected with Salmonella enteritidis. J. Vet. Sci. Anim. Husb. 4, 307.

Rishi, P., Preet, S., Kaur, P., 2011. Effect of L. plantarum cell-free extract and co-trimoxazole against Salmonella Typhimurium: a possible adjunct therapy. Ann. Clin. Microbiol. Antimicrob. 10, 9.

Rokana, N., Singh, R., Mallappa, R.H., Batish, V.K., Grover, S., 2016. Modulation of intestinal barrier function to ameliorate Salmonella infection in mice by oral administration of fermented milks produced with Lactobacillus plantarum MTCC 5690—a probiotic strain of Indian gut origin. J. Med. Microbiol. 65, 1482–1493.

Sabag-Daigle, A., Blunk, H.M., Gonzalez, J.F., Steidley, B.L., Boyaka, P.N., Ahmer, B.M., 2016. Use of attenuated but metabolically competent Salmonella as a probiotic to prevent or treat Salmonella infection. Infect. Immun. 84, 2131–2140.

Tanner, S.A., Chassard, C., Rigozzi, E., Lacroix, C., Stevens, M.J., 2016. Bifidobacterium thermophilum RBL67 impacts on growth and virulence gene expression of *Salmonella enterica* subsp. enterica serovar Typhimurium. BMC Microbiol. 16, 46.

Truusalu, K., Mikelsaar, R.H., Naaber, P., Karki, T., Kullisaar, T., Zilmer, M., Mikelsaar, M., 2008. Eradication of Salmonella Typhimurium infection in a murine model of typhoid fever with the combination of probiotic Lactobacillus fermentum ME-3 and ofloxacin. BMC Microbiol. 8, 132.

Umar, M., Nisar, S., Sheikh, B.A., Syed, S., HinaQayoom, Mir, M.A., 2020. Chapter-4, therapeutic cytokines. In: Book; Cytokines and Their Therapeutic Potential. Nova Biomedical Publishers, New York, USA, ISBN 978-1-53617-017-7.

Umar, M., Nisar, S., Sheikh, B.A., Suhail, S., HinaQayoom, Mir, M.A., 2020a. Chapter 5. Chemokines and cytokines in infectious diseases. In: Book; Cytokines and Their Therapeutic Potential. Nova Biomedical Publishers, New York, USA, ISBN 978-1-53617-017-7.

Umar, M., Nisar, S., Qayoom, H., Mir, M.A., 2020b. Chapter 6. Monoclonal antibodies in therapeutics. In: Book; Immunoglobulins, Magic Bullets and Therapeutic Antibodies. Nova Biomedical Publishers, New York, USA, ISBN 978-1-53616-903-4.

Yoshimura, K., Matsui, T., Itoh, K., 2010. Prevention of *Escherichia coli* O157:H7 infection in gnotobiotic mice associated with Bifidobacterium strains. Antonie Leeuwenhoek 97, 107–117.

Zihler, A., Gagnon, M., Chassard, C., Lacroix, C., 2011. Protective effect of probiotics on Salmonella infectivity assessed with combined in vitro gut fermentation-cellular models. BMC Microbiol. 11, 1–13.

CHAPTER

5

Combating human fungal infections: need for new antifungal drugs and therapies

Manzoor Ahmad Mir, Hafsa Qadri, Shariqa Aisha and Abdul Haseeb Shah

Department of Bioresources, School of Biological Sciences, University of Kashmir, Srinagar, Jammu and Kashmir, India

5.1 Introduction

The antimicrobial drug resistance (AMR) phenomenon represents a global concern associated with the growing spectrum of drug-tolerance mechanisms (Brown and Wright, 2016). This growing AMR phenomenon is required to be confronted immediately. The agents with antimicrobial properties as the therapeutic options for such rising MDR microbial infections are restricted, and there are various adverse causes responsible for the occurrence of such a growing AMR crisis (Qadri et al., 2021a). This growing concern is as burdensome as other global problems such as global warming, etc., and requires proper control and intervention (Ayukekbong et al., 2017). Moreover, the rapidly growing infectious diseases are continuously causing serious issues in various health sectors (Sheikh et al., 2021).

The Kingdom fungi is predicted to have about six million species that are broadly spread in the surroundings. Several hundred pathogenic yeast and mold species are among them, causing harmful infectious diseases in humans in a variety of ways (Firacative, 2020). Infections related to fungal pathogens represent a grave concern around the world (Cowen et al., 2015). Moreover, a sharp increase in the occurrence of tolerant fungal infections produced by multidrug-resistant fungal organisms mostly because of the vigorous usage of different classes of antifungal drugs in both agriculture and modern medicine has been reported (Kontoyiannis, 2017). The emergence of human fungal infections chiefly in immunocompromised persons with diseases like cancer, AIDS, diabetes, etc., and/or undergoing therapies

Human Pathogenic Microbes
https://doi.org/10.1016/B978-0-323-96127-1.00010-3

(corticosteroids/antibiotics) and other such related conditions have been found related to growing mortality and morbidity rates (Sardi et al., 2013; Qadri et al., 2021b). Moreover, invasive fungal infections have been reported to be a chief public health concern creating a serious obstacle in the process of appropriate treatment and control strategies (Firacative, 2020). Causing around >90% fungal-related deaths *Candida, Cryptococcus,* and *Aspergillus* species (Fig. 5.1) are known to influence millions around the world (Schmiedel and Zimmerli, 2016). Bongomin et al. (2017) reported that about 1.6 million individuals die yearly due to the invasive fungal diseases produced by different human pathogenic fungal species including *Candida, Aspergillus, Cryptococcus,* etc. (Bongomin et al., 2017).

The prevalence of resistant fungal diseases has grown rapidly as different antifungal agents are being vigorously employed for agricultural and modern medicinal usage (Kontoyiannis, 2017). The available treatment and control strategies for the existing fungal infections have become very restricted, as there are only a few classes of antifungals currently accessible (Groll et al., 1998; Kathiravan et al., 2012). For the appropriate treatment of the growing infectious diseases, there is a need for the establishment of multiple molecular, genetic, and immunological procedures (Mir and Agrewala, 2007; Mir and Al-baradie, 2013; Mir and Albaradie, 2014; Mir, 2015; Mir et al., 2020; Mir et al., 2021). The establishment of novel antibacterial agents in contrast to antifungal agents is more burdensome since fungi are eukaryotic organisms and multiple promising therapeutic targets are also present in human hosts with considerable toxicity threat (Roemer and Krysan, 2014; Denning and Bromley, 2015). Presently, multiple antifungal drug classes viz. polyenes, azoles, echinocandins, etc., which are applied orally/topically/intravenously are frequently employed for

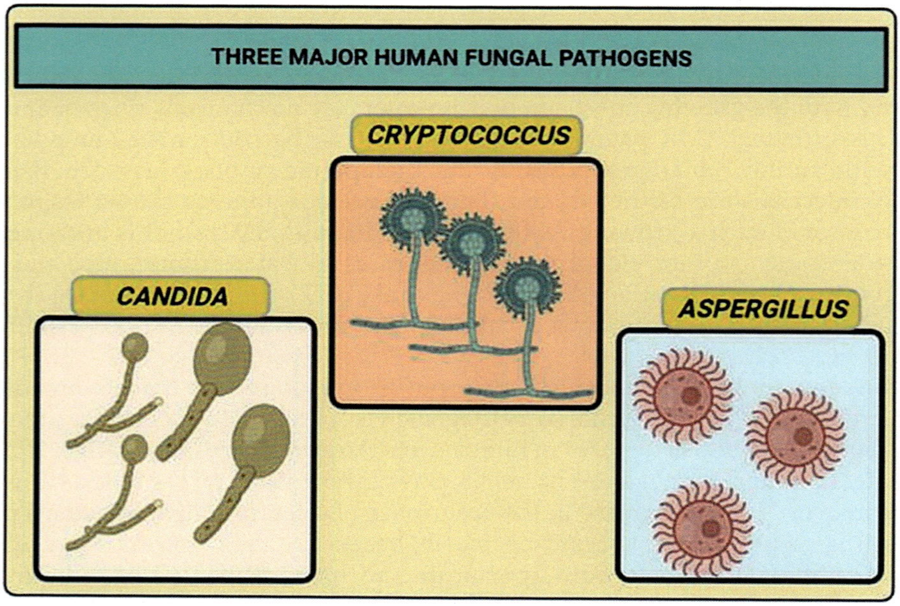

FIGURE 5.1 Diagrammatic illustration of the three major human fungal pathogens.

fungal disease treatment (Qadri et al., 2021a; Castelli et al., 2014; Perfect, 2016). Following the cure of fungal-associated infections, the antifungal agents fall under five main categories, depending on the structure of each group. Since fungi are eukaryotic organisms, just like human cells, antifungals are much related to antitumor drugs in comparison to antibiotics which target the radically different prokaryotic subcellular structures and enzymes present in bacterial organisms (De Pauw, 2000). Nevertheless, all these antifungal drug classes have been found to have multiple disadvantages regarding the toxicity, safety, range of activity, and pharmacokinetic characteristics (Pianalto and Alspaugh, 2016; Perfect, 2016). The development of tolerant fungal strains against the currently available antifungal drugs has guided toward the establishment of novel antifungals with multiple modes of action targeting the biosynthesis of the cell wall, lipid, as well as fungal proteins (Perfect, 2016). However, many novel antifungal drug targets are at present being explored; fungal RNA synthesis and cell wall/membrane constituents serve as the most frequent antifungal targets. The pathogenic fungi have evolved enormous processes (Fig. 5.2) of antifungal drug resistance involving overexpression of drug-efflux pump proteins, drug-target modification, biofilm generation, and alternative associated drug-resistance mechanisms to be identified and explored (Scorzoni et al., 2017).

The mechanism of transport alterations is one of the most significant drug-resistance processes identified in the fungal pathogenic organisms in response to the impact of various antifungal drug categories. In such pathogenic fungi, transport alterations providing antifungal resistance are commonly regulated via multiple kinds of fungal transporters namely ABC

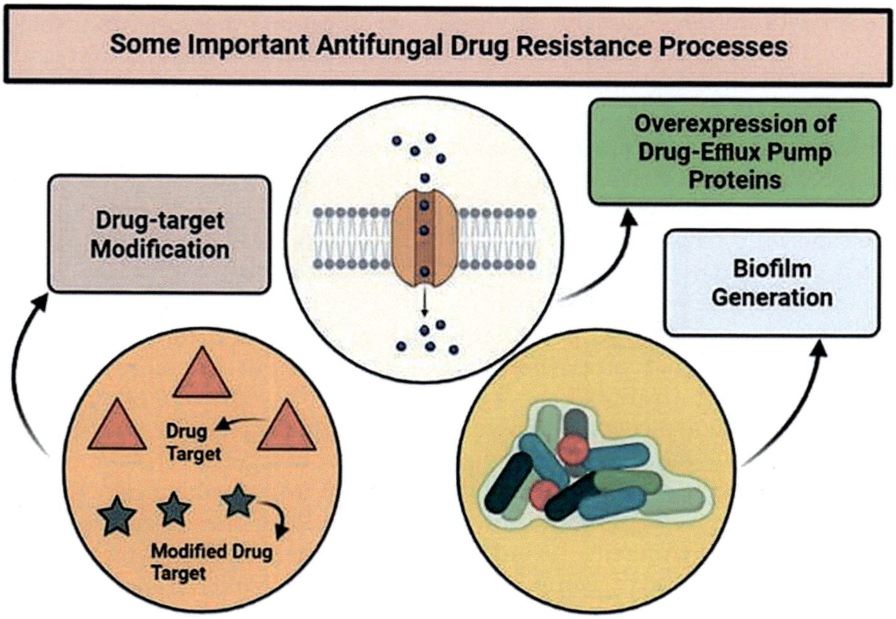

FIGURE 5.2 Representation of some of the processes of antifungal drug resistance involving overexpression of drug-efflux pump proteins, drug-target modification, and biofilm generation in different human pathogenic fungi.

and MFS transporters. *C. albicans* comprises two extremely homologous ABC transporters, viz. *Cdr1* and *Cdr2* (Sanglard et al., 2009). In different pathogenic *Candida* species, the process of overexpression of *CDR-1* and *CDR-2* genes has been identified to be involved in the establishment of tolerance against the most frequently utilized antifungal agents (i.e., azoles) having*CDR1* gene acting as a primary resistance contributor (Ramage et al., 2002; White et al., 2002). Another crucial resistance process employed by such pathogenic fungal organisms is target mutations and gene upregulation. As a resistance mechanism, mutations or upregulation of *Erg11/Cyp51* (azole drug targets) has been extensively observed as a resistance system (Sanglard et al., 2009). Variations in the ergosterol biosynthetic mechanism, such as enhanced activity of the *ERG11* gene, coding14a-demethylase enzyme, or point mutations in the target enzymes, were identified as important contributing factors for reduced fluconazole affinity (Prasad and Kapoor, 2005). Biofilm formation represents another intriguing resistance process used by these fungal pathogens. Biofilm development in fungal pathogens has long been thought to play a major part in their pathogenicity (Martinez and Fries, 2010). Biofilms are three-dimensional structures made up of a complex system of yeast/filamentous cells implanted in an exopolymeric network primarily made up of sugars, etc. (Cowen et al., 2015). The inherent tolerance of *Candida* biofilms is attributed to a variety of processes, such as large biofilm cell density, biofilm matrix effect, the occurrence of persister cells, and so on (Silva et al., 2017). The leading human fungal pathogen *C. albicans* forms properly organized biofilms that are made up of a variety of cell kinds encased in an ECM (Chandra et al., 2001; Ramage et al., 2005, 2009). Biofilms are considered to be well-known for developing on healthcare appliances like pacemakers, catheters, dentures, and other such instruments and systems that provide a surface and a safe place for biofilm development. Biofilm-based infectious diseases are a major clinical challenge because biofilms produced by *C. albicans* are innately resistant to traditional antifungal therapeutic strategies, the host immune response, as well as other environmental disturbances (Nobile and Johnson, 2015).

To tackle the growing number of fungal pathogens and the growing issue of fungal drug resistance and cross-resistance, novel antifungal drugs particularly novel antifungal drug classes which show very little cross-resistance to the already available antifungal agents are especially required (De Pauw, 2000). Presently, various novel antifungal drugs are reported to be undergoing preclinical/clinical studies which might help in the process of controlling the rising emergence of antifungal drug tolerance (Wiederhold, 2017). Peptides have been reported to be potential agents in the establishment of possible compounds with antimicrobial properties via high-throughput screening, and successive development following a rational strategy (Ciociola et al., 2016). Certain peptides (isolated from different body sources) possess antimicrobial activities and have also been found promising in the development of novel antimycotic agents. These molecules (small cationic peptides) having antifungal activities obtained from the host cells (large proteins) could be utilized in the control of multiple fungal infections. These peptide molecules like lysozyme, defensins, lactoferrin, histatin, etc., exert their antifungal action mainly by intensifying the movement of compounds via fungal cell membrane thereby enabling the process of permeabilization (Mehra et al., 2012; De Oliveira Santos et al., 2018). Various plant-based natural substances are identified as potential antifungals (De Andrade Monteiro and Dos Santos, 2019).

Nevertheless the establishment of novel drug delivery procedures along with novel antifungal agents is an urgent requirement, concerning the below-mentioned points (Perfect, 2016; Sousa et al., 2020):

- The process of antifungal drug tolerance in all therapeutic antifungal groups displays an alarming increase.
- The mortality rates associated with invasive mycoses are increasing rapidly (20%–40%) and there is a growing need to improve these figures.
- There is a growing need to establish more desirable fungicidal agents in order to minimize treatment duration and treatment cost associated with the increasing number of patients having prolonged antifungal treatment/therapies.
- Novel therapeutic antifungal classes with diverse mechanisms of action are required, in order to use such novel antifungals in combination with currently available ones for more desirable responses.
- Much awareness is required to be specified to host toxicities and associations between drugs of present antifungal treatment/therapies in order to eradicate or reduce their possible effects.
- To minimize the occurrence of drug usage, the system of pharmacokinetics and pharmacodynamics could also be advanced (Perfect, 2016; Sousa et al., 2020).

Here, we will present a detailed summary of different kinds of antifungals along with novel antifungal drugs/therapies to combat the process of drug resistance in different pathogenic fungi.

5.2 Different classes of antifungals

An antifungal compound represents a drug agent which specifically eradicates the pathogenic fungal organisms from the host having minimal toxic effects on the host organism. The establishment of antifungal drugs has somehow lagged behind that of the development of antibacterial agents, which is an expected impact on the cellular structure of the organisms involved. Bacterial organisms being prokaryotic offer various structural and metabolic targets differing from that of the human host. Comparatively, fungal organisms are eukaryotic, and inevitably almost all drugs toxic to fungal organisms show toxicity in the host as well. Moreover, since fungal organisms exhibit multicellular forms and generally grow slowly, it is way harder to quantify than bacterial organisms. Therefore the experiments devised to assess the in vitro/in vivo characteristics of a possible antifungal drug are greatly halted (Dixon and Walsh, 1996).

Given the scarcity of antifungals in the utilization and advancement of MDR in human fungal diseases, an ongoing necessity of innovative wide-range antifungals with improved potency is required (Prasad et al., 2016). Antifungal drugs are generally grouped into different types (Fig. 5.3) on the basis of their targets for antifungal therapy:

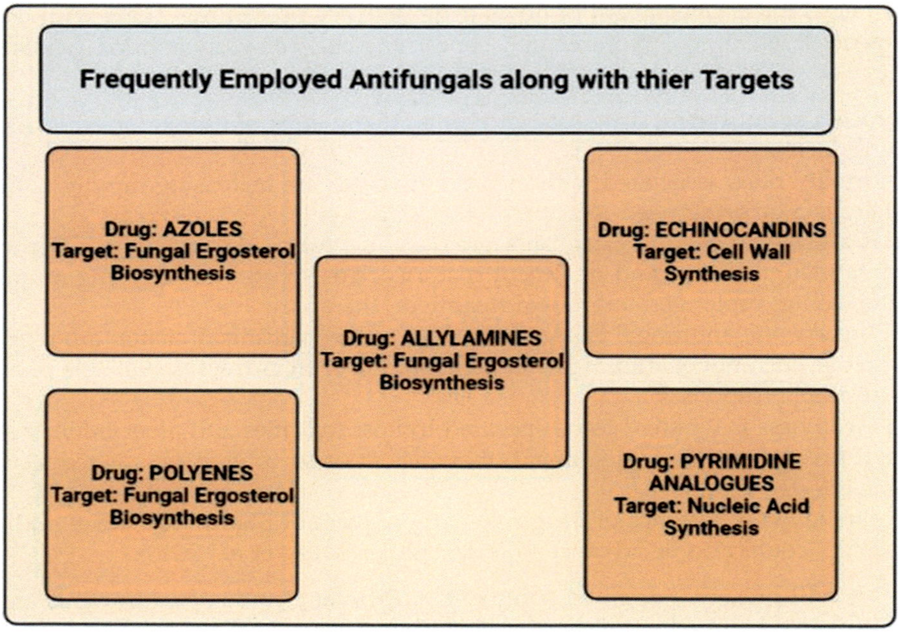

FIGURE 5.3 Some frequently utilized antifungal drugs and their targets.

5.2.1 Ergosterol biosynthesis inhibitors

5.2.1.1 Azoles

The drug azoles represent the most common antifungal drugs available for clinical usage. These drugs hinder the ergosterol synthesis. In fungi, ergosterol represents the primary constituent of the cell membrane and is involved in various cell-based processes like membrane fluidity, membrane integrity, etc. (Campoy and Adrio, 2017). Azoles inhibit the 14a-lanosterol demethylase (*CYP51*) enzyme encoding *ERG11* gene which transforms lanosterol into ergosterol in the plasma membrane, thereby restraining the proliferation and multiplication of the organism (Ghannoum, 1997; Sheehan et al., 1999; Carrillo-Munoz et al., 2006). The antifungal azoles are grouped into the following types: Group (I) imidazoles and Group (II) triazoles. Group-I azoles including ketoconazole, miconazole, and clotrimazole constitute the first developed azoles (Shukla et al., 2016). Group-II azoles including itraconazole and fluconazole (1st generation triazoles) show a wide antifungal activity range in comparison to the group-I azoles, i.e., imidazoles. Antifungal fluconazole is the most active drug against different pathogenic species belonging to *Cryptococcus*, *Candida*, *Histoplasma*, etc. (Pappas et al., 2004; Pardasani, 2000). Alternately, itraconazole display higher potency in the case of *Aspergillus* and additional pathogenic yeast species (Andriole, 2000; Denning and Hope, 2010).

5.2.1.2 *Allylamines*

The antifungals allylamines represent the synthetic fungicidal compounds that inhibit the process of ergosterol biosynthesis, since these compounds are reversible, noncompetitive inhibitors of squalene epoxidase coded by the *ERG1* gene. The enzyme catalyzes the conversion of squalene to 2,3-squalene epoxide. The blockage of such enzymatic potency results in squalene accumulation causing cell disruption by increasing the cell permeability (Andriole, 2000; Denning and Hope, 2010). Terbinafine and naftifine are the two main antifungals of this class (Abdel-Kader and Muharram, 2017).

5.2.1.3 *Morpholines*

The antifungal "Amorolfine" having both in-vitro fungistatic and fungicidal property represents a synthetic water-soluble compound derived from morpholine. It blocks two main enzymes (*ERG2* and *ERG24*) of the ergosterol biosynthetic pathway (Campoy and Adrio, 2017).

5.2.2 Fungal cell membrane disruptors

The antifungals "Polyenes" represent the macrocyclic natural particles called "macrolides" having both fungicidal and a wide range of antifungal properties. These drugs are the first antifungal agents available for clinical usage. Amphotericin B, nystatin, natamycin represent the three main polyenes available for clinical purposes (Farid et al., 2000; Lalitha et al., 2008; Sklenár et al., 2013). Because of their amphiphilic structure, polyenes envelop the lipid bilayer forming a network with the ergosterol creating pores. The process of pore generation causes membrane disarrangement, resulting in the discharge of the cytoplasmic constituents and finally oxidative impairment resulting in cellular destruction (Andes, 2003; Hossain and Ghannoum, 2000; Hossain and Ghannoum, 2001). Amphotericin B shows antifungal activity against most infectious and filamentous yeast organisms including *Candida*, *Cryptococcus*, *Aspergillus*, etc. (Caffrey et al., 2001; Laniado-Laborín and Cabrales-Vargas, 2009). Alternately, nystatin and natamycin also display antifungal potency in the case of the pathogenic species belonging to *Fusarium*, *Aspergillus*, *Cryptococcus*, and *Candida* species (Zotchev, 2003).

5.2.3 Fungal cell wall synthesis

The polysaccharides glucans contain D-glucose monomers connected by Beta-(1,3)/Beta-(1,6)-glucan linkages (Lorand and Kocsis, 2007). Beta-(1,3)-D-glucan comprises over 50% of the cell wall and is the primary structural polysaccharide binding additional cell wall constituents. Presently, there are three antifungals known as echinocandins which target the fungal cell wall viz. micafungin, anidulafungin, and caspofungin (Mukherjee et al., 2011; Sucher et al., 2009). All these three antifungal agents possess both in vitro/in vivo fungicidal properties in the case of different *Candida* strains and also fungistatic properties in the case of different *Aspergillus* species. Micafungin and anidulafungin are reported to be recommended for the cure and control of invasive candidiasis, etc., and caspofungin for invasive aspergillosis also (Pappas et al., 2004; Arévalo et al., 2003; Akins, 2005; Gershkovich et al., 2009).

5.2.4 Chitin synthesis inhibitors

Chitin represents a secondary cell wall constituent containing a linear homopolymer of Beta-(1,4)-linked N-acetylglucosamine covalently attached to Beta-(1,3)-D-glucan (Vicente et al., 2003; Akins, 2005; Lorand and Kocsis, 2007). Chitin synthesis is regulated by different enzymes known as chitin synthases. nikkomycins and polyoxins represent the broadly reported inhibitors of chitin synthase. Nikkomycin has been found to possess antifungal activity against the extremely chitinous dimorphic fungi while polyoxins exhibit potency in the case of different phytopathogens (Lorand and Kocsis, 2007; Akins, 2005).

5.2.5 Nucleic acid synthesis inhibitors

Flucytosine (5-FC; 5-fluorocytosine), a fluorinated pyrimidine analog having fungistatic property, impedes pyrimidine metabolism, and also the process of Dna/Rna and protein biosynthesis (Lewis and Fothergill, 2015). The compound shows both in vitro/in vivo potency in the case of numerous *Candida* and *Cryptococcus* organisms (Zhanel et al., 1997). The majority of the filamentous yeast species are devoid of the enzyme thymidylate synthase; as such the essential scope of flucytosine is restricted to fungal pathogens. Since resistance is frequently observed, therefore 5-FC is commonly employed as an adjunctive instead of primary therapy (Zhanel et al., 1997).

5.2.6 Protein biosynthesis inhibitors

The US Food and Drug Administration (FDA) in the year 2014 recommended "Tavaborole" an oxaborole antifungal for the topical treatment of toenail onychomycosis caused by *Trichophyton rubrum* and *T. mentagrophytes*. The compound possesses antifungal properties in the case of pathogenic yeast, molds, and dermatophytes (Gupta and Versteeg, 2016). The antifungal "Tavaborole" has been reported to block the enzyme leucyl-tRNA synthetase, important for protein synthesis. "Cispentacin" and its synthetic derivative "Icofungipen" having better antifungal property against the leading human fungal pathogen represents other beta-amino acids which target the process of protein biosynthesis by blocking isoleucyl tRNA synthetase (Petraitis et al., 2004; Rock et al., 2007).

5.3 Need for new antifungal drugs and therapies

The present available antifungal drugs and therapies are restricted in terms of their capacity to deal with fungal infections, mostly systemic fungal diseases, and no significant development in the concerned area has been observed recently. Hence there is a growing requirement for the advancement of new treatment strategies against pathogenic fungi (Tavanti et al., 2012; Bermas and Geddes-McAlister, 2020). However, over the past years, multiple strategies have been established in order to identify novel treatments/solutions/strategies (Tavanti et al., 2012). The researchers in the concerned field are focusing to identify novel antifungal agents in different ways such as either by testing the already available medical agents, naturally derived products, e.g., from microorganisms, plants, sea, or by the process of systematic testing of different chemical compound libraries. Investigators are also striving to explain the basic biology of fungal pathogenic organisms both in vitro as well as in vivo. Host-fungal associations have been found to perform a significant part in the process of virulence in fungal pathogens. Thus, targeting such host-fungal interaction paves way for the advancement of new remedial therapies that may be applied as single or combined with the already available antifungal agents. That novel drug combination could also help in determining the establishment of the process of antifungal drug resistance (Tavanti et al., 2012).

For the analysis of antifungal actions of different natural bioactive agents derived from marine organisms, microbes, and plants, major efforts have been made (Arif et al., 2009; Di Santo, 2010; Mayer et al., 2011). Many of such compounds were evaluated due to their known role in the process of triggering mechanisms essential for the fungal organisms, and many additional compounds were evaluated for their unknown antifungal activities directly. However, presently, not many research studies have provided any compound significant for undergoing clinical trials even though many fascinating results were derived (Tavanti et al., 2012).

Additionally, many research studies emphasized in vitro screening of various drugs presently applied for clinical usage concerning their enhancement of the antifungal activity of the fungi-static fungal compound fluconazole (FLC) in *C. albicans*. This enabled the recognition of various compounds, for example, the inhibitors of the calcineurin or Tor pathways (Cruz et al., 2000, 2001), inhibitors of the efflux transporter (milbemycin-derived compounds) (Dryden and Payne, 2005; Sharma et al., 2009), and also some antibodies in the case of heat-shock 90 protein (HSP90) (Pachl et al., 2006). Specifically, the calcineurin pathway inhibitors are found to be completely active in vivo in the enhancement of FLC and have also contributed toward a significant reduction in the phenomenon of fungal pathogenesis (Bader et al., 2006; Sanglard et al., 2003; Steinbach et al., 2006). Moreover, many industrial-based clinical labs to recognize and produce innovative antifungal agents have undergone the process of systematic screening of different chemical compound libraries. A novel glucan synthase inhibitor having antifungal activity against two important species of *Candida,* i.e., *C. albicans* and *C. glabrata,* has been determined by high performance testing of the legacy Schering-Plough compound collection (Ting et al., 2011; Zhou et al., 2011; Walker et al., 2011). Few analyses utilizing the reverse genetic assay concerning *C. albicans* heterozygous deletion/transposon disruption mutants collection have been evaluated for proliferation under treatment with chemical compound collections (Xu et al., 2007). The strategy has paved way for the analysis of both antifungal agents as well as the genes in association with the system of activity of the concerned chemical agents (Tavanti et al., 2012).

Furthermore, for the successful establishment of drugs, the genes important for the survival of fungal pathogens serve as suitable targets (Tavanti et al., 2012). Researchers have evaluated the significance of *C. albicans* and *A. fumigatus* genes utilizing the GRACE (gene replacement and conditional expression)/CPR (conditional promoter replacement) techniques (Roemer et al., 2003; Hu et al., 2007). Around 567 suitable genes in *C. albicans* have been successfully described by a research investigation (Roemer et al., 2003). Similarly about 54 genes of *A. fumigatus* have been successfully screened on the basis of ortholog role and importance in *C. albicans* and *S. cerevisiae* (Hu et al., 2007), and among them, 35 genes have been considered important for *A. fumigatus*. Researchers found that although the *ERG11* gene family (*CYP51A* and *ERG11B*) is important in *A. fumigatus*, the individual genes themselves were not. Such studies gave proper information and data regarding antifungal drug design and upgraded the existing in-silico studies which while utilizing *S. cerevisiae* data verified only 61% of homologous genes identified by the study done by Roemer et al. (2003).

Comprehending the relation between the fungal pathogen and its host during infection could disseminate additional significant data and information for the proper establishment and advancement of antifungal drugs, treatment/strategies. For the proper analysis of the communication/cross-talk which takes place among the fungal pathogen and host during the stage of infection, researchers have been screening the colonization characteristics of mutants directly in their hosts. A research investigation was done on around 1201 gene knockout mutants of *Cryptococcus neoformans* by examining their in vivo growth account in the murine lung; the study successfully reported 40 infectivity mutants (Liu et al., 2008).

5.4 Potential alternative antifungals to combat drug resistance in pathogenic fungi

The occurrence of fungal diseases is growing drastically, and unfortunately, the resistance mechanisms adopted by these pathogenic species are not fully understood. Among such pathogenic fungi, *Candida* species represents a grave concern usually connected with growing death and disease rates around the world causing severe public healthcare threats (De Andrade Monteiro and Dos Santos, 2019). Unfortunately, the therapeutic alternatives for fungal diseases are very finite, due to the availability of fewer categories of antifungals viz. polyenes, echinocandins, azoles, allylamines, and flucytosine (Groll et al., 1998; Kathiravan et al., 2012). Although the existence of drug resistance against azoles has been widely reported, the reoccurrence of traditional fungal diseases arises because of the increasing development of antifungal drug tolerance. Therefore the establishment of novel therapeutic approaches against the growing fungal diseases represents a major medical challenge (De Andrade Monteiro and Dos Santos, 2019). Hence, there is a growing necessity for the advancement of alternative antifungal agents having more potential and efficiency than the traditional ones. Here, some of the alternative compounds having the ability to be utilized as antifungals with favorable antifungal activities in case of the fungal pathogens are described (Fig. 5.4).

5.4.1 Novel triazoles as potential antifungals

Due to the scarcity and lack of efficiency of antifungals, some advanced triazoles as antifungal compounds are currently being investigated and developed including, ravuconazole, albaconazole, and isavuconazole (Table 5.1). As per the preliminary analysis, these drugs have been found to exert better pharmacokinetic profiles and low toxicity, and in vitro action in the case of different *Candida* species. Albaconazole represents a wide spectrum antifungal with outstanding tolerance ability, and the drug has been found to show good antifungal potency in the case of different *Candida* species, both in vitro and in vivo, having comparatively good activities as compared to frequently used antifungal, fluconazole (Bartroli et al., 2011;

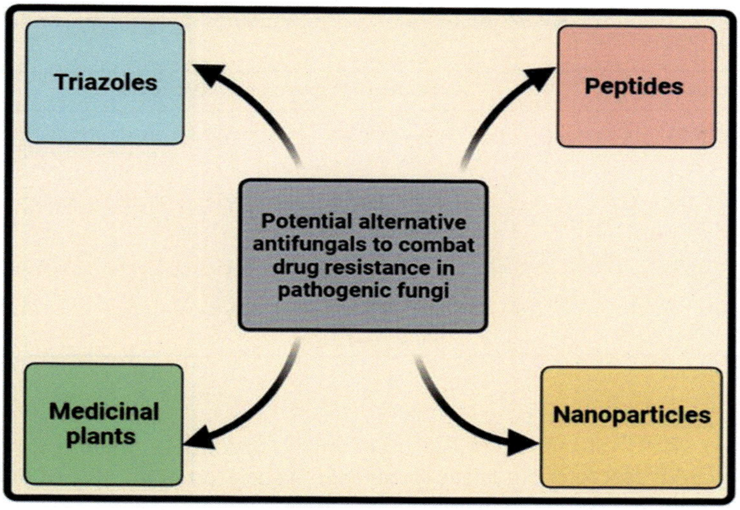

FIGURE 5.4 Schematic outline of the potential alternative antifungals to combat drug resistance in pathogenic fungi.

TABLE 5.1 Representation of some novel triazoles as potential antifungal agents, their mechanism of action, and stage of development (Pasqualotto et al., 2010; Girmenia and Finolezzi, 2011; Allen et al., 2015; De Oliveira Santos et al., 2018).

Novel triazoles	Mechanism of action	Stage of development	Administration via
Albaconazole	Ergosterol biosynthetic pathway inhibition	In phase II trials	Oral
Isavuconazole	Ergosterol biosynthetic pathway inhibition	In phase III trials	Oral and IV
Ravuconazole	Ergosterol biosynthetic pathway inhibition	In phase II trials	Oral and IV

De Oliveira Santos et al., 2018). An investigation described that the majority of the *C. albicans* and *C. glabrata* isolates are vulnerable to the drug albaconazole (Pasqualotto et al., 2010). Isavuconazole represents another novel, second-generation triazole compound with broadspectrum antifungal properties (Pasqualotto et al., 2010). The drug has shown promising antifungal activity related to that of other triazoles against different resistant *Candida* species. Alternately, the antifungal ravuconazole shares structural similarity with isavuconazole and has also been found to show better antimycotic activity in many fluconazole-resistant *Candida* isolates (Pasqualotto et al., 2010). Fortunately, many new antifungal drugs are undergoing clinical tests (Table 5.2, Jacobs et al., 2021).

Lysozyme represents an enzyme present in a variety of bodily fluids, including saliva and respiratory productions. It is best known for its muramidase activity, which allows it to destroy microbes like bacteria. Furthermore, lysozyme is effective against a variety of pathogenic *Candida* species clinical strains, and also in the case of other pathogenic species like *Aspergillus fumigatus* and *Penicillium* species as well (Papini et al., 1981). The antifungal modes

TABLE 5.2 Representation of some of the new exploratory antifungal drugs under clinical tests (Jacobs et al., 2021).

S.No	New antifungal agent	Concerned pathogenic fungal species
1.	Fosmanogepix (APX001); (GPI) inhibitors	Pathogenic species of *Cryptococcus, Aspergillus, Candida* (except *C. krusei)*,etc.
2.	Ibrexafungerp (SCY-078); (Triterpenoid)	Different pathogenic species of *Candida, Aspergillus,* etc.
3.	Rezafungin (CD101); (Echinocandin)	Different pathogenic species of *Candida, Aspergillus,* etc.
4.	Oteseconazole (VT-1161), VT-1598; (Tetrazole)	Different pathogenic species of *Cryptococcus, Coccidioides, Candida, Aspergillus,* etc.

of activity linked with lysozyme, on the other hand, are still a mystery. In *C. albicans*, the enzyme inhibits SAP (secreted aspartic protease; associated with the process of virulence in *Candida*) action and secretion (Wu et al., 1999); its fungicidal action at large concentrations is likely due to the destruction to the cell membrane/wall, resulting in the impairment of osmotic balance (Wu et al., 1999). Human Lactoferrin (hLF) is a peptide which links to iron and functions as a protease in the case of humans. It is found in saliva and other bodily secretions. It was discovered to be effective against the pathogenic species of *Candida* viz. *C. albicans* and *C. krusei* (Samaranayake et al., 1997). Lactoferrin's modes of activity are most likely linked to the formation of a cationic peptide with a wide range of antimicrobial properties (Orsi, 2004). It became confirmed that hLF1-11, a synthetic peptide containing the first cationic domain of lactoferricin H (produced by pepsinolysis of hLF), has a strong antifungal property, in addition to adding to the removal of diseases by elevating the formation of macrophages and dendritic cells (Lupetti et al., 2007; Van Der Does et al., 2012). In *C. albicans*, the peptide hLF1-11 suppressed biofilm development at an initial phase, interrupting biofilm cellular intensity and metabolic functions, and causing biofilm and hyphal-linked genes to be downregulated (Morici et al., 2016). Histatin-5 represents a piece of the salivary protein histatin-3, containing an N-terminal fragment having 24 amino acids. Histatin-5 can destroy both yeast and filamentous types of *Candida* species at very small doses; the peptide could also exhibit its fungicidal action by linking to a Candidate protein (67 kilodalton) and then interrupting with nonlytic ATP efflux (Edgerton and Koshlukova, 2000). Human defensins (HBD) are little cationic peptides related to the defensin protein family. Human β-defensin-1 (HBD-1), HBD-2, and HBD-3 represent the three kinds of human β-defensin with fungicidal action in *C. albicans* that all work through the similar mechanism(Tomalka et al., 2015; De Oliveira Santos et al., 2018).

5.4.2 Peptides as promising antifungals

Because of their intriguing features, the identification of naturally existing antimicrobial peptides (AMPs) has gotten a lot of attention among various strategies investigated in the hunt for innovative, efficient, and efficacious antifungal drugs (Cruz et al., 2014). Tolerance to AMPs is extremely unlikely to evolve, based on their mode of activities. Moreover, the majority of AMPs are nontoxic to human tissues, and a few are persistent across a broad-spectrum of temperatures (Wei et al., 2007).

Several AMPs of various origins have been shown to produce intracellular ROS inside the target cells, start signaling cascades when they connect with particular cell receptors, and affect membrane stability and/or ion channel activity (Vriens et al., 2014; Van Der Weerden et al., 2013). Furthermore, as evidenced by phosphatidylserine externalization, enhanced ROS generation, and DNA breakage, certain AMPs have the capacity to trigger destruction of the concerned fungal organism (De Brucker et al., 2011). The capacity of certain AMPs to impose their effect on sessile species and/or to disrupt biofilm development is of special relevance. The ability of certain fungal organisms, especially different *Candida* species to attach to biotic/abiotic materials has clinical implications for mucosal ailments (e.g., vaginal candidiasis) as well as invasive and systemic mycoses (e.g., implantation/catheter colonization). Microbes implanted in a self-produced polysaccharide-rich extracellular matrix are shielded from mechanical/chemical damage, are resistant to host defenses, and, very importantly, are resistant to major antimicrobial treatments (Nett, 2016; Ramage et al., 2012).

Several AMPs were reported in a variety of microbial, plants, and vertebrate/invertebrate species (Tam et al., 2015; Shishido et al., 2015; Faruck et al., 2016). The Antimicrobial Peptide Database (APD3) has been upgraded to include 2712 peptides, among which 988 have the antifungal property (Wang et al., 2016). Furthermore, the library of biofilm-active antimicrobial peptides (BaAMPs) presently contains about 200 organic, altered, or newly generated peptides, with about 25 of them being effective toward fungal biofilms (Di Luca et al., 2015).

Peptides with antifungal properties have been discovered in both invertebrate/vertebrate species as part of their innate immune responses. Certain (e.g., defensins) have sequence similarity across species, whereas others are species-specific (Van Der Weerden et al., 2013). Considering the paucity of the adaptive immune system in different invertebrate species, AMPs serve an important part in the defense process. Antifungal peptides having antifungal properties are abundant in insects, sea invertebrates, etc. In response to infectious diseases, AMPs are synthesized in the fat body and deposited into the hemolymph in the case of insects (Faruck et al., 2016). On the other hand, these peptides are generated in hemocytes in the case of sea invertebrates (Kang et al., 2015).

It has been reported that some of the AMPs discovered in nonmammalian vertebrate species like fishes, reptiles, etc., have antifungal properties. Peptides with antifungal properties, primarily temporins and brevinins, are primarily found in skin fluids, especially those of frogs (Van Der Weerden et al., 2013). The peptides displaying antifungal properties are largely processed by epithelial cells and neutrophils in the case of mammalian vertebrate species. Such peptides in mammals are classified into three categories: (1) defensins (2) histatins, and (3) cathelicidins. Mammalian defensins are split into three classes based on form and dimensions viz: Alpha, Beta, and Theta defensins, which all include three disulfide links. Defensins have a significant immunoregulatory action in addition to their fungicidal function. Such compounds can trigger a proinflammatory reaction by chemoattracting macrophages and additional immune cells to infection regions or interfering in the inflammatory process based on their expression level (Van Der Weerden et al., 2013; Matejuk et al., 2010). Apart from mankind, primates and rodents possess alpha-defensins, and numerous mammals possess beta-defensins, but only Rhesus macaques have theta-defensins. Even though various cathelicidins were separated in different mammalian species, only a member of such category was detected in human beings, cathelicidin LL-37. Human saliva contains histatins, which are histidine-rich peptides. Mucin segments were discovered as additional salivary peptides with antifungal properties (Ciociola et al., 2016).

Plants represent the most abundant bioactive chemical sources on the planet. Plants produce a huge amount of peptides with antifungal activities, which are grouped into distinct categories depending on their sequence homology, the occurrence of cysteine motifs, and tertiary form, in relation to the constant danger of fungal diseases (Tam et al., 2015; Van Der Weerden et al., 2013). The majority of such peptides are from the defensin category and have a CSAlpha/Beta form. The radish defensins RsAFP1 and RsAFP2 are the fairly marked peptides having a CSAlpha/Beta form. RsAFP1 and the very potent RsAFP2 have been found to cause caspase-regulated death and ROS generation in *C. albicans* as a result of their association with sphingolipids present in the membrane. RsAFP2 also suppressed *C. albicans* biofilm generation by preventing the yeast-hyphal conversion, and a biofilm-specified interaction among both peptides and caspofungin was discovered (Vriens et al., 2014, 2016).

Different antifungal peptides are produced by eukaryotic and prokaryotic microbial organisms, and they may provide a comparative benefit in certain ecosystems. While certain fungal defensins were discovered (Zhu, 2008), many of such peptides have no homologs in additional species. AF, a cysteine-rich polypeptide derived from *Penicillium chrysogenum* with a sheet shape, is one of the fungal organic compounds to consider. PAF (*P. chrysogenum* antifungal protein) was found to suppress a variety of infectious molds while having no harmful impacts on mammalian cells in vitro or in vivo in mouse models. PAF therapy has been shown to affect cellular membranes, interfere with G-protein signal transduction, disrupt Ca^{2+} regulation, induce ROS generation, and induce death in fungal cells (Palicz et al., 2013). Human localized microorganisms may also synthesize antifungal peptides with antifungal characteristics, which may perform a part in limiting the growth of fungal pathogen *C. albicans* (Swidergall and Ernst, 2014). Others were derived from cyanobacteria, which are universal photosynthetic prokaryotic organisms that are found in both land and aquatic habitats (Shishido et al., 2015).

5.4.3 Medicinal plants as potential antifungals

Plants are applied as medicine for very long all across the world. Herbs, herbal components, and objects containing various portions of plants or other plant products which have conventionally been utilized to address different health problems are all examples of medicinal plants (Petrovska, 2012). Since their phenolic classes operate as the principal antimicrobial bioactive ingredient, essential oils and herbal extracts exhibit enormous antifungal activities (Natu and Tatke, 2019; Fogarasi et al., 2019; Semeniuc et al., 2018; Socaciu et al., 2020).

In *Candida* species, the utilization of medicinal plants and their bioactive substances has risen as a possible substitute to conventional antifungal agents. The bioactive compounds/antifungal substances obtained from plants possess the ability to act on diverse particular targets with negative impacts (De Oliveira Santos et al., 2018). Moreover, medicinal plants are most frequently utilized due to the availability of a wide spectrum of functional groups, low-cost extracts, and other related properties. Many bioactive substances like phytochemicals, essential oils have been found in abundance as alternatives to currently available antifungals (De Andrade Monteiro and Dos Santos, 2019).

Different kinds of plant synthesized products are continuously evaluated for their antifungal potency against different pathogenic fungi. For instance, the ethanolic extract of the medicinal plant *Lonicera japonica* (aerial parts), has been reported to show better antifungal potency in the case of different *Candida* organisms and also with strong wound healing abilities (Chen et al., 2012). The extract made from *Piper betel* leaves restricted the proliferation of different *Candida* strains, and also that four distinct *Strychnos spinosa* extracts displayed good activities against *Candida* (Harun et al., 2014; Isa et al., 2014). Hydromethanolic extracts of leaves from *Juglans regia* and *Eucalyptus globulus* and methanol extract of *Cynomorium coccineum* were also found to harbor very good antimycotic actions against different *Candida* isolates (Martins et al., 2015). Akroum and colleagues reported the antifungal activity of acetylic extract of *Vicia faba* in the case of *C. albicans* in vitro and also the minimum death frequency in *Candida*-infected mice treated with the same plant extract (Akroum, 2017).

Phenolic groups represent intricate, volatile, aromatic chemicals with a variety of chemical forms that are deposited in a variety of plant parts, including glandular hairs, oil cells, and oil ducts (Fajinmi et al., 2019). Their antibacterial, pathogen-killing, antiinflammatory, and antioxidant activities have made them well-known (Swamy et al., 2016). Antifungal action is found in several essential oils and their parts, including lemongrass oil, cinnamon oil, tea tree oil, etc. (Kaur et al., 2021).

The polyphenolic compounds obtained from different natural sources such as flavonoids, quinones, tannins, coumarins, lignans, and neolignans are reported to harbor good antifungal properties (Lopes et al., 2017). In *C. albicans*, the action of flavonols like quercetin, myricetin, and kaempferol is reported, and also these compounds from propolis have demonstrated better antifungal potency in the case of various species of *Candida* (Herrera et al., 2010). The flavanol subclass (flavan-3-ol) and gallotannin, isolated from *Syzygium cordatum*, have also been reported to display inhibitory activities on *C. albican's* growth (Vickers, 2017). The organic substances, phenylpropanoids categorized as coumarins, lignans, etc., have also been investigated for the desired antifungal activities against many *Candida* species (Lu et al., 2017).

Multiple types of essential oils were reported to exhibit good antifungal action in the case of many pathogenic *Candida* species. In a study by Sharifzadeh et al., essential oils from *Trachyspermum ammi* were reported to possess antifungal activities against different *Candida* isolates displaying tolerance to FLC (Sharifzadeh et al., 2015). Herbal essences from *Foeniculum vulgare, Satureja hortensis, C. cyminum,* and *Zataria multiflora* have been investigated for their antifungal potency in *C. albicans and that the* essential oils from *Z. multiflora* displayed a potent antimicrobial action comparatively (Gavanji et al., 2015). Moreover, many terpenoids have been found to possess excellent antimycotic potency against *C. albicans* at nontoxic concentrations (Zore et al., 2011). Terpenes' anti-biofilm action, as well as the potency of thymol, geraniol, and carvacrol for the cure of different *Candida*-based diseases linked to hospital devices have all been connected (Dalleau et al., 2008). Changes in the cellular cytoplasmic membrane and introduction of cell death are among the processes underlying carvacrol's impacts, according to an in vitro macrodilution investigation in different species of *Candida* (Mulaudzi et al., 2012). The above findings suggest that plants possess substances with large bioactive potency. The separation, recognition, and enhancement of pharmacokinetic and pharmacodynamic activities, and also the choice of lead compounds for further drug establishment, are all part of the process of discovering bioactive compounds (De Oliveira Santos et al., 2018).

Moreover, the combinatorial impact of different plant extracts or bioactive molecules in association with traditional antimicrobials (or another extract or bioactive compound) in the case of clinical MDR microbes offers a promising therapeutic strategy (Mukherjee et al., 2005). Combinatorial therapies have the benefit of small antifungal doses, potential additive antifungal effects, and lower drug resistance advancement. The goal of such an interactive approach is to increase the antifungal impacts as much as possible (De Oliveira Santos et al., 2018). When employed as agents in the case of the *C. albicans*, Tangarife-Castao et al. found an additive interaction between essential oils/plant extracts and antifungal agents. The combo of itraconazole and *P. bredemeyeri* extract had the perfect additive impacts in the case of *C. albicans* (Tangarife-Castaño et al., 2011). Similarly, in a research investigation, methanolic extract of *T. catappa* leaves blended with nystatin/AMP-B in the case of *Candida albicans*, *Candida glabrata* and some other pathogens confirmed a synergistic potential (Chanda et al., 2013). The antimycotic activities of an ethanolic extract of *Hyptis martiusii* (EEHM) in the case of *C. albicans*, *C. tropicalis* and, *C. krusei* were also studied (Santos et al., 2013). When employed against *C. tropicalis*, it was found that EEHM combined with metronidazole had an additive antifungal action. Other researchers found that an *Echinophora platyloba* ethanol extract and various azoles had an additive impact against different strains of *C. albicans* isolated from the vaginal fluid of patients having recurrent vulvovaginitis. The MIC and MFC values of the *E. platyloba* ethanolic extract in association with itraconazole and fluconazole successfully indicated potent additive impacts (Avijgan et al., 2014). Despite the fact that various in vitro research reports investigating the additive impacts between potential antifungal biological molecules and conventional antifungals have been published, the actual mechanisms responsible for such synergistic impacts are incompletely defined. As a result, in order to gain extra insight, it is critical to continue looking for novel Phytocompounds and to meticulously investigate any potential synergistic effect between them and traditional antifungal drugs (De Oliveira Santos et al., 2018).

5.4.4 Nanoparticles as potential antifungal agents

It has been reported that metallic nanoparticles have been utilized to eradicate human as well as plant pathogenic fungal organisms due to their inherent antimicrobial property (Mashitah et al., 2016). Three major pathways Fig. 5.5 have been reported to explain the precise mechanisms underlying such property viz. (a) direct uptake of nanoparticles (b) indirect property of nanoparticles by the generation of ROS (c) cell wall/membrane damage via cumulation (Mashitah et al., 2016). It could be extremely possible that the fusion of all these mechanistic pathways is responsible for such antimicrobial activity (Slavin et al., 2017, Table 5.3).

Due to their electrochemical potential, nanoparticles go through dissolution (Qidwai et al., 2018). This causes them to separate into ions in the microbiological fluid or culture media. Such ions can also build up on the inside or outside of microtubules, creating an antagonistic response. The buildup of nanoparticles exterior to the microtubules results in the creation of sheets which obstruct the cellular respiratory chain and damage microtubules (Mashitah et al., 2016). The relationship between the nanoparticle and the drug it carries is dependent on its electrical charge. The electrostatic system explains why silver nanoparticles were the first to show antimicrobial action. The electrostatic interaction among the negatively

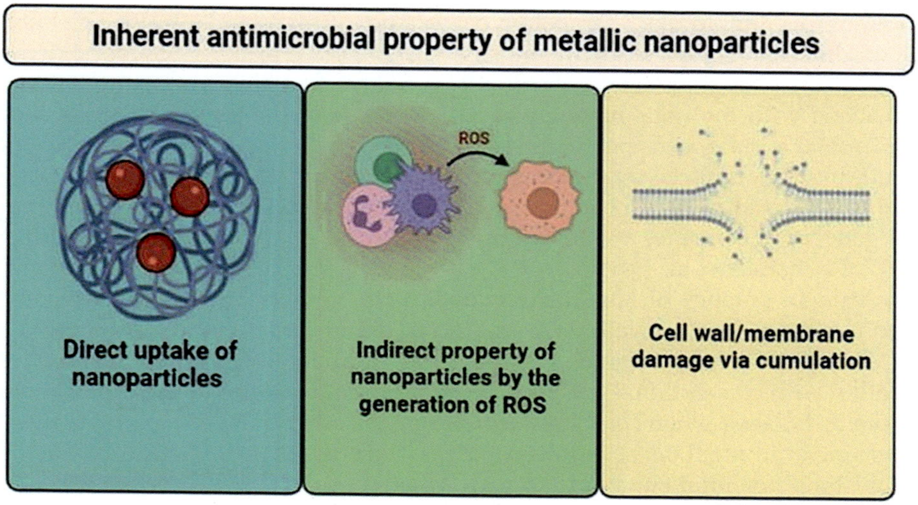

FIGURE 5.5 Representation of the three distinct pathways underlying the inherent antimicrobial property of metallic nanoparticles (Mashitah et al., 2016).

TABLE 5.3 List of certain ongoing clinical investigations in the area of myco-nanotechnology (Sousa et al., 2020).

	Concerned antifungal	Nano-formulation	Clinical phase
1.	Amphotericin-B	Cochleate lipid-crystal nanoparticle	II-phase
2.	Amphotericin-B	Cochleate lipid-crystal nanoparticle	II-phase
3.	Terbinafine (TDT067)	Transfersome	III-phase
4.	Itraconazole	Nanoemulsion gel	II-phase

charged cellular microbial membrane and the positively charged nanoparticle membrane is commonly acknowledged as being critical for the antimicrobial properties of such nanoparticles (Mashitah et al., 2016). Silver nanoparticles possess vigorous antifungal action in the case of *C. albicans* with an MIC of 25 μg/mL (Zhang et al., 2016). Silver nanoparticles have antifungal properties in the case of *Aspergillus niger*, impeding spore formation and restricting biofilm development; in combination with simvastatin, the antifungal impact is enhanced, possibly since simvastatin, as an ergosterol synthesis inhibitor, destroys the fungal cellular membrane, allowing the nanoparticles to enter (Bocate et al., 2019). On exposure to acidic lysosomal state or reacting with oxidative organelles, metals found in nanoparticles could serve as catalysts, combining with biomolecules and triggering the direct generation of free radicals (Slavin et al., 2017; Qidwai et al., 2018). Because of their biocompatibility, biodegradability, and mucoadhesivity, chitosan and its chemical byproducts are utilized as basic components for drug delivery nanoformulations, with benefits like in situ gelling, mucoadhesive characteristics, and the capacity to extend the discharge of low-molecular mass substances to macromolecular therapeutic compounds (Calvo et al., 2019). The antimicrobial efficacy of chitosan nanoparticles in the case of *Candida* infections has been demonstrated (Rai et al., 2017). This antimicrobial property is linked to amino acid units having positive charge reacting with the units having a negative charge of lipopolysaccharides and proteins on the microbial surface, causing cell membrane breakdown, as per several previous studies. As a result, the nanoparticles are capable of binding with DNA strands and blocking the formation of mRNA and proteins. Chitosan works against pathogenic fungi by limiting sporulation and generation of spores via impeding the function of growth-promoting enzymes (Rai et al., 2017; Kucharska et al., 2019).

The antifungal potency of zinc oxide nanoparticles (ZnONPs) has been demonstrated in dermatophyte infections as well as in additional other infectious pathogens like *Candida*, etc. (Rai et al., 2017). However, the combinatorial antifungal action of ZnONPs was tested in association with conventional antifungal agents, and it was discovered that their inhibitory efficacy can be boosted when combined with ZnONPs, potentially reducing overuse, toxicity, and increasing antifungal effect (Sun et al., 2018). Furthermore, in the future, such nanoparticles could be a potential substitute to conventional preservatives in cosmetics (Singh and Nanda, 2013).

As far as the utilization of metallic nanoparticles for antifungal purposes is concerned, the antifungal agents can be vectorized using three types of metallic nanoparticles: magnetic, gold, and silver nanoparticles. Biologically, chemically, and physically, metallic nanoparticles are made in three modes. Chemical synthesis has been linked to a number of negative impacts, including the accumulation of harmful chemicals on nanoparticle surfaces (Hasan, 2015), and as a result, biological synthesis is gaining traction (Abbasi et al., 2016). In immunochemical research, Gold nanoparticles are used to determine protein associations, and in DNA fingerprinting to identify DNA molecules in a specimen. Aminoglycosides such as streptomycin, gentamicin, and neomycin are also detected using them (Hasan, 2015). On the other hand, because of their antimicrobial activities against various microbial pathogens like bacteria, viruses, and other such pathogenic microbes, silver nanoparticles are the most effective. They are, in turn, the most widely utilized materials, with antimicrobial drugs employed in the textile sector, water treatment, solar protection, and other applications (Hasan, 2015).

5.5 Future perspectives

A tremendous rise in the frequency of infectious fungal diseases possibly due to the spread in the population of persons with a compromised immune defense mechanism is often reported. The constant application of various antifungals has ultimately caused the formation of resistant fungal strains (Qadri et al., 2021b). An extremely complicated form of drug resistance currently emerging in the major fungal pathogenic organisms is described as the phenomenon of multidrug resistance (MDR). The MDR phenomenon refers to the simultaneous acquirement of resistance/tolerance to a wide range of drugs via a finite/single genetic change.

The existing antifungal treatments and therapies are very restricted regarding their capacity to handle different pathogenic fungal infections, mostly systemic fungal infections. The pathogenic fungal organisms are tolerant to currently available antifungal drugs and can adjust to a variety of host niches, posing a significant threat to public health. As of now, there has been no serious contribution toward the development of potential and novel antifungal drugs and therapies in the concerned area recently. As such, there is a growing necessity for the establishment of innovative remedial strategies and therapies against pathogenic fungi. In summary, the chapter has been framed to give a detailed account of different types of clinically used antifungal agents along with potential alternative antifungals to combat the process of drug tolerance in different fungal pathogens and eventually help in the process of identification and establishment of novel antifungal agents and therapies. The phenomenon of antifungal resistance is caused by a complex set of systems comprising various mechanisms and genes. Furthermore, these processes constantly change and develop, posing a challenge to the health sector and exacerbating the requirement for novel antifungal treatments. In this manner, the discovery of novel bioactive components, as well as the advancement of novel antifungal preparations and combos involving bioactive molecules and traditional agents, can pave the way for a successful treatment strategy.

References

Abbasi, E., Milani, M., Fekri Aval, S., Kouhi, M., Akbarzadeh, A., Tayefi Nasrabadi, H., Nikasa, P., Joo, S.W., Hanifehpour, Y., Nejati-Koshki, K., 2016. Silver nanoparticles: synthesis methods, bio-applications and properties. Crit. Rev. Microbiol. 42, 173–180.

Abdel-Kader, M.S., Muharram, M.M., 2017. New microbial source of the antifungal allylamine "Terbinafine". Saudi Pharmaceut. J. 25, 440–442.

Akins, R.A., 2005. An update on antifungal targets and mechanisms of resistance in *Candida albicans*. Med. Mycol. 43, 285–318.

Akroum, S.J., 2017. Antifungal activity of acetone extracts from *Punica granatum* L., *Quercus suber* L. and *Vicia faba* L. J. de Mycol. Med. 27, 83–89.

Allen, D., Wilson, D., Drew, R., Perfect, J., 2015. Azole antifungals: 35 years of invasive fungal infection management. Exper. Rev. Anti-infect. Ther. 13, 787–798.

Andes, D., 2003. In vivo pharmacodynamics of antifungal drugs in treatment of candidiasis. Antimicrob. Agents & Chemother. 47, 1179–1186.

Andriole, V.T., 2000. Current and future antifungal therapy: new targets for antifungal therapy. Int. J. Antimicrob. Agents 16, 317–321.

Arévalo, M.P., Carrillo-Muñoz, A.-J., Salgado, J., Cardenes, D., Brió, S., Quindós, G., Espinel-Ingroff, A., 2003. Antifungal activity of the echinocandin anidulafungin (VER002, LY-303366) against yeast pathogens: a comparative study with M27-A microdilution method. J. Antimicrob. Chemother. 51, 163–166.

Arif, T., Bhosale, J., Kumar, N., Mandal, T., Bendre, R., Lavekar, G., Dabur, R., 2009. Natural products–antifungal agents derived from plants. J. Asian Natur. Prod. Res. 11, 621–638.

Avijgan, M., Mahboubi, M., Nasab, M.M., Nia, E.A., Yousefi, H., 2014. Synergistic activity between *Echinophora platyloba* DC ethanolic extract and azole drugs against clinical isolates of *Candida albicans* from women suffering chronic recurrent vaginitis. J. de Mycol. Méd. 24, 112–116.

Ayukekbong, J.A., Ntemgwa, M., Atabe, A.N., 2017. The threat of antimicrobial resistance in developing countries: causes and control strategies. Antimicrob. Resist. & Infect. Contr. 6, 1–8.

Bader, T., Schröppel, K., Bentink, S., Agabian, N., Köhler, G., Morschhäuser, J., 2006. Role of calcineurin in stress resistance, morphogenesis, and virulence of a *Candida albicans* wild-type strain. Infect. & Immun. 74, 4366–4369.

Bartroli, J., Merlos, M., Sisniega, H., 2011. Overview of albaconazole. Eur. Infect. Dis. 5, 88–91.

Bermas, A., Geddes-Mcalister, J., 2020. Combatting the evolution of antifungal resistance in *Cryptococcus neoformans*. Mol. Microbiol. 114, 721–734.

Bocate, K.P., Reis, G.F., De Souza, P.C., Junior, A.G.O., Durán, N., Nakazato, G., Furlaneto, M.C., De Almeida, R.S., Panagio, L.A., 2019. Antifungal activity of silver nanoparticles and simvastatin against toxigenic species of *Aspergillus*. Int. J. Food Microbiol. 291, 79–86.

Bongomin, F., Gago, S., Oladele, R.O., Denning, D.W., 2017. Global and multi-national prevalence of fungal diseases—estimate precision. J. Fungi 3, 57.

Brown, E.D., Wright, G.D., 2016. Antibacterial drug discovery in the resistance era. Nature 529, 336–343.

Caffrey, P., Lynch, S., Flood, E., Finnan, S., Oliynyk, M., 2001. Amphotericin biosynthesis in *Streptomyces nodosus*: deductions from analysis of polyketide synthase and late genes. Chem. & Biol. 8, 713–723.

Calvo, N.L., Sreekumar, S., Svetaz, L.A., Lamas, M.C., Moerschbacher, B.M., Leonardi, D., 2019. Design and characterization of chitosan nanoformulations for the delivery of antifungal agents. International J. Mol. Sci. 20, 3686.

Campoy, S., Adrio, J.L., 2017. Antifungals. Biochem. Pharm. 133, 86–96.

Carrillo-Munoz, A., Giusiano, G., Ezkurra, P., Quindós, G., 2006. Antifungal agents: mode of action in yeast cells. Rev. Esp. Quimioter. 19, 130–139.

Castelli, M.V., Butassi, E., Monteiro, M.C., Svetaz, L.A., Vicente, F., Zacchino, S.A., 2014. Novel antifungal agents: a patent review (2011–present). Exper. Opin. Ther. Patents 24, 323–338.

Chanda, S., Rakholiya, K., Dholakia, K., Baravalia, Y., 2013. Antimicrobial, antioxidant, and synergistic properties of two nutraceutical plants: *Terminalia catappa* L. and *Colocasia esculenta* L. Turkish J. Biol. 37, 81–91.

Chandra, J., Kuhn, D.M., Mukherjee, P.K., Hoyer, L.L., Mccormick, T., Ghannoum, M.A., 2001. Biofilm formation by the fungal pathogen *Candida albicans*: development, architecture, and drug resistance. J. Bacteriol. 183, 5385–5394.

Chen, W.-C., Liou, S.-S., Tzeng, T.-F., Lee, S.-L., Liu, I.-M., 2012. Wound repair and anti-inflammatory potential of *Lonicera japonica* in excision wound-induced rats. BMC Compl. & Alter. Med. 12, 1–9.

Ciociola, T., Giovati, L., Conti, S., Magliani, W., Santinoli, C., Polonelli, L., 2016. Natural and synthetic peptides with antifungal activity. Futur. Med. Chem. 8, 1413–1433.

Cowen, L.E., Sanglard, D., Howard, S.J., Rogers, P.D., Perlin, D.S., 2015. Mechanisms of antifungal drug resistance. Cold Spr. Harb. Perspect. Med. 5, a019752.

Cruz, J., Ortiz, C., Guzman, F., Fernandez-Lafuente, R., Torres, R., 2014. Antimicrobial peptides: promising compounds against pathogenic microorganisms. Curr. Med. Chem. 21, 2299–2321.

Cruz, M.C., DEL Poeta, M., Wang, P., Wenger, R., Zenke, G., Quesniaux, V.F., Movva, N.R., Perfect, J.R., Cardenas, M.E., Heitman, J., 2000. Immunosuppressive and nonimmunosuppressive cyclosporine analogs are toxic to the opportunistic fungal pathogen *Cryptococcus neoformans* via cyclophilin-dependent inhibition of calcineurin. Antimicrob. Agents & Chemother. 44, 143–149.

Cruz, M.C., Goldstein, A.L., Blankenship, J., DEL Poeta, M., Perfect, J.R., Mccusker, J.H., Bennani, Y.L., Cardenas, M.E., Heitman, J., 2001. Rapamycin and less immunosuppressive analogs are toxic to *Candida albicans* and *Cryptococcus neoformans* via FKBP12-dependent inhibition of TOR. Antimicrob. Agents & Chemother. 45, 3162–3170.

Dalleau, S., Cateau, E., Bergès, T., Berjeaud, J.-M., Imbert, C., 2008. In vitro activity of terpenes against *Candida biofilms*. Int. J. Antimicrob. Agents 31, 572–576.

De Andrade Monteiro, C., Dos Santos, J.R.A., 2019. Phytochemicals and their antifungal potential against pathogenic yeasts. Phytochem. Human Health 1–31.

De Brucker, K., Cammue, B.P., Thevissen, K., 2011. Apoptosis-inducing antifungal peptides and proteins. Biochem. Soc. Transac. 39, 1527–1532.

De Oliveira Santos, G.C., Vasconcelos, C.C., Lopes, A.J., De Sousa Cartágenes, M.D.S., Do Nascimento, F.R., Ramos, R.M., Pires, E.R., De Andrade, M.S., Rocha, F.M., De Andrade Monteiro, C., 2018. Candida infections and therapeutic strategies: mechanisms of action for traditional and alternative agents. Front. Microbiol. 9, 1351.

De Pauw, B., 2000. Is there a need for new antifungal agents? Clin. Microbiol. & Infect. 6, 23–28.

Denning, D.W., Bromley, M.J., 2015. How to bolster the antifungal pipeline. Science 347, 1414–1416.

Denning, D.W., Hope, W.W., 2010. Therapy for fungal diseases: opportunities and priorities. Trends Microbiol. 18, 195–204.

DI Luca, M., Maccari, G., Maisetta, G., Batoni, G., 2015. BaAMPs: the database of biofilm-active antimicrobial peptides. Biofouling 31, 193–199.

Di Santo, R., 2010. Natural products as antifungal agents against clinically relevant pathogens. Nat. Product Rep. 27, 1084–1098.

Dixon, D.M., Walsh, T.J., 1996. Antifungal agents. Med. Microbiol.

Dryden, M.W., Payne, P.A., 2005. Preventing parasites in cats. Veter. Ther. 6, 260.

Edgerton, M., Koshlukova, S., 2000. Salivary histatin 5 and its similarities to the other antimicrobial proteins in human saliva. Adv. Dental Res. 14, 16–21.

Fajinmi, O., Kulkarni, M., Benická, S., Zeljković, S.Ć., Doležal, K., Tarkowski, P., Finnie, J., van Staden, J., 2019. Antifungal activity of the volatiles of *Agathosma betulina* and *Coleonema album* commercial essential oil and their effect on the morphology of fungal strains *Trichophyton rubrum* and *T. Mentagrophytes*. South Afr. J. Botany 122, 492–497.

Farid, M.A., EL-Enshasy, H.A., EL-Diwany, A.I., EL-Sayed, E.S.A., 2000. Optimization of the cultivation medium for natamycin production by *Streptomyces natalensis*. J. Basic Microbiol. Int. J. Biochem. Physiol. Genet. Morphol. & Ecol. Microorgan. 40, 157–166.

Faruck, M.O., Yusof, F., Chowdhury, S., 2016. An overview of antifungal peptides derived from insect. Peptides 80, 80–88.

Firacative, C., 2020. Invasive fungal disease in humans: are we aware of the real impact? Memórias do Instit. Oswaldo Cruz 115.

Fogarasi, M., Socaci, S.A., Fogarasi, S., Jimborean, M., Pop, C., Tofană, M., Rotar, A., Tibulca, D., Salagean, D., Salanta, L., 2019. Evaluation of biochemical and microbiological changes occurring in fresh cheese with essential oils during storage time. Studia UBB Chem. 64, 527–537.

Gavanji, S., Zaker, S.R., Nejad, Z.G., Bakhtari, A., Bidabadi, E.S., Larki, B., 2015. Comparative efficacy of herbal essences with amphotricin B and ketoconazole on *Candida albicans* in the in vitro condition. Integr. Med. Res. 4, 112–118.

Gershkovich, P., Wasan, E.K., Lin, M., Sivak, O., Leon, C.G., Clement, J.G., Wasan, K.M., 2009. Pharmacokinetics and biodistribution of amphotericin B in rats following oral administration in a novel lipid-based formulation. J. Antimicrob. Chemother. 64, 101–108.

Ghannoum, M., 1997. Future of antimycotic therapy. Der. Ther. 3, 104–111.

Girmenia, C., Finolezzi, E., 2011. New-generation triazole antifungal drugs: review of the phase II and III trials. Clin. Invest. 1, 1577–1594.

Groll, A.H., Piscitelli, S.C., Walsh, T.J., 1998. Clinical pharmacology of systemic antifungal agents: a comprehensive review of agents in clinical use, current investigational compounds, and putative targets for antifungal drug development. Adv. Pharmaco. 343–500.

Gupta, A.K., Versteeg, S.G., 2016. Tavaborole—a treatment for onychomycosis of the toenails. Exper. Rev. Clin. Pharmacol. 9, 1145–1152.

Harun, W.H.-A.W., Razak, F.A., Musa, M.Y., 2014. Growth inhibitory response and ultrastructural modification of oral-associated candidal reference strains (ATCC) by *Piper betle* L. extract. Int. J. Oral Sci. 6, 15–21.

Hasan, S., 2015. A review on nanoparticles: their synthesis and types. Res. J. Recent Sci. 2277, 2502.

Herrera, C.L., Alvear, M., Barrientos, L., Montenegro, G., Salazar, L.A., 2010. The antifungal effect of six commercial extracts of *Chilean propolis* on *Candida* spp. Ciencia e Invest. Agrari 37, 75–84.

Hossain, M.A., Ghannoum, M.A., 2000. New investigational antifungal agents for treating invasive fungal infections. Exper. Opin. Investig. Drugs 9, 1797–1813.

Hossain, M.A., Ghannoum, M.A., 2001. New developments in chemotherapy for non-invasive fungal infections. Exper. Opin. Investig. Drugs 10, 1501–1511.

Hu, W., Sillaots, S., Lemieux, S., Davison, J., Kauffman, S., Breton, A., Linteau, A., Xin, C., Bowman, J., Becker, J., 2007. Essential gene identification and drug target prioritization in *Aspergillus fumigatus*. PLoS Pathogen. 3, e24.

Isa, A.I., Awouafack, M.D., Dzoyem, J.P., Aliyu, M., Magaji, R.A., Ayo, J.O., Eloff, J.N., 2014. Some *Strychnos spinosa* (Loganiaceae) leaf extracts and fractions have good antimicrobial activities and low cytotoxicities. BMC Complem. Alter. Med. 14, 1–8.

Jacobs, S.E., Zagaliotis, P., Walsh, T., 2021. Novel antifungal agents in clinical trials. F1000Research 10, 507.

Kang, H.K., Seo, C.H., Park, Y., 2015. Marine peptides and their anti-infective activities. Mar. Drugs 13, 618–654.

Kathiravan, M.K., Salake, A.B., Chothe, A.S., Dudhe, P.B., Watode, R.P., Mukta, M.S., Gadhwe, S., 2012. The biology and chemistry of antifungal agents: a review. Bioorg. & Med. Chem. 20, 5678–5698.

Kaur, N., Bains, A., Kaushik, R., Dhull, S.B., Melinda, F., Chawla, P., 2021. A review on antifungal efficiency of plant extracts entrenched polysaccharide-based nanohydrogels. Nutrients 13, 2055.

Kontoyiannis, D.P., 2017. Antifungal resistance: an emerging reality and a global challenge. J. Infect. Dis. 216, S431–S435.

Kucharska, M., Sikora, M., Brzoza-Malczewska, K., Owczarek, M., 2019. Antimicrobial properties of chitin and chitosan. Chitin & Chitosan Proper. & Appl. 169–187.

Lalitha, P., Vijaykumar, R., Prajna, N., Fothergill, A., 2008. In vitro natamycin susceptibility of ocular isolates of *Fusarium* and *Aspergillus* species: comparison of commercially formulated natamycin eye drops to pharmaceutical-grade powder. J. Clin. Microbiol. 46, 3477–3478.

Laniado-Laborín, R., Cabrales-Vargas, M.N., 2009. Amphotericin B: side effects and toxicity. Revista Iberoamer. de Micol. 26, 223–227.

Lewis, RE., Fothergill, A.W., 2015. Antifungal agents. InDiagnosis and Treatment of Fungal Infections. Springer, Cham. 79–97.

Liu, O.W., Chun, C.D., Chow, E.D., Chen, C., Madhani, H.D., Noble, S.M., 2008. Systematic genetic analysis of virulence in the human fungal pathogen *Cryptococcus neoformans*. Cell 135, 174–188.

Lopes, G., Pinto, E., Salgueiro, L., 2017. Natural products: an alternative to conventional therapy for dermatophytosis? Mycopathologia 182, 143–167.

Lorand, T., Kocsis, B., 2007. Recent advances in antifungal agents. Mini Rev. Med. Chem. 7, 900–911.

Lu, M., Li, T., Wan, J., Li, X., Yuan, L., Sun, S., 2017. Antifungal effects of phytocompounds on *Candida* species alone and in combination with fluconazole. Int. J. Antimicrob. Agents 49, 125–136.

Lupetti, A., Brouwer, C.P., Bogaards, S.J., Welling, M.M., De Heer, E., Campa, M., Van Dissel, J.T., Friesen, R.H., Nibbering, P.H., 2007. Human lactoferrin-derived peptide's antifungal activities against disseminated *Candida albicans* infection. J. Infect. Dis. 196, 1416–1424.

Martinez, L.R., Fries, B.C., 2010. Fungal biofilms: relevance in the setting of human disease. Curr. Fung. Infect. Rep. 4, 266–275.

Martins, N., Barros, L., Santos-Buelga, C., Henriques, M., Silva, S., Ferreira, I.C., 2015. Evaluation of bioactive properties and phenolic compounds in different extracts prepared from *Salvia officinalis* L. Food Chem. 170, 378–385.

Mashitah, M.D., San Chan, Y., Jason, J., 2016. Antimicrobial properties of nanobiomaterials and the mechanism. Nanobiomater. Antimicr. Ther. 261–312 (Elsevier).

Matejuk, A., Leng, Q., Begum, M., Woodle, M., Scaria, P., Chou, S., Mixson, A., 2010. Peptide-based antifungal therapies against emerging infections. Drugs Futur. 35, 197.

Mayer, A.M., Rodríguez, A.D., Berlinck, R.G., Fusetani, N., 2011. Marine pharmacology in 2007–8: marine compounds with antibacterial, anticoagulant, antifungal, anti-inflammatory, antimalarial, antiprotozoal, antituberculosis, and antiviral activities; affecting the immune and nervous system, and other miscellaneous mechanisms of action. Compar. Biochem. & Physiol. Part C: Toxicol. & Pharmacol. 153, 191–222.

Mehra, T., Köberle, M., Braunsdorf, C., Mailänder-Sanchez, D., Borelli, C., Schaller, M., 2012. Alternative approaches to antifungal therapies. Exper. Dermatol. 21, 778–782.

Mir, M.A., 2015. Developing Costimulatory Molecules for Immunotherapy of Diseases. Academic Press.

Mir, M.A., Agrewala, J.N., 2007. Influence of CD80 and CD86 co-stimulation in the modulation of the activation of antigen presenting cells. Curr. Immunol. Rev. 3, 160–169.

Mir, M.A., Al-Baradie, R., 2013. Tuberculosis time bomb-A global emergency: need for alternative vaccines. Majmaah J. Health Sci. 1, 77–82.

Mir, M.A., Albaradie, R.S., 2014. Inflammatory mechanisms as potential therapeutic targets in stroke. Adv. Neuroimmune Biol. 5, 199–216.

Mir, M.A., Bhat, B.A., Sheikh, B.A., Rather, G.A., Mehraj, S., Mir, W.R., 2021. Nanomedicine in human health therapeutics and drug delivery: nanobiotechnology and nanobiomedicine. Appl. Nanomater. Agric. Food Sci. & Med. (IGI Global).

Mir, M.A., Mehraj, U., Sheikh, B.A., Hamdani, S.S., 2020. Nanobodies: the "magic bullets" in therapeutics, drug delivery and diagnostics. Human Antibod. 28, 29–51.

Morici, P., Fais, R., Rizzato, C., Tavanti, A., Lupetti, A., 2016. Inhibition of *Candida albicans* biofilm formation by the synthetic lactoferricin derived peptide hLF1-11. PLoS one 11, e0167470.

Mukherjee, P., Sheehan, D., Puzniak, L., Schlamm, H., Ghannoum, M., 2011. Echinocandins: are they all the same? J. Chemother. 23, 319–325.

Mukherjee, P.K., Sheehan, D.J., Hitchcock, C.A., Ghannoum, M.A., 2005. Combination treatment of invasive fungal infections. Clin. Microbiol. Rev. 18, 163–194.

Mulaudzi, R., Ndhlala, A., Kulkarni, M., van Staden, J., 2012. Pharmacological properties and protein binding capacity of phenolic extracts of some Venda medicinal plants used against cough and fever. J. Ethnopharmacol. 143, 185–193.

Natu, K.N., Tatke, P.A., 2019. Essential oils—prospective candidates for antifungal treatment? J. Essen. Oil Res. 31, 347–360.

Nett, J.E., 2016. The host's reply to *Candida biofilm*. Pathogens 5, 33.

Nobile, C.J., Johnson, A.D., 2015. *Candida albicans* biofilms and human disease. Ann. Rev. Microbiol. 69, 71–92.

Orsi, N., 2004. The antimicrobial activity of lactoferrin: current status and perspectives. Biometals 17, 189–196.

Pachl, J., Svoboda, P., Jacobs, F., Vandewoude, K., van der Hoven, B., Spronk, P., Masterson, G., Malbrain, M., Aoun, M., Garbino, J., 2006. A randomized, blinded, multicenter trial of lipid-associated amphotericin B alone versus in combination with an antibody-based inhibitor of heat shock protein 90 in patients with invasive candidiasis. Clin. Infect. Dis. 42, 1404–1413.

Palicz, Z., Jenes, Á., Gáll, T., Miszti-Blasius, K., Kollár, S., Kovács, I., Emri, M., Márián, T., Leiter, É., Pócsi, I., 2013. In vivo application of a small molecular weight antifungal protein of Penicillium chrysogenum (PAF). Toxicol. & Appl. Pharmacol. 269, 8–16.

Papini, M., Simonetti, S., Franceschini, S., Scaringi, L., Binazzi, M., 1981. Lysozyme distribution in healthy human skin. Archiv. Dermatol. Res. 272, 167–170.

Pappas, P.G., Rex, J.H., Sobel, J.D., Filler, S.G., Dismukes, W.E., Walsh, T.J., Edwards, J., 2004. Guidelines for treatment of candidiasis. Clin. Infect. Dis. 38, 161–189.

Pardasani, A., 2000. Oral antifungal agents used in dermatology. J. Amer. Acad. Dermatol. 12, 270–275.

Pasqualotto, A.C., Thiele, K.O., Goldani, L.Z., 2010. Novel triazole antifungal drugs: focus on isavuconazole, ravuconazole and albaconazole. Curr. Opin. Investig. Drugs 11, 165–174.

Perfect, J.R., 2016. Is there an emerging need for new antifungals? Exper. Opin. Emerg. Drugs 21, 129–131.

Petraitis, V., Petraitiene, R., Kelaher, A.M., Sarafandi, A.A., Sein, T., Mickiene, D., Bacher, J., Groll, A.H., Walsh, T., 2004. Efficacy of PLD-118, a novel inhibitor of *Candida* isoleucyl-tRNA synthetase, against experimental oropharyngeal and esophageal candidiasis caused by fluconazole-resistant *C. albicans*. Antimicrob. Agents & Chemother. 48, 3959–3967.

Petrovska, B.B., 2012. Historical review of medicinal plants' usage. Pharmac. Rev. 6, 1.

Pianalto, K.M., Alspaugh, J., 2016. New horizons in antifungal therapy. J. Fungi 2, 26.

Prasad, R., Kapoor, K., 2005. Multidrug resistance in yeast *Candida*. Int. Rev. Cytol. 242, 215–248.

Prasad, R., Shah, A.H., Rawal, M.K., 2016. Antifungals: mechanism of action and drug resistance. Yeast Membr. Transp. 327–349.

Qadri, H., Haseeb, A., Mir, M., 2021a. Novel strategies to combat the emerging drug resistance in human pathogenic microbes. Curr. Drug Targets 22, 1–13.

Qadri, H., Qureshi, M.F., Mir, M.A., Shah, A.H., 2021b. Glucose-the X Factor for the survival of human fungal pathogens and disease progression in the host. Microbiol. Res. 126725.

Qidwai, A., Kumar, R., Shukla, S., Dikshit, A., 2018. Advances in biogenic nanoparticles and the mechanisms of antimicrobial effects. Indian J.Pharmaceut. Sci. 80, 592–603.

Rai, M., Ingle, A., Pandit, R., Paralikar, P., Gupta, I., Anasane, N., Dolenc-Voljč, M., 2017. Nanotechnology for the treatment of fungal infections on human skin. Microbiol. Skin Soft Tissue Bone & Joint Infect. 169–184 (Elsevier).

Ramage, G., Bachmann, S., Patterson, T.F., Wickes, B.L., López-Ribot, J.L., 2002. Investigation of multidrug efflux pumps in relation to fluconazole resistance in *Candida albicans* biofilms. J. Antimicrob. Chemother. 49, 973–980.

Ramage, G., Mowat, E., Jones, B., Williams, C., Lopez-Ribot, J., 2009. Our current understanding of fungal biofilms. Crit. Rev. Microbiol. 35, 340–355.

Ramage, G., Rajendran, R., Sherry, L., Williams, C., 2012. Fungal biofilm resistance. Crit. Rev. Microbiol. 2012.

Ramage, G., Saville, S.P., Thomas, D.P., Lopez-Ribot, J.L., 2005. Candida biofilms: an update. Eukaryot. Cell 4, 633–638.

Rock, F.L., Mao, W., Yaremchuk, A., Tukalo, M., Crépin, T., Zhou, H., Zhang, Y.-K., Hernandez, V., Akama, T., Baker, S., 2007. An antifungal agent inhibits an aminoacyl-tRNA synthetase by trapping tRNA in the editing site. Science 316, 1759–1761.

Roemer, T., Jiang, B., Davison, J., Ketela, T., Veillette, K., Breton, A., Tandia, F., Linteau, A., Sillaots, S., Marta, C., 2003. Large-scale essential gene identification in *Candida albicans* and applications to antifungal drug discovery. Mol. Microbiol. 50, 167–181.

Roemer, T., Krysan, D., 2014. Antifungal drug development: challenges, unmet clinical needs, and new approaches. Cold Spr. Harb. Perspect. in Med. 4, a019703.

Samaranayake, Y., Samaranayake, L., Wu, P., So, M., 1997. The antifungal effect of lactoferrin and lysozyme on *Candida krusei* and *Candida albicans*. Apmis 105, 875–883.

Sanglard, D., Coste, A., Ferrari, S., 2009. Antifungal drug resistance mechanisms in fungal pathogens from the perspective of transcriptional gene regulation. FEMS Yeast Res. 9, 1029–1050.

Sanglard, D., Ischer, F., Marchetti, O., Entenza, J., Bille, J., 2003. Calcineurin A of *Candida albicans*: involvement in antifungal tolerance, cell morphogenesis and virulence. Mol. Microbiol. 48, 959–976.

Santos, K.K., Matias, E.F., Sobral-Souza, C.E., Tintino, S.R., Morais-Braga, M.F., Guedes, G.M., Rolón, M., Coronel, C., Alfonso, J., Vega, C., 2013. Trypanocide, cytotoxic, and anti-*Candida* activities of natural products: *Hyptis martiusii* Benth. Eur. J. Integ. Med. 5, 427–431.

Sardi, J., Scorzoni, L., Bernardi, T., Fusco-Almeida, A., Giannini, M.M., 2013. Candida species: current epidemiology, pathogenicity, biofilm formation, natural antifungal products and new therapeutic options. J. Med. Microbiol. 62, 10–24.

Schmiedel, Y., Zimmerli, S., 2016. Common invasive fungal diseases: an overview of invasive candidiasis, aspergillosis, cryptococcosis, and *Pneumocystis pneumonia*. Swiss Med. Weekly 146, w14281.

Scorzoni, L., De Paula e Silva, A.C., Marcos, C.M., Assato, P.A., De Melo, W.C., De Oliveira, H.C., Costa-Orlandi, C.B., Mendes-Giannini, M.J., Fusco-Almeida, A.M., 2017. Antifungal therapy: new advances in the understanding and treatment of mycosis. Front. Microbiol. 8, 36.

Semeniuc, C.A., Socaciu, M.-I., Socaci, S.A., Mureşan, V., Fogarasi, M., Rotar, A.M., 2018. Chemometric comparison and classification of some essential oils extracted from plants belonging to Apiaceae and Lamiaceae families based on their chemical composition and biological activities. Molecules 23, 2261.

Sharifzadeh, A., Khosravi, A., Shokri, H., Sharafi, G., 2015. Antifungal effect of *Trachyspermum ammi* against susceptible and fluconazole-resistant strains of *Candida albicans*. J. de Mycol. Med. 25, 143–150.

Sharma, M., Manoharlal, R., Shukla, S., Puri, N., Prasad, T., Ambudkar, S.V., Prasad, R., 2009. Curcumin modulates efflux mediated by yeast ABC multidrug transporters and is synergistic with antifungals. Antimicrob. Agents & Chemother. 53, 3256–3265.

Sheehan, D.J., Hitchcock, C.A., Sibley, C.M., 1999. Current and emerging azole antifungal agents. Clin. Microbiol. Rev. 12, 40–79.

Sheikh, B.A., Bhat, B.A., Mehraj, U., Mir, W., Hamadani, S., Mir, M.A., 2021. Development of new therapeutics to meet the current challenge of drug resistant tuberculosis. Curr. Pharmaceut. Biotechnol. 22, 480–500.

Shishido, T.K., Humisto, A., Jokela, J., Liu, L., Wahlsten, M., Tamrakar, A., Fewer, D.P., Permi, P., Andreote, A.P., Fiore, M.F., 2015. Antifungal compounds from cyanobacteria. Mar. Drugs 13, 2124–2140.

Shukla, P., Singh, P., Yadav, R.K., Pandey, S., Bhunia, S.S., 2016. Past, present, and future of antifungal drug development. Commun. Dis. Devel. World 125–167 (Springer).

Silva, S., Rodrigues, C.F., Araújo, D., Rodrigues, M.E., Henriques, M., 2017. *Candida* species biofilms' antifungal resistance. J. Fungi 3, 8.

Singh, P., Nanda, A., 2013. Antimicrobial and antifungal potential of zinc oxide nanoparticles in comparison to conventional zinc oxide particles. J. Chem. Pharm. Res. 5, 457–463.

Sklenár, Z., Scigel, V., Horácková, K., Slanar, O., 2013. Compounded preparations with nystatin for oral and oromucosal administration. Acta Poloniae Pharmaceut. 70, 759–762.

Slavin, Y.N., Asnis, J., Häfeli, U.O., Bach, H., 2017. Metal nanoparticles: understanding the mechanisms behind antibacterial activity. J. Nanobiotechnol. 15, 1–20.

Socaciu, M.-I., Fogarasi, M., Semeniuc, C.A., Socaci, S.A., Rotar, M.A., Muresan, V., Pop, O.L., Vodnar, D.C., 2020. Formulation and characterization of antimicrobial edible films based on whey protein isolate and tarragon essential oil. Polymers 12, 1748.

Sousa, F., Ferreira, D., Reis, S., Costa, P., 2020. Current insights on antifungal therapy: novel nanotechnology approaches for drug delivery systems and new drugs from natural sources. Pharmaceuticals 13, 248.

Steinbach, W.J., Cramer Jr., R.A., Perfect, B.Z., Asfaw, Y.G., Sauer, T.C., Najvar, L.K., Kirkpatrick, W.R., Patterson, T.F., Benjamin Jr., D.K., Heitman, J., 2006. Calcineurin controls growth, morphology, and pathogenicity in *Aspergillus fumigatus*. Eukaryot. Cell 5, 1091–1103.

Sucher, A.J., Chahine, E.B., Balcer, H.E., 2009. Echinocandins: the newest class of antifungals. Annal. Pharmacother. 43, 1647–1657.

Sun, Q., Li, J., Le, T., 2018. Zinc oxide nanoparticle as a novel class of antifungal agents: current advances and future perspectives. J. Agric. & Food Chem. 66, 11209–11220.

Swamy, M.K., Akhtar, M.S., Sinniah, U.R., 2016. Antimicrobial properties of plant essential oils against human pathogens and their mode of action: an updated review. Evid. Based Complem. & Alter. Med. 2016.

Swidergall, M., Ernst, J.F., 2014. Interplay between *Candida albicans* and the antimicrobial peptide armory. Eukaryot. Cell 13, 950–957.

Tam, J.P., Wang, S., Wong, K.H., Tan, W.L., 2015. Antimicrobial peptides from plants. Pharmaceuticals 8, 711–757.

Tangarife-Castaño, V., Correa-Royero, J., Zapata-Londoño, B., Durán, C., Stanshenko, E., Mesa-Arango, A.C., 2011. Anti- *Candida albicans* activity, cytotoxicity and interaction with antifungal drugs of essential oils and extracts from aromatic and medicinal plants. Infection 15, 160–167.

Tavanti, A., Naglik, J.R., Osherov, N., 2012. Host-fungal Interactions: Pathogenicity versus Immunity. Hindawi.

Ting, P.C., Kuang, R., Wu, H., Aslanian, R.G., Cao, J., Kim, D.W., Lee, J.F., Schwerdt, J., Zhou, G., Wainhaus, S., 2011. The synthesis and structure–activity relationship of pyridazinones as glucan synthase inhibitors. Bioorg. & Med. Chem. Lett. 21, 1819–1822.

Tomalka, J., Azodi, E., Narra, H.P., Patel, K., O'neill, S., Cardwell, C., Hall, B.A., Wilson, J.M., Hise, A.G., 2015. β-Defensin 1 plays a role in acute mucosal defense against *Candida albicans*. J. Immunol. 194, 1788–1795.

van der Does, A.M., Joosten, S.A., Vroomans, E., Bogaards, S.J., Van Meijgaarden, K.E., Ottenhoff, T.H., Van Dissel, J.T., Nibbering, P.H., 2012. The antimicrobial peptide hLF1–11 drives monocyte-dendritic cell differentiation toward dendritic cells that promote antifungal responses and enhance Th17 polarization. J. Innate Immun. 4, 284–292.

Van Der Weerden, N.L., Bleackley, M.R., Anderson, M.A., 2013. Properties and mechanisms of action of naturally occurring antifungal peptides. Cell. & Mol. Life Sci. 70, 3545–3570.

Vicente, M.F., Basilio, A., Cabello, A., Peláez, F., 2003. Microbial natural products as a source of antifungals. Clin. Microbiol. Infect. 9, 15–32.

Vickers, N., 2017. Animal communication: when i'm calling you, will you answer too? Curr. Biol. 27, R713–R715.

Vriens, K., Cammue, B., Thevissen, K., 2014. Antifungal plant defensins: mechanisms of action and production. Molecules 19, 12280–12303.

Vriens, K., Cools, T.L., Harvey, P.J., Craik, D.J., Braem, A., Vleugels, J., De Coninck, B., Cammue, B.P., Thevissen, K., 2016. The radish defensins RsAFP1 and RsAFP2 act synergistically with caspofungin against *Candida albicans* biofilms. Peptides 75, 71–79.

Walker, S.S., Xu, Y., Triantafyllou, I., Waldman, M.F., Mendrick, C., Brown, N., Mann, P., Chau, A., Patel, R., Bauman, N., 2011. Discovery of a novel class of orally active antifungal β-1, 3-d-glucan synthase inhibitors. Antimicrob. Agents & Chemother. 55, 5099–5106.

Wang, G., Li, X., Wang, Z., 2016. APD3: the antimicrobial peptide database as a tool for research and education. Nucl. Acids Res. 44, D1087–D1093.

Wei, G.-X., Campagna, A.N., Bobek, L.A., 2007. Factors affecting antimicrobial activity of MUC7 12-mer, a human salivary mucin-derived peptide. Ann. Clin. Microbiol. & Antimicrob. 6, 1–10.

White, T.C., Holleman, S., Dy, F., Mirels, L.F., Stevens, D.A., 2002. Resistance mechanisms in clinical isolates of *Candida albicans*. Antimicrob. Agents & Chemother. 46, 1704–1713.

Wiederhold, N.P., 2017. Antifungal resistance: current trends and future strategies to combat. Infect. Drug Resist. 10, 249.

Wu, T., Samaranayake, L., Leung, W., Sullivan, P.J., 1999. Inhibition of growth and secreted aspartyl proteinase production in *Candida albicans* by lysozyme. J. Med. Microbiol. 48, 721–730.

Xu, D., Jiang, B., Ketela, T., Lemieux, S., Veillette, K., Martel, N., Davison, J., Sillaots, S., Trosok, S., Bachewich, C., 2007. Genome-wide fitness test and mechanism-of-action studies of inhibitory compounds in *Candida albicans*. PLoS Pathog. 3, e92.

Zhanel, G.G., Karlowsky, J.A., Harding, G., Balko, T.V., Zelenitsky, S.A., Friesen, M., Kabani, A., Turik, M., Hoban, D.J., 1997. In vitro activity of a new semisynthetic echinocandin, LY-303366, against systemic isolates of *Candida* species, *Cryptococcus neoformans*, *Blastomyces dermatitidis*, and *Aspergillus* species. Antimicrob. Agents & Chemother. 41, 863–865.

Zhang, X.-F., Liu, Z.-G., Shen, W., Gurunathan, S.J.I., 2016. Silver nanoparticles: synthesis, characterization, properties, applications, and therapeutic approaches. Int. J. Mol. Sci. 17, 1534.

Zhou, G., Ting, P.C., Aslanian, R., Cao, J., Kim, D.W., Kuang, R., Lee, J.F., Schwerdt, J., Wu, H., Herr, R., 2011. SAR studies of pyridazinone derivatives as novel glucan synthase inhibitors. Bioorg. & Med. 21, 2890–2893.

Zhu, S., 2008. Discovery of six families of fungal defensin-like peptides provides insights into origin and evolution of the CSαβ defensins. Mol. Immunol. 45, 828–838.

Zore, G.B., Thakre, A.D., Jadhav, S., Karuppayil, S.M., 2011. Terpenoids inhibit *Candida albicans* growth by affecting membrane integrity and arrest of cell cycle. Phytomedicine 18, 1181–1190.

Zotchev, S.B., 2003. Polyene macrolide antibiotics and their applications in human therapy. Curr. Med. Chem. 10, 211–223.

CHAPTER

6

Significance of immunotherapy for human bacterial diseases and antibacterial drug discovery

Manzoor Ahmad Mir, Syed Suhail Hamdani and Hafsa Qadri

Department of Bioresources, School of Biological Sciences, University of Kashmir, Srinagar, Jammu and Kashmir, India

6.1 Introduction

Immunotherapy is a broad term that refers to any treatment intervention that aims to manipulate and target the immune system in order to protect the body from deadly diseases (Papaioannou et al., 2016). Immunotherapies are disease-management approaches that target or modulate specific immune system components. Immunotherapies exert control over immune system components, allowing them to locate and eliminate infections and infected cells, thus relieving symptoms and providing protection against diseases. Finally, immunotherapy tries to harness the host's adaptive and innate immune responses in order to accomplish long-lasting elimination of sick cells. It is roughly classified into active (vaccine therapy and allergen-specific) and passive (adoptive and antibody-based) approaches. Active immunotherapy stimulates the patient's immune response, resulting in the development of specific immunological effectors (T cells and antibodies), whereas passive immunotherapy gives patients ex vivo-generated immune elements (immune cells, antibodies) without activating the host immune response (Tur and Barth, 2014).

People's health is at risk from infectious diseases, as governments around the world continue to deal with new and reemerging threats like the SARS-CoV2 pandemic. As a result, a variety of immunotherapeutic techniques are being examined as potential alternative therapeutics for infectious disorders, resulting in major advancements in our understanding of immunological interactions between pathogen and the host. To address the limitations of presently available infectious disease prevention and control procedures, including insufficient efficacy, medication toxicity, and the emergence of drug resistance, novel and inventive therapeutic options are necessary. When it comes to overcoming these limitations,

Human Pathogenic Microbes
https://doi.org/10.1016/B978-0-323-96127-1.00004-8

immunotherapies already show great promise, as seen by recent breakthroughs and successes with drugs like monoclonal antibodies (mAbs) (Ramamurthy et al., 2021). This chapter summarizes some of the most significant advances in the fight against infectious disease over the previous 5 years, utilizing the advantages of immunotherapies such as monoclonal antibodies, vaccinations, checkpoint inhibition, and cytokine level manipulation. Since immunotherapy is widely used in cancer treatment, recent developments have been made in the treatment of human immunodeficiency virus (HIV), zika virus, malaria, TB, and, most recently in COVID-19 thus highlighting the importance of immunotherapeutic techniques to control bacterial and other infections. Ultimately, immunotherapeutics' safety, specificity, and affordability will have an effect on their broad application.

Infectious diseases are still the primary cause of mortality and morbidity in humans. While significant progress has been achieved in diagnosing and understanding the biology of infectious diseases, current antimicrobial chemotherapy has been shown to be ineffective against a large number of current pathogens. Emerging and reemerging microorganisms create new infectious diseases, have developed resistance to available antimicrobial agents, or are unable to be treated due to a lack of treatment options or are generally ineffective due to underlying host immunological weakness. The number of people with a compromised immune system has increased dramatically in recent years as a result of the HIV epidemic and new advances in cancer, transplantation, and autoimmune disease treatment. As a result, novel approaches are required for the treatment of infectious diseases. Almost certainly, such therapies will be utilized as adjunctive therapy, i.e., to improve the efficacy of currently available antibacterial chemotherapy and to restore or redirect the immune response in response to microbiological infections. In this regard, therapies aiming at lowering the sometimes-destructive host inflammatory response to infections should also be mentioned, for example, the use of glucocorticosteroids to minimize mortality and unfavorable sequelae in pneumococcal meningitis (van Dissel).

Historically, immunotherapeutic measures such as vaccination or serum therapy have been extremely efficient in preventing or treating infectious diseases caused by microbial pathogens that induce an acute infectious disease followed by lifelong natural immunity (Qayoom et al., 2021; Ramamurthy et al., 2021). For example, following the discovery of diphtheria toxin in the 1890s, von Behring and Kitasato pioneered passive immunotherapy by delivering toxin-antiserum to diphtheria patients. Serum therapy acquired recognition as a treatment for diphtheria, pneumococcal pneumonia, malignant scarlatina, and meningococcal meningitis throughout the subsequent decades. It became clear that serum therapy (i.e., antibodies) may work by neutralizing toxins, inhibiting microorganism adhesion or lysis, or by improving immune cell activity in innate immunity, for example, by stimulating phagocytosis. Simultaneously, bacteriophages and nonspecific immune stimulants (e.g., BCG) were investigated as potential therapeutic agents for infectious diseases (van Dissel).

By contrast, vaccination as a preventative measure has become widespread and has had a profound effect on human health. Furthermore, multiple new examples of vaccination success are visible today, including those against human papillomavirus infection and related cervical cancer, varicella, genital herpes in women, and *Staphylococcus aureus*, while the first tuberculosis vaccine candidates are in phase I-II research. Numerous breakthroughs have rekindled interest in infectious disease immunotherapy. Various developments have been made to understand host–microbe interaction, and protective immunity should enable a more rational approach to the development of immunotherapeutic therapies. While such an intervention may include a pathogen-specific modality, it is frequently used in an adjunctive setting and is nonpathogen specific, aimed at modulating host effector cells and immune response. Chemotherapy has a direct inhibitory effect on the microbial pathogen in this scenario. Because nonspecificity enables broad application, it will be more cost-effective than earlier pathogen-specific techniques. In spite of this, there are still prospects for pathogen-specific therapies such as antibody therapy against multidrug-resistant microbes and techniques for HIV and tuberculosis postinfection vaccination. Various modalities are likely to incorporate humanized monoclonal antibody and cytokine technologies, as well as advanced vaccination and cellular therapies. The cytokines M-CSF, GM-CSF,G-CSF, and IFN-gamma being nonspecific immunomodulatory agents have gained attention and are utilized as adjuvant treatment for a variety of infectious disorders due to their potential to boost macrophages, neutrophils, and monocyte activity. IFN-gamma is approved for use in chronic granulomatous disorders as a preventative measure. Immunotherapeutic interventions as adjuncts to chemotherapy are gaining popularity in the treatment of viral, bacterial, and fungal infections, for example, IFN-alpha and ribavirin in hepatitis C, intravenous immunoglobulin in necrotizing fasciitis, various interventions in sepsis, and IL-12, IFN-gamma, or G-CSF in combination with azoles or amphotericin B in immunocompromised hosts with fungal disease. Additionally, alternative options are being studied, such as the development of antiadhesion therapy for bacterial disorders, particularly urinary tract infections.

Immunotherapies show tremendous potential for the treatment of autoimmunity, infectious diseases, and cancer. Classical vaccination is the most effective and oldest immunotherapy modality; additionally, it is the most cost-effective technique for public health to battle infectious diseases. Antibiotic resistance is on the rise, especially among Gram-negative bacteria, and may soon represent a major medical and economical burden. Antibiotic resistance is becoming increasingly common, posing a serious danger to modern medicine as we know it (Chan, 2012; Hamdani et al., 2020). As a result, alternative strategies are critical. As a result, we should begin to understand the molecular mechanisms underlying innate immune responses and develop innovative antimicrobial immunotherapies to cope with various diseases. The evident hazards of immunotherapies, such as autoinflammation, autoimmunity, and subsequent organ damage, must be considered carefully as they remain the main roadblocks to clinical use. Despite the situation's seriousness, precise ligands and signaling pathways capable of mediating host defense while conserving healthy tissues must be discovered. To preserve immunological homeostasis, antimicrobial immune responses are rigorously scaled to the level of infectious danger (Blander and Sander, 2012).

The field of antibacterial drug discovery has evolved significantly over the previous decade, including the primary bacterial organisms studied, the discovery methodologies used, and the pharmaceutical companies involved. Antibiotic research had been frustratingly ineffective with target-based high-throughput screening. The lessons learned about target characteristics, screening libraries, and screening procedures have significantly increased our understanding of this failure. Most people no longer use the term "genomics" when talking about drug discovery. Instead, they use terms like "structure-based design and analysis of focused libraries" and "engineering resistance-breaking properties into established antibiotics" to find novel natural product leads among previously abandoned compounds or novel microbial sources. Additionally, different therapy modalities like antivirulence methods and immunotherapeutic approaches are being explored (Brötz-Oesterhelt and Sass, 2010).

Clinical experience with immunotherapeutic treatments is still limited and frequently exploratory. The last decade has demonstrated how difficult it is to forecast the success and demand for such treatments (e.g., failure of anticytokine strategies in sepsis). As a result, corporations will be hesitant to create such drugs. The outlook for immunization is more optimistic. Comparative controlled clinical trials will be required to assess the immunomodulatory drugs' (long-term) safety and efficacy, as well as to determine particular indications for their administration in therapy regimens.

6.2 Immunotherapy and its types

Immunotherapeutic approaches have developed over the past two decades in pathogenesis and have revolutionized the treatment of cancer and other infections, moving away from nonselective harmful drugs and other traditional therapies. Despite substantial global efforts, infectious diseases continue to be the largest cause of mortality and morbidity worldwide, necessitating the development of novel innovative therapies that will address current antimicrobial-resistance concerns. Infectious pathogens have been successful in setting up an inviting environment within the host and adjusting metabolic activity to fulfill their nutritional needs while suppressing host defenses by manipulating regulatory mechanisms. Numerous immune system host characteristics influence therapy outcomes and play a role in disease progression or regression. immunotherapy encompasses a broad array of treatment options that work by targeting or altering the immune system (Papaioannou et al., 2016). Finally, immunotherapy attempts to harness the host's adaptive and innate immune responses in order to eliminate diseased cells permanently. It is broadly classified into passive (antibody-based and adoptive) and active approaches (allergen-specific and vaccine therapy). Active immunotherapy triggers the patient's immune response, resulting in the creation of specific immune effectors (T cells and antibodies), whereas passive immunotherapy injects ex vivo generated immunological components into patients without triggering the host immune response (Naran et al., 2018).

Immunotherapy can take place in active or passive manner. Active immunotherapy entails priming the patient's immune system to combat various infections. Passive immunotherapy involves the administration of immunological molecules to patients who are unable to make them on their own. Both techniques might be either specific or general (Fig. 6.1).

6.2.1 Specific active immunotherapy

One method the immune system can distinguish between cells is by the appearance of antigens on the cell surface. Active immunotherapy involves the analysis of infected cells in the laboratory and the identification of their antigens. Following that, a treatment is developed to elicit a specific immune response against them. This is referred to as *specific active immunotherapy*. Despite its theoretical validity, this approach has shown unsatisfactory outcomes, and practitioners have typically abandoned it (Mir et al., 2020; Waldman et al., 2020).

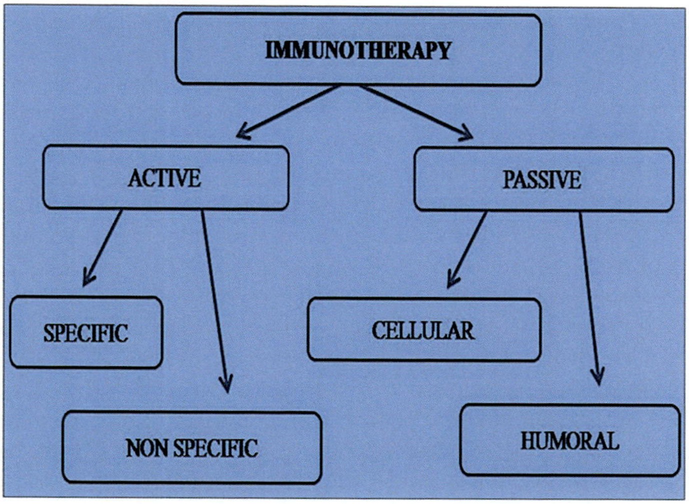

FIGURE 6.1 Immunotherapy can take place in active or passive manner. Active immunotherapy entails priming the patient's immune system to combat various infections. Passive immunotherapy involves the administration of immunological molecules to patients who are unable to make them on their own. Both techniques might be either specific or general.

6.2.2 Nonspecific active immunotherapy

The *nonspecific active method* is the use of drugs that initiate a broad immune response by activating a diverse array of immune cells. The method tries to boost the immune system's overall activity rather than focusing on the elements most capable of fighting infections. This strategy has had some success. Interleukins (IL-2 and IL-12) and interferons are employed in this procedure, as well as Bacillus Calmette-Guerin (BCG), the bacterium that causes tuberculosis. Certain cancers may regress as a result of the immunological response to tuberculosis. Researchers have had some success by directly injecting nonspecific drugs into body cavities containing tumors.

6.2.3 Passive immunotherapy

Passive immunotherapy, also known as *adoptive immunotherapy*, involves the administration of antibodies and other medicines to a cancer patient in order for the patient to adopt an immune response produced in a test tube. Passive approaches might be either specific or nonspecific. Monoclonal antibodies (mAbs), also known as "magic bullets," can be created to specifically target infected cells. One nonspecific strategy is removing part of the patient's T cells, cultivating them in the laboratory, and then injecting expanded populations of the treated cells into the patient. An alternate strategy is to activate lymphocytes isolated from a patient's tumor to create yet another agent termed tumor-infiltrating lymphocytes (TILs) and then return these "foreign" cells to the patient to fight the cancer cells (Janeway et al., 2001; Mir et al., 2020).

The difference between active and passive immunotherapy is in how they affect the patient's immune system. An active immunotherapy agent is one that establishes immunological memory and so induces a long-lasting, sustained response. This is the most accurate representation of a typical immunological response. However, just as immune system function differs among healthy individuals, so does the extent of response to an active immunotherapy agent. A passive immunotherapy agent, on the other hand, generates an immediate effect by the introduction of immune-cell components such as cytokines or antibodies. Because the effects of passive immunotherapy are temporally related to the agent's delivery, continuous doses may be required for a lasting response. This is because the immune system's memory is not engaged (Mehraj et al., 2020; Waldmann, 2003).

Vaccines are the most established and effective method of immunotherapy (Rappuoli et al., 2014). Vaccines provide protection in two ways: (i) an increase in the precision and efficiency of the host immune system's response to previously encountered pathogens by inducing clonal proliferation of antigen-specific lymphocytes; and (ii) delivering neutralizing agents like antibodies to provide passive protection after infection, for example, bacterial toxins/antigens (Naran et al., 2018). Smallpox and polio have been nearly or completely eradicated thanks to the success of immunization initiatives. Past three decades have seen scientific advancements which have aided in the development of novel vaccination platforms based on nucleic acid—based vaccines, recombinant antibodies, and adjuvant optimization (Rappuoli et al., 2014).

Since 1986, monoclonal antibodies (mAbs) have become the most extensively used immunotherapies (Hooks et al., 1991; Naran et al., 2018). mAbs work in three ways, all of which rely on the antigen's specificity and selectivity. The first method is to attach to cell surface receptors and activate a signaling cascade that results in cell death. The second way is by interfering with ligand—receptor interactions, which are essential for continuing cell growth or viability. The third method is antibody-dependent cytotoxicity, in which the antibody's Fc region aids in the recruitment of cells involved in cell-mediated immunity [like NK cells, macrophages, and monocytes]. Complement-dependent cytotoxicity is the final approach, which happens when the complement cascade is activated following attaching to the target structure. The targeting domain of a monoclonal antibody is fused to a toxic payload, like small compounds or apoptosis-inducing toxins, in antibody conjugates that target disease-associated antigens. They internalize and release their payload once hooked to their target, causing cell death with the least amount of injury to healthy tissues. Bispecific antibodies that possess two binding domains, one for an antigen and the other for an effector cell, have also been created. Bispecific antibodies can affect molecules engaged in cell proliferation or inflammation by interfering with multiple surface receptors/ligands and thus bring targets closer together to facilitate protein complexation during the clotting cascades, or recruiting immune cells to the diseased site without requiring MHC engagement (Suurs et al., 2019).

Checkpoint blockade therapy, one of the most well-known immunotherapies in recent years, includes breaking the connection between immunological suppressive receptor-ligand pairs utilizing mAbs. Immunological checkpoints are biological processes that block the immune system from going on the offensive and thus preventing healthy cells being destroyed by the immune system. The immune system returns to normal function by suppressing disease-associated aberrant immune checkpoint activation, allowing for increased immunological responses to elevated ligands. Checkpoint blockade mAbs specifically target programmed cell death 1 ligand 1 (PD-L1), T-cell immunoglobulin, mucin domain-containing protein 3 (TIM3), cytotoxic T-lymphocyte-associated protein 4 (CTLA4), programmed cell death protein 1 (PD1), and OX40, which all help to prevent T-cell inhibition and thus promote effector T-cell activation. Since their initial regulatory approval in 2011, a number of medications have been developed and approved by the US Food and Drug Administration for the treatment of various cancers using checkpoint inhibitors. Their use in combination with other medications has shown promising results in the treatment of malaria, HIV, and tuberculosis (TB), and is explored in further depth under the heading "Checkpoint inhibition." Cytokines are soluble proteins that play a role in inflammation, cell proliferation, wound healing, angiogenesis, immunity, and repair, among other biological processes. Since 1986, the use of cytokines for therapeutic use has been approved. They mediate signaling that is critical for disease spread and management (Wykes and Lewin, 2018).

An immunotherapeutic technique that was recently approved involves increasing T-cell activity using a chimeric antigen receptor (CAR) (Ramamurthy et al., 2021). Using CAR-T cells, researchers hope to construct an entirely new type of cell receptor by merging a T-cell-specific antibody derivative with a transmembrane signaling domain expressed on a target cell. This eliminates the need for MHC and enables T-cell activation without it. There are several ways to treat cancer using CAR-T-cell therapy, but the most common is by using the patient's own T cells to make the modified CAR-T cells, which can then be injected back into the patient's body. Early CARs had a tendency to be anergic, which made it possible to design targeting and transmembrane signaling domains like a costimulatory receptorlike CD28 and CD3 chain using genetic engineering. CARs of the third and fourth generations contain an inducible gene and a second costimulatory protein that expresses proliferative and proinflammatory cytokines, respectively (Ramamurthy et al., 2021) (Fig. 6.2).

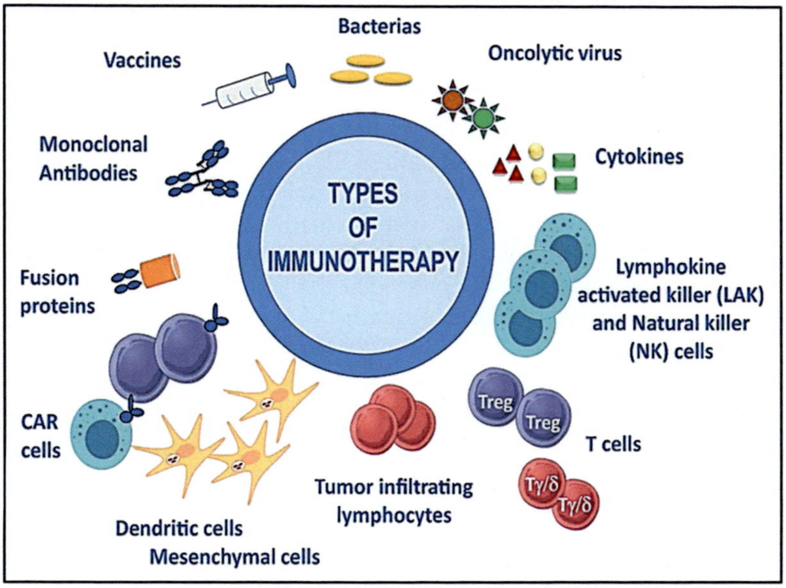

FIGURE 6.2 Various immunotherapies, including vaccines, monoclonal antibodies, pathogens, recombinant proteins, cellular immunotherapies, CAR cells, and cytokines are used in the treatment of a wide range of diseases.

6.3 Immunotherapies for bacterial infections

6.3.1 Vaccines

Vaccination was the first sort of host-directed immunotherapy to be developed, and it includes a wide range of treatments. Most vaccinations work by injecting a noninfectious version of a disease-causing microbe into a person, giving a more favorable stimulus for disease-specific T-cell activation and immunological memory formation. Immune memory cells have the ability to quickly destroy infections and prevent infection. While these types of therapeutic strategies were effective at removing smallpox and greatly lowering the burden of diseases of infectious pathogens like typhoid, rabies, hepatitis, and cholera, it was ineffective against cancer and chronic diseases like the human immunodeficiency virus (HIV). The use of therapeutic vaccines in conjunction with various immune-based treatments has the potential to boost efficacy (Waldmann, 2003).

Immunization against infectious diseases is one of the most cost-effective health interventions ever devised, benefiting individuals and society far more than the vast majority of other healthcare interventions we provide patients, such as diabetes or heart disease care or dialysis for end-stage renal disease. Indeed, vaccinations have been so effective in eradicating numerous chronic diseases that many members of the public have forgotten how dreadful those diseases were, culminating in the recent appearance of an antivaccine movement in industrialized countries (Laxminarayan et al., 2014). Active immunization is a technique that stimulates the immune system to create antibodies by exposing it to antigens found in vaccines and toxoids. Vaccines contain live, attenuated, or dead microorganisms. A toxoid is a bacterial toxin that has been genetically modified to be safe. Various types of vaccines used for various diseases are given in Table 6.1. Active immunization is intended to produce an adaptive immune response that occurs several days after vaccine injection and confers long-term immunity (Kaufmann et al., 2010). Among vaccinations, live attenuated vaccines, which reproduce in the patient, are believed to elicit the most identical immune response to the actual disease. Those manufactured from inactivated organisms, a component of the bacterial capsule, a toxoid, recombinant proteins, and chemically purified components are examples of killed vaccines. Vaccines that contain killed components may require additional scheduled doses to have an impact. After vaccination with inactivated antigens, immunoglobulin M (IgM) and immunoglobulin G are formed. Typically, live vaccinations require at least one incubation period and many weeks for a more effective response.

Mycobacterium tuberculosis (Mtb) is the pathogen that causes tuberculosis (TB), which is the leading infectious disease—related cause of death (Mir and Al-baradie, 2013; Organization, 2013). Bacillus Calmette-Guérin (BCG) is the only licensed tuberculosis vaccine that consistently protects children against severe extrapulmonary tuberculosis (TB). It affords relatively moderate protection against pulmonary tuberculosis (TB) in adults (Sable et al., 2019). Additionally, despite its broad use as a tuberculosis vaccine, it has failed to prevent active tuberculosis infection and thus highlights the need for innovative techniques. A successful tuberculosis vaccine should ideally outperform BCG in terms of protective efficacy and disease prevention, hence stopping *Mtb* transmission. Regrettably, several previous tuberculosis vaccinations failed to accomplish this. Notably, the MVA85A vaccination had no effect on

TABLE 6.1 Various types of vaccines and diseases against which they are used.

S. No.	Type of vaccine	Description	Disease
1.	Inactivated vaccine	Vaccines containing pathogens killed through chemical treatment or heat	Cholera, hooping cough, plague, typhus
2.	Live attenuated vaccine	Vaccines containing less pathogenic strain of microbe	MMR, smallpox, typhoid fever, tuberculosis, tularemia
3.	Toxoid	*T* ***oxoids*** are inactivated bacterial toxin elements that are capable of instructing the immune system to develop antibodies to activated toxins. Induces an immune response to the pathogen's toxin	Diphtheria, tetanus
4.	Subunit	A vaccination intended to generate immune responses against a pathogen's most prevalent epitopes	Hepatitis B
5.	DNA vaccine	Vaccines comprising fragments of DNA encoding antigens for certain diseases are administered in order to stimulate endogenous synthesis	Tuberculosis
6.	Conjugate	A powerful antigen (often a protein) that is covalently linked to a weaker antigen (often a bacterial polysaccharide)	Bacterial meningitis, pneumonia
7.	Recombinant	Recombinant DNA delivered through bacterial vaccine vector	HPV

BCG's protective effectiveness in newborns or HIV-1-infected adults (Mir et al., 1993; Tameris et al., 2013). Numerous new vaccine candidates are presently undergoing or have just finished clinical trials, with varied degrees of success. An immunogenic fusion protein (M72) derived from two *Mycobacterium tuberculosis* antigens and the GlaxoSmithKline adjuvant AS01E demonstrated 49.7% effectiveness in producing tuberculosis disease protection in HIV-negative individuals with latent tuberculosis (Tait et al., 2019). Due to inadequate study designs that ignore specific bacteria strains, patient heterogeneity, disease progression, and hospital epidemiology, there has been limited success to develop vaccines against infections caused by antibiotic-resistant *S. aureus* and *P. aeruginosa* in clinical trials. This highlights the importance of fully characterizing these parameters to ensure meaningful results (Bekeredjian-Ding, 2020).

6.3.2 Monoclonal antibody therapy

C. Milstein and G. Köhler's 1975 discovery of monoclonal antibodies (mAbs) revolutionized medicine and immunology, as well as a variety of other disciplines (Mehraj et al., 2020; Köhler and Milstein, 1975). mAbs are produced when two cells unite to form a hybrid cell or hybridoma that possesses two different characteristics, i.e., the ability to produce a single particular antibody and immortality (Fig. 6.3). Dr. Milstein is widely regarded as the "Father of Modern Immunology" for his seminal contributions to the field (Springer, 2002). Numerous mAbs have enabled the identification of novel molecules and the development

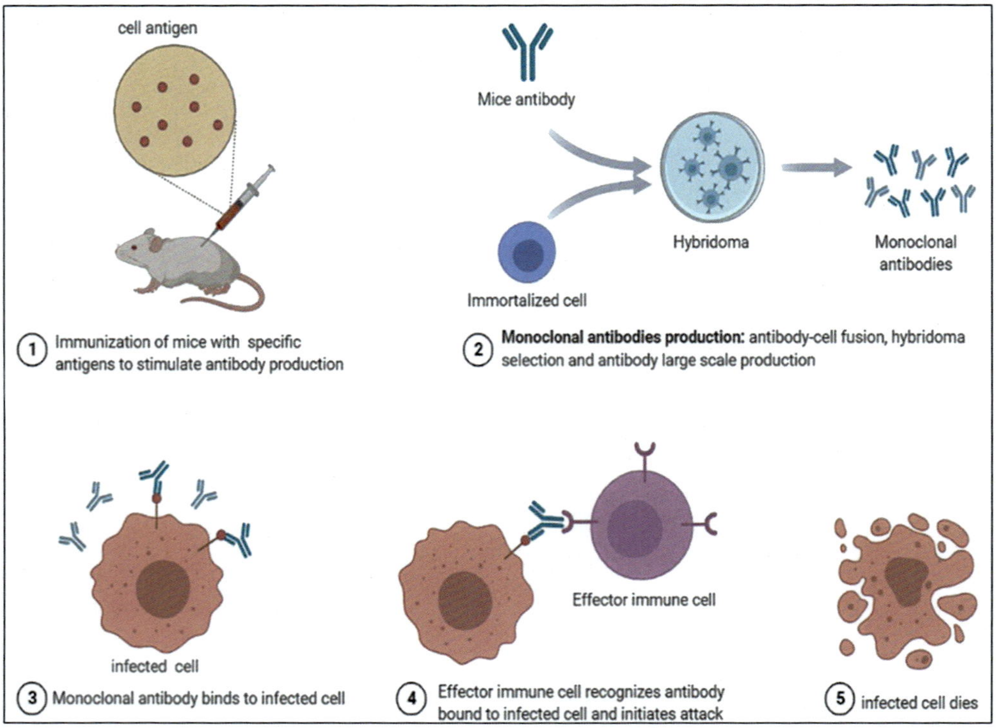

FIGURE 6.3 Diagrammatic demonstration showing the production of monoclonal antibodies by B cells and specifically target antigens.

of more precise diagnostic methods; quick, specific, and cost-effective technologies; purification/concentration processes for substances; and more effective and targeted therapies. Since 1975, the number of monoclonal antibodies licensed for human therapy has increased steadily, and many more are currently being studied in clinical trials.

More than any other disease, mAb treatment has been explored and produced for cancer treatment. The cancer field's subsequent experiments and errors, as well as the solutions that resulted, paved the way for sophisticated antibody-based therapy. The first challenges of providing mouse hybridoma-derived antibodies to humans, as well as the subsequent resolution of the human anti-mouse antibody (HAMA) reaction employing recombinant DNA technology via chimerization and humanization, are among them (Mehraj et al., 2020). mAbs exert their inhibitory effect by inhibiting soluble mediators such as cytokines and growth factors from binding to their target receptors. This is accomplished through the mediator's or its cognate receptor's binding. Antibodies are directed against a certain cell population in which Fc receptor functions are triggered or effector moieties are delivered. The crosslinking of receptors linked with programmed cell death or the usage of specific receptors that function as agonists for the activation of specific cell types is then used to activate antibodies. Surface antigens, like the majority of antibodies, are appealing targets for antimicrobial monoclonal antibodies. Complement attachment for immune system engagement and opsonophagocytic death are the two main properties of surface-specific antibacterial monoclonal antibodies (OPK). Panobacumab is a human anti-*Pseudomonas aeruginosa* LPS serotype O11 monoclonal antibody that is one of the most sophisticated antibacterial mAbs. In vitro, this mAb demonstrated OPK activity and significantly reduced bacterial burden and enhanced survival in infection models (Horn et al., 2010). In a phase II clinical trial, panobacumab was taken in combination with conventional antibiotics and resulted in a shorter time to disease resolution in people with *P. aeruginosa* 011 pneumonia (Que et al., 2014). While strain-specific monoclonal antibodies have shown promise, their efficacy is limited to strains that express the receptors they are designed to target. A protective monoclonal antibody directed against the *P. aeruginosa* type III secretion protein, PcrV, on the other hand, is essential for the transfer of virulence components into host cells. As a result, KB001, a phase II monoclonal antibody against *P. aeruginosa* for the prevention of ventilator-associated pneumonia, was developed (Baer et al., 2009).

MAbs are being looked into again to see if they could be utilized to treat bacterial infections. Antibodies are important in tuberculosis immunomodulation, as demonstrated by antibody profiles obtained during latent tuberculosis infection, which show improved Fc-mediated immune effector function and macrophage clearance of intracellular pathogens, highlighting antibodies' protective role (Lu et al., 2016). However, protective monoclonal antibodies against Mtb have not been developed so far. In comparison, several designed monoclonal antibodies against *P. aeruginosa* and *S. aureus* have advanced to clinical testing. MEDI3902, a bispecific IgG1 antibody directed against *P. aeruginosa's* Psl exopolysaccharide (colonization and tissue adhesion) and PcrV protein (host cell cytotoxicity) is being researched as a potential treatment for pneumonia in high-risk patients (NCT02696902) (Ali et al., 2019). Additionally, the targets are conserved across *P. aeruginosa* samples from around the world, suggesting that they may mediate broad coverage. The mAb AR-301 is capable of neutralizing alpha toxin and hence protects against host cell damage caused by alpha toxin in patients with methicillin-resistant *Staphylococcus aureus* (MRSA) pneumonia (François et al., 2018). The long-acting monoclonal antibody MEDI4893 is also capable of neutralizing alpha toxin and has shown efficacy in immune prophylaxis against *S. aureus* infection, as well as maintaining serum levels in healthy individuals after intravenous administration, and is presently in phase II clinical trials (Ruzin et al., 2018).

6.3.3 Checkpoint inhibition

Checkpoint blockade therapy, one of the most well-known immunotherapies in recent years, includes breaking the connection between immunological suppressive receptor-ligand pairs utilizing mAbs. Immune checkpoints are biological systems that prevent the host immune system from indiscriminately attacking healthy cells. The immune system returns to normal function by suppressing disease-associated aberrant immune checkpoint activation, allowing for enhanced immunological responses against higher ligands. Negative immunoregulation mechanisms, which are employed in immune checkpoint inhibition, provide a variety of inhibitory pathways necessary for the maintenance of self-tolerance and the adjustment of the severity and duration of immune responses in peripheral tissues (Pardoll, 2012). These pathways are essentially linked to T cell exhaustion and involve the development of inhibitory receptors on the cell surface that govern autoreactivity and immunopathology (Sharpe et al., 2007). Although inhibitory receptors are only produced transiently in functional effector T cells, their elevated and sustained expression may be a defining feature of exhausted T cells (Wherry, 2011). Pathogens and tumors increase inhibitory immune checkpoint interactions to elude immune regulation. Infections such as cancer, HIV, TB, and malaria have all been linked to checkpoint proteins.

Programmed cell death 1 ligand 1 (PD-L1), mucin domain-containing protein 3 (TIM3), T-cell immunoglobulin, and cytotoxic T-lymphocyte-associated protein 4 (CTLA4) are specifically targeted by prominent checkpoint blockade mAbs) and thus promote effector T-cell activation by preventing T-cell inhibition. Since their initial regulatory approval in 2011, various drugs have been developed for the treatment of various cancers by the utilization of checkpoint inhibitors and have been approved by the US FDA. Their combination with other therapies has demonstrated promising results in the treatment of tuberculosis (TB), malaria, and human immunodeficiency virus (HIV) (Wykes and Lewin, 2018; Qayoom et al., 2021).

While immune checkpoint inhibitors have transformed cancer therapy, there is conflicting evidence on their efficacy in the treatment of tuberculosis. Despite $CD8^+$ and $CD4^+$ T cells having the protective roles in *Mtb* containment, mounting evidence suggests that they gradually deteriorate in patients with active tuberculosis, frequently as a result of the expression of inhibitory receptors (LAG3, CTLA-4, TIM3, and PD-1), which results in T-cell exhaustion (Day et al., 2018). While mABs targeting PD-1 and its ligand (PD-L1) have been demonstrated to restore tumor-specific T-cell activity, it is unknown if this would be effective in the treatment of human tuberculosis (Barber et al., 2019). For example, PD-1 mutant mice infected with Mtb are significantly more prone to developing new tuberculosis infections, which are characterized by increased mycobacterial loads and mortality. Similarly, Tezera et al. (2020), report that inhibiting PD-1 promotes *Mtb* growth via increased tumor necrosis factor-alpha (TNF-α) release in a 3D cell culture model of human tuberculosis. Elevated IFN-production may also be achieved by inhibiting the PD-1/PD-L1 pathway in vitro, but this may not be sufficient to restore the proliferative ability of *Mtb*-specific $CD4^+$ T cells. Those findings are backed up by an increase in checkpoint blockade-associated tuberculosis and unusual *Mycobacterium tuberculosis* infections in cancer patients treated with anti-PD-1/PD-L1 monoclonal antibodies (Anand et al., 2020). Other inhibitory receptors were shown to coexpress with functionally compromised TIM3+ T cells following infection with *M. tuberculosis*, indicating a role of TIM3 (Jayaraman et al., 2016). Notably, anti-TIM3 mAb therapy increased T-cell activity and improved bacterial load control in chronically infected mice. As a result, inhibiting LAG3's activity stimulates T-cells while also removing the suppressive activity induced by regulatory T-cells, making it a more appealing option than PD-1 (Phillips et al., 2017). This can be summarized as immune checkpoint expression in tuberculosis and may be considered as a physiological response to the persistent *M. tuberculosis* pathogen, and its inhibition may enhance infection and pathogenesis, as indicated by cellular and epidemiological studies of PD-1 knockout mice. A variety of factors will likely influence immune checkpoint inhibition for tuberculosis treatment, including the host (immunocompetence and HIV status) and mycobacterial-specific characteristics (*Mtb* strain and medication resistance) (Langan et al., 2020) (Fig. 6.4).

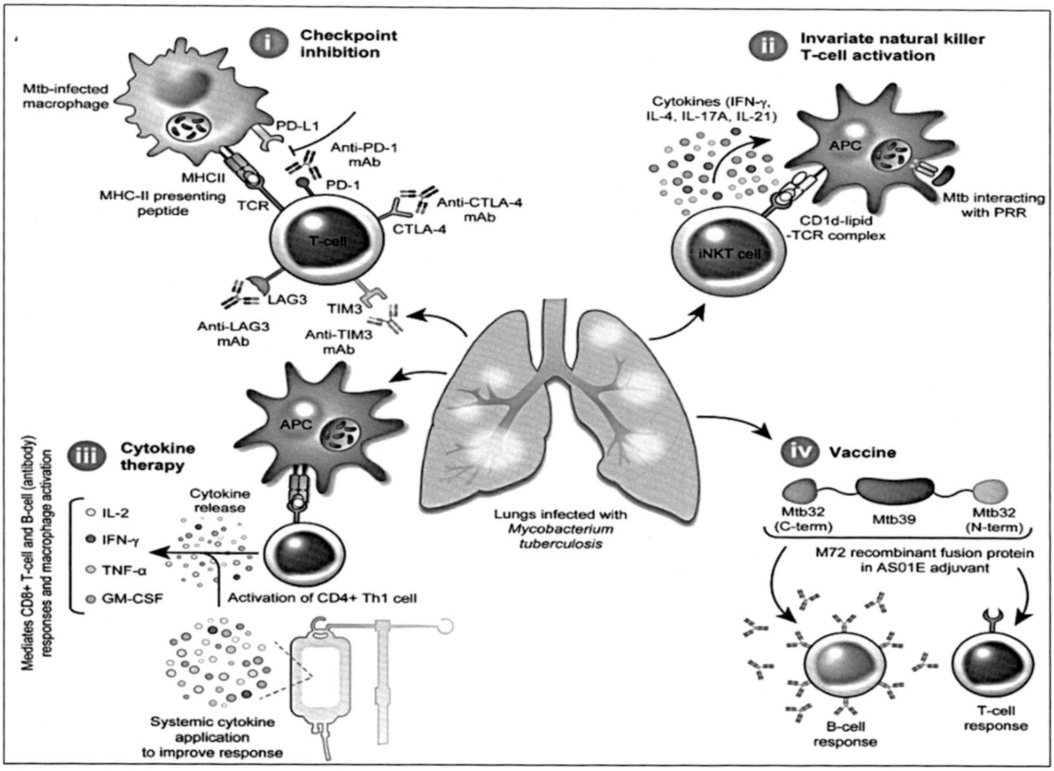

FIGURE 6.4 (i) mAbs target T-cell receptors or APC ligands, disrupting the interaction of receptors and ligands, resulting in T-cell activation. (ii) To initiate an immune response, invariant natural killer T-cells identify mycobacterial lipids on APCs and produce cytokines. (iii) Systemically administered cytokines (IL-2, IFN-γ, TNF-α, GM-CSF) can enhance the Th1 response to Mtb infection. The M72/AS01E vaccine, developed by fusing two Mtb antigens (Mtb32 and Mtb39) in the AS01E adjuvant induces both humoral (B-cell) and cellular (T-cell) responses, protecting against active TB infection.

6.3.3.1 *Programmed cell death protein 1 (PD-1)*

A variety of immune cells, such as innate lymphoid cells (ILCs) and innate lymphocytes (natural killer (NK) cells and adaptive lymphocytes (e.g., B and T cells)) play important roles in the management and elimination of infection. They accomplish this by killing sick cells and releasing inflammatory chemicals that activate or stimulate myeloid bactericidal responses. Various triggers induced during infection, on the other hand, can cause an increase in immunological regulatory checkpoint molecules. When the evolutionary significance of these mechanisms, like the important T-cell checkpoint inhibitory pathway PD-1/PD-L1, is explored, it is theorized that they prevent immune response hyperactivation, which can lead to negative outcomes like increased tissue damage and inflammation (Qayoom et al., 2021). Overexpression of PD-1 on T cells, for example, appears to be critical for avoiding immune-mediated pathology during infection with *Mycobacterium tuberculosis*. Other infectious organisms, particularly pathogenic bacteria, can subvert the immune system by using immune-regulatory mechanisms (Wykes and Lewin, 2018), and blocking these pathways may represent feasible possibilities for restoring immune cell function in the battle against infection. One such bacteria, *Burkholderia pseudomallei,* causes chronic infection that is intrinsically resistant to a wide range of antibiotics, making it very difficult to treat and associated with a high risk of death. In mouse infection models, several small-colony variants of *B. pseudomallei* upregulate PD-1 on adaptive and innate immune cells (See, Samudi et al., 2016). In an in vitro infection model capable of suppressing interferon production and proliferation of PD-1^+ T cells, *B. pseudomallei* increases the expression of PD-L1, a PD-1 ligand on human neutrophils. The PD-1/PD-L1 pathway may be inhibited to aid bacterial clearance in this severe infection. Infection with *Helicobacter pylori* increased the expression of PD-1 and PD-L1 on T cells and epithelial cells, respectively. In an in vitro *H. pylori* infection model, inhibiting PD-L1 led to a boost in CD4+ T cell-mediated IL-2 production. Innate and adaptive lymphocytes both benefit from IL-2 for proliferation and survival. This PD-1/PD-L1 upregulation has been proposed as a significant starter of *H. pylori*-induced gastric cancer via premalignant lesions (Shen et al., 2019). Through improved Kupffer cell activity, inhibition of the PD-1/PD-L1 axis restores susceptibility to bacterial infection linked with chronic liver disease and acute liver injury. To cure immunological paralysis following septic shock, targeting this route is gaining attention. To date, the most well-characterized checkpoint in infectious diseases and cancer is overexpression and inhibition of PD-1. However, evidence is accumulating suggesting other immunological checkpoints, such as mucin domain-containing protein 3 (Tim-3) and T-cell immunoglobulin may potentially play a role in disease (McCulloch et al., 2021).

6.3.3.2 T-cell immunoglobulin and mucin domain-containing protein 3 (Tim-3)

While anti-PD1 medication is almost certainly contraindicated during tuberculosis infection, blocking Tim-3 looks to be more promising. Tim-3, like PD-1, is an immunological checkpoint receptor found on T and NK cells. When bound to its ligand, galectin-9, it transmits an inhibitory signal (Pose et al., 2021). Infection with *Mycobacterium TB* has been demonstrated to boost Tim-3 expression in human and mouse T cells (Jayaraman et al., 2016). While the research focused on T cells, it did not rule out the possibility that Tim-3 signaling is also involved in NK cells. Tim-3 expression is increased in Natural Killer cells from tuberculosis patients with latent or active tuberculosis, and this negatively correlates with IFN-ɣ production in response to IL-12. IL-12 is a well-known IFN-ɣ inducer. This inhibition may be restored in vitro once more in the presence of a Tim-3-blocking antibody (Wang et al., 2015). The impact of Tim-3 overexpression on NK cells in a mouse model of tuberculosis has yet to be identified, but it is an attractive subject for further research. While checkpoint inhibitors like Tim-3 and PD-1 are being examined widely in the context of bacterial infection, scientific prospects indicate that other well-known checkpoints studied in cancer research could also be useful in infection research. One of the targets of attention is the T cell immunoreceptor with Ig and ITIM domains (TIGIT).

6.3.3.3 TIGIT

TIGIT is another critical immune checkpoint that is expressed mostly on exhausted NK cells and T cells. TIGIT interacts to the antigen-presenting cell ligands CD155 and CD112, which are expressed on macrophages and dendritic cells (DCs). The interaction directly suppresses TIGIT-expressing lymphocytes while also competing with the activating receptor DNAM-1 for costimulation (Mir, 2015; Harjunpää and Guillerey, 2020). TIGIT is well characterized in cancer situations and may potentially be used to treat bacterial infection. TIGIT expression is upregulated in T cells from septic patients, as well as in a mouse polymicrobial sepsis caecal ligation puncture model (Washburn et al., 2019). While these findings were intriguing, they were solely descriptive and left open the question of whether TIGIT was beneficial in terms of lowering immunological pathology or deleterious in terms of increasing immune exhaustion and impeding microbiological clearance in the case of sepsis. Interestingly, the role of TIGIT in either a protective or pathogenic manner has been found to be cell type and disease model dependent.

6.3.4 T-cell-based immunotherapies

T cells mount highly effective and targeted immune responses against foreign antigens. T-cell treatments are defined broadly as strategies that trigger these immune responses and are frequently explored in combination with mAbs. T-cell coreceptors binding to APC counter-receptors and TCR binding to the peptide-MHC complex are both essential for effective T-cell activation. T-cell exhaustion is a situation in which T cells become dysfunctional as a result of extended antigen exposure and/or inflammation, and it's linked to a range of chronic infections and cancers. Inhibitory receptors that persist and are diverse, the progressive and hierarchical loss of effector cytokines, metabolic dysregulation, altered transcription factor expression and function, and the inability to convert to quiescence and the inability to achieve antigen-independent memory T-cell homeostasis are all characteristics of this condition (Schietinger and Greenberg, 2014). As a result, T-cell depletion is an immune evasion strategy that essentially results in inefficient infection and tumor suppression. Exhausted T cells are not inert; rather, they continue to execute insufficient, vital activities that obstruct pathogen infection or tumor development (Wherry and Kurachi, 2015).

Since its discovery in the LCMV mouse model, this condition of T-cell dysfunction has been reported in animal and human models of persistent viral infections, as well as different malaria and *Mycobacterium TB* infections (Khan et al., 2017). Significant breakthroughs have been achieved in three critical areas: (i) the formation and homeostasis of exhausted T cells; (ii) T-cell properties and maintenance; and (iii) the absence of conventional memory. This dysfunctional state of exhaustion has therefore attracted researchers' curiosity, as has restoring or strengthening immune responses to effectively combat infection or cancers (Pauken and Wherry, 2015).

The efficacy and use of T-cell-based immunotherapies are being extensively explored with the goal of creating more effective treatment options for tuberculosis (with or without coinfection with HIV). Unconventional T-cells [HLA-E-restricted T-cells, mucosal-associated invariant T-cells (MAIT), γδT-cells, and natural killer T-cells (NKT)] are a diverse population of T lymphocytes that are not restricted to antigen recognition via the classical MHC and thus could be promising candidates for developing TB-directed T-cell-based therapeutics (La Manna et al., 2020). Invariant NKT (iNKT) cells identify a variety of lipids associated with mycobacterial cells and release a variety of cytokines (IL-21, IL-17A, IL-4 and IFN-γ) capable of mounting an immune response against *M. tuberculosis* (Langan et al., 2020).

6.3.5 Cytokine therapy

Cytokines play a part in a variety of activities, including immunity, wound healing and repair, cell proliferation and inflammation, angiogenesis, fibrosis, and cell migration. Given that cytokines are important signals for fundamental processes in a wide range of disorders, it appears conceivable that manipulating them can influence sick states in both positive and negative ways. Since 1986, cytokines have been approved for therapeutic use. They mediate signaling that is critical for disease spread and management (Riley, 2019). The antitumor properties of cytokines were revealed in a number of mouse experiments, leading to the development of cytokine-based cancer treatments. Interleukin-1, interleukin-12, interleukin-15, interleukin-18, interleukin-21, and GM-CSF are only a few of the cytokines that have been studied in clinical studies in advanced cancer patients (Feldmann, 2008). Treatment with proinflammatory cytokines IL-12 and IL-18 was recently reported to repair tumor-induced NK cell anergy in MHC-1 knockout mice. Cytokine therapy raised NK cell activity, which aided anticancer responses in mice challenged with MHC-1-deficient tumors, resulting in improved survival rate in mouse (Ardolino and Raulet, 2016).

It is the goal of this department to continue to investigate and develop antisuppressive agents such as TGF and IL-10 because their results have been promising. Overexpression of IL-10 in a variety of cell types, including dendritic cells (DCs), monocytes, and CD4+ T cells, has been linked to disease progression in chronic infections and a bad prognosis in cancer patients (Said et al., 2010). Furthermore, IL-10 has been linked to the progression of human visceral leishmaniasis by inhibiting Th1 cell responses and/or neutralizing parasitized tissue macrophages, hence reducing therapeutic response. When *M. tuberculosis* is infected in vivo, it has been discovered that IL-10 acts as a negative modulator of the immune response, hence leading to the development of chronic infection. Because blocking antibodies to IL-10R specifically promotes in vivo IFN-ɣ and IL-17A production and Th1 and Th17 responses during *M. tuberculosis* infection, it provides much greater protection against aerogenic challenge with *M. tuberculosis* than BCG vaccination alone (Pitt et al., 2012).

6.3.5.1 TNF-α

TNF-α is a highly pleiotropic inflammatory cytokine that plays a vital role in infectious events. Immune cells and other organs have shown that it can induce apoptosis or necroptosis and activate the immune system; however, the dichotomy between these two pathways is still unclear (Kalliolias and Ivashkiv, 2016). When bacterial patterns are detected via the Toll-like receptor (TLR) system, macrophages release high amounts of TNF-α in response. Macrophages and other myeloid cells become more active against bacteria as a result of this. When infected with *Escherichia coli*, intestinal epithelial cells express higher levels of PD-L1, suggesting that TNF-α may be involved in the enhanced expression of PD-L1 (Lee et al., 2020). TNF-α is necessary for tuberculosis prevention by maintaining the granuloma, but excessive levels also aggravate disease by increasing macrophage necrosis. As a result, TNF-α action can be harmful, and further research should be done to determine the pleiotropic effect of TNF-α in bacterial infections in order to identify individual patients in whom this cytokine operates pathologically. Personalized medicine advances may enable the identification of immunological markers that separate patients who would benefit from TNF-α inhibition from those who would benefit from infection augmentation (Roca and Ramakrishnan, 2013).

6.3.5.2 TGF-β

TGF-β is a versatile cytokine that plays a crucial function in cellular defenses. TGF-β1, TGF-β2, and TGF-β3 are the three isoforms of TGF-β. All of these isoforms bind to the TGF-β receptor, a heterodimeric complex comprising TGF-βRI and TGF-βRII that begins a typical signaling cascade through SMAD2/3 phosphorylation. TGF-β signaling has a number of implications, but the most important for the immune response is that it reduces lymphocyte proliferation, cytokine synthesis, metabolism, and killing power, and it is essential for regulatory T-cell growth (Batlle and Massagué, 2019). In the event of tuberculosis infection, reversing immune suppression by TGF-inhibition rather than checkpoint inhibition may be beneficial to the patient. Tuberculosis lung granulomas in nonhuman primates are linked with relatively modest levels of T-cell fatigue, as shown by immune checkpoint expression (Warsinske et al., 2017). It was thought that this was due to the granuloma's restricted interaction with infected macrophages, minimal antigen activation, and the presence of inhibitory chemicals like TGF-β. During tuberculosis infection, TGF-β expression was found to be localized to infected granulomas, and so blocking it may limit the improvement of immune responses to these infected granulomas (Gern et al., 2021), whereas checkpoint inhibition may result in a more systemic immune enhancement, which may account for the increased pathology observed with anti-PD-1 therapy.

6.3.6 Cellular therapies

There is still research being done on how cell treatments can be used to treat bacterial infections, which is why they are still relatively new. In one case, naive T cells were transduced with a TCR specific for an *S. typhimurium* antigen and then adoptively transplanted into *S. Typhimurium*-infected animals. T cells were given either therapeutically or prophylactically, and transferred T cells cleared more germs than untransduced naive T cells in both instances (Kalinina et al., 2020). To combat bacterial infection, adoptive transplantation of macrophages can also be performed. Tacke et al. demonstrated a strategy for creating monocyte-derived macrophages (MDMs) that, when administered intravenously, drastically reduced bacterial loads in mice models of *Klebsiella pneumoniae* and *P. aeruginosa* (Tacke et al., 2019). Ackermann et al. recently developed a stirred bioreactor-based approach for producing macrophages from human-induced pluripotent stem cells (iPSCs). Stirred bioreactor production is advantageous because it can scale up to the volumes necessary to manufacture therapeutic macrophages, which is challenging to achieve with 'off-the-shelf' MDMs. These iPSC-derived macrophages displayed functional characteristics to primitive macrophages when breathed and were capable of rescuing animals from an acute *P. aeruginosa* lung infection (McCulloch et al., 2021).

While intriguing new uses for macrophage and T-cell adoptive transfer are being discovered, the effectiveness of NK cellular therapy to treat bacterial diseases is yet unknown. This may be especially effective in the current era of genetic modification, when such approaches offer near-limitless possibility for modifying cells to improve their persistence and function (Lin et al., 2020). As we know bacterial infections can employ a variety of ways to directly or indirectly inhibit NK cell activity. NK cells can be created from iPSCs that lack particular genes that make them generally sensitive to bacterial-induced dysregulation, like the

TABLE 6.2 Various types of immunotherapies involved in bacterial infections.

S.No.	Disease/pathogen	Type of immunotherapy	Target	Cell type involved
1.	Tuberculosis	Checkpoint inhibition	Tim-3	NK cells, $CD4^+$ and $CD8^+$ T cells
2.	*H. pylori*		PD-L1	Epithelial cells
3.	*P. aeruginosa*		PD-L1	Macrophage, dendritic cells
4.	Sepsis		PD-1	Macrophages, $CD4^+$ and $CD8^+$ T cells
5.	*L. monocytogenes*		Tim-3	Macrophage
6.	*K. pneumoniae, P. aeruginosa*, MRSA	Cellular therapies	N/A	Macrophages
7.	*S. typhimurium*		N/A	T cells
8.	Tuberculosis	Cytokine modulation	Type I IFNs/PEG2, TGF-β	Macrophages, CD4+ T cells
9.	*K. pneumoniae, S. pneumoniae*, MRSA		IL-20	$CD4^+$ and $CD8^+$ T cells, NK cells, macrophages
10.	*S. pneumoniae*		Type I IFNs	Macrophages
11.	Sepsis		TNF-α, IL-7, GM-CSF	$CD4^+$ and $CD8^+$ T cells, macrophages, NK cells, lymphocytes

immunological checkpoint molecules TGF-α or TNF-β or Tim-3 or TIGIT receptors. These increased NK cells may be resistant to immunosuppression and capable of eradicating infection reservoirs by the death of infected cells, as well as increasing long-term adaptive and innate immunity through the release of important cytokines like as IFN-γ and GM-CSF. NK cell therapy may be especially successful in the treatment of tuberculosis, as the granuloma environment has been demonstrated to be highly immunosuppressive, and NK cells are critical for disease control.

Various types of immunotherapies involved in bacterial infections are shown in Table 6.2.

6.4 Emerging technologies against bacterial pathogens

An immunotherapeutic technique that was recently approved involves increasing T-cell activity using a chimeric antigen receptor (CAR) (Ramamurthy et al., 2021). Recombinant receptor-engineered T cells (CAR-Ts) express a recombinant receptor, which is usually an antibody derivative binding to a specific target cell-surface receptor and fused to the transmembrane signaling domain of the cell. This allows activation of T cells without the need for the MHC to be present on the surface of the target cell. CAR T-cell therapy is a sort of adoptive cell therapy that involves extracting peripheral blood T cells from patients, genetically altering them to express CAR ex vivo, and then injecting the altered cells back into the patients. Given that many first-generation CARs were anergic, future modifications allowed for the incorporation of costimulatory receptorlike CD28 as well as targeting and transmembrane signaling domains like a CD3 chain. Third and fourth-generation CARs were created by including a second costimulatory protein and an inducible gene that produces proinflammatory or proliferative cytokines, respectively (Naran et al., 2018).

To treat bacterial infections, numerous novel technologies are being developed. Antibody–antibiotic conjugates make it easier to deliver antibiotics precisely where they're needed while yet maintaining their bactericidal activity. In phase I clinical studies, Genentech's DSTA4637S, an anti-*S. aureus* antibody–antibiotic conjugate constituted of a monoclonal antibody directed against *S. aureus*-specific wall teichoic acids conjugated to an antibiotic, displayed favorable safety and pharmacokinetic properties (Peck et al., 2019). In addition, research on synthetic immunobiotics is being done. Polymyxin B (an antibiotic that binds to the surface of Gram-negative bacteria) and antibody-recruiting antigenic epitopes that elicit a specific immunological response are used to create these (Feigman et al., 2018). These findings are encouraging for future bacterial infection treatment techniques.

6.5 Adjuvant immunotherapies as novel strategies to bacterial infection

One of the most serious medical issues of the 21st century is the increasing spread of multidrug-resistant diseases, notably mycobacteria and Gram-negative bacteria. The impending extinction of many efficient conventional antibiotics for bacterial infections, along with rising population size and mobility, necessitates the creation of innovative medications. It is critical to find ligand and receptors pathways that precisely manage protective responses at the infection site in order to stimulate optimal immune responses while minimizing undesired inflammatory tissue damage. Antibiotic resistance is a global problem that is quickly spreading; nevertheless, it has a wide range of geographic distribution patterns (Helbig et al., 2013). Eastern Europe and Asia are seeing disturbingly increasing rates of multidrug-resistant and highly drug-resistant *M. tuberculosis* infections which, when paired with rising population mobility, will almost certainly lead to a tuberculosis renaissance in Western nations (Jean and Hsueh, 2011; Sheikh et al., 2021). Gram-positive bacteria that are resistant to antibiotics, such as vancomycin-resistant *Enterococci* and methicillin-resistant *S. aureus* have garnered considerable public attention (Calfee, 2012). Fortunately, infection rates have been gradually declining over the last few years, and the majority of isolates remain susceptible to second- and third-line antibiotics like linezolid. In contrast, the dramatic rise in multidrug resistance in Gram-negative bacteria, particularly Enterobacteriaceae like *E. coli* and *Klebsiella* spp., as well as nonfermenters like *Acinetobacter* and *Pseudomonas* spp., is cause for grave concern (Arias and Murray, 2009; Livermore, 2012; Qadri et al., 2021).

Increasing worldwide population density and the rapid emergence of multidrug resistance require for novel, widely efficient, and cost-effective therapeutics for infectious diseases. While new antibiotic development is critical, the pipelines may still be insufficient to keep up with the growth of resistance. Therefore, it will be critical to develop ways that reduce our reliance on antibiotics as the sole treatment choice for bacterial illnesses. Rather than focusing exclusively on pathogens, we address the potential for host-focused approaches to trigger protective antimicrobial immune responses selectively. Adjuvant immunotherapies include those treatments that could be administered in place of or in addition to standard antibiotics (Kak et al., 2012; Mir, 2015). Intentional induction of antimicrobial immunity is a well-established and extremely effective preventive idea that was initially developed by Edward Jenner in 1798 and is commonly referred to as *vaccination*. Specific vaccines against diseases that are frequently associated with antibiotic resistance could significantly limit their spread. Apart from technological barriers to developing effective targeted vaccinations, there is growing public aversion to immunization due to safety concerns (Detmer and Glenting, 2006). Furthermore, acute bacterial infections necessitate rapid-acting treatments. Vaccines in the conventional sense give protection by eliciting adaptive immune responses that typically take many days or even weeks to completely develop. In comparison, the innate immune system includes a vast array of antimicrobial effectors that could be triggered in a matter of minutes or hours. Additionally, because innate immune responses are antigen-independent, adjuvant immunotherapies can be used more broadly, at least against the similar class of pathogens (e.g., Gram-negative extracellular bacteria). Additionally, immunotherapies have been indicated to be beneficial for controlling persistent infections such as *M. tuberculosis* infections (Grange et al., 2011).

6.5.1 Innate immune detection of microbial infections

The innate immune system detects microorganisms through the use of pattern recognition receptors (PRRs), which are a diverse array of immune receptors that are encoded in the germline. Pathogen-associated molecular patterns (PAMPs), which are regarded to be integral components of all bacteria, which are as highly conserved and act as a signal for the immune system to identify them as foreign, are recognized by PRRs (Medzhitov and Janeway). When the PRR is ligated, intracellular signaling cascades are activated, frequently convergent on NF-kB, MAPKs, and/or interferon regulatory factors, resulting in the expression of cytokines, increased phagocytosis, and the generation of ROS or the formation of antimicrobial peptides (Medzhitov, 2007). When activated, several members of the nucleotide oligomerization domain (NOD)-like receptor (NLR) family of PRRs can form large cytosolic protein complexes called *inflammasomes* that act as platforms for caspase-1 autocatalysis and subsequent maturation and release of IL-1b and IL-18. In APCs, activation of the PRR enhances antigen processing and presentation, costimulatory molecule upregulation, and cytokine generation. Thus, PRR ligation establishes a link between innate and adaptive immunity; a fact that is commonly utilized due to the fact that a large number of known vaccine adjuvants act as PRR ligands (Iwasaki and Medzhitov, 2010; Helbig et al., 2013).

6.5.2 Broad innate immune stimulation: targeting TLRs and NODs, among other PRRs

PPR ligands are clear candidates for vaccine adjuvants and treatments for disorders ranging from cancer to viral infections due to their prominent role in microorganism recognition and capacity to swiftly stimulate innate immune responses (Hennessy et al., 2010). We will discuss several prior studies in this section, with an emphasis on bacterial infections.

6.5.2.1 Tolllike receptors

TLR ligation induces proinflammatory responses, and while the majority of TLRs have significant similarities in terms of intracellular signaling, specific TLRs or different combinations of TLRs have unique impacts on subsequent immunological responses. Pretreatment of mice with soluble TLR agonist like Pam3-CSK4 or MALP2 (TLR-2), poly-inosinic: polycytidinic acid (polyI:C, TLR-3), CpG DNA (TLR-9), or lipopolysaccharide (LPS, TLR-4) has been demonstrated to augment host defense by increasing bacterial death and phagocytosis (Ribes et al., 2009). Additionally, synthetic TLR-4 agonists can bolster host defense. Prophylactic treatment of the AGP CRX-524 or a combination of CRX-524 and CRX-527 enhances resistance to infection with *Yersinia pestis* and *Listeria monocytogenes* in mice (Airhart et al., 2008). Flagellin enhances pulmonary immunity and protects mice against bacterial pneumonia caused by *S. pneumoniae* and *P. aeruginosa*, when delivered mucosally (Munoz et al., 2010). TLRs, however, remain appealing targets, particularly for infections caused by bacteria that are poorly TLR stimulatory, such as *Francisella tularensis*, which expresses a modified version of LPS (Lembo et al., 2008). On the other hand, some infections take advantage of TLR-mediated cellular responses. *Salmonella enterica serovar typhimurium* requires TLR signaling to activate virulence genes encoded by SPI2, which is required for the development of the salmonella-containing vacuole and bacterial reproduction, implying that wide TLR activation may not be uniformly effective (Arpaia et al., 2011).

6.5.2.2 Nucleotide oligomerization domains (NODs)

An extensive collection of proteins known as NLRs have a three-domain structure, with a leucine-rich repeat domain, a nucleotide oligomerization domain (NOD), and a varied effector domain (Ye and Ting, 2008). NLRs are ideal sensors for invading infections or translocated microbial products due to their cytosolic location. The possible use of NLR ligands to cure acute infections has been primarily investigated in many mouse infection models using NOD1 and NOD2 ligands. The identification of *H. influenzae* peptidoglycan by NOD1 improved neutrophils' opsonophagocytic killing of *S. pneumoniae*, a result that can be reproduced by a synthetic muropeptide (FK-156) (Lysenko et al., 2007). Apart from its potential therapeutic implications, this study neatly depicts how bacteria compete with one another by utilizing the host's immune system. These findings were later validated using aerosolized *Haemophilus influenzae* lysates (Evans et al., 2010). The NOD2 ligand muramyl dipeptide stimulates autophagy and production of the antimicrobial peptide LL37 in human alveolar macrophages, resulting in enhanced mycobacterial death; on the other hand mice lacking NOD2 are highly susceptible to *M. tuberculosis* infections. These results collectively suggest that NOD activation may be beneficial as an adjunct to tuberculostatic therapy (Juárez et al., 2012).

6.5.2.3 Other PRRs

Poly(I:C), a synthetic dsRNA mimic, is undoubtedly one of the most studied immunostimulants. It activates numerous PRRs and was long recognized as a powerful inducer of type-I IFNs prior to the discovery of PRRs. Poly(I:C) and its derivatives have demonstrated interesting adjuvant properties in a variety of mice models (Coffman et al., 2010). Nevertheless, its therapeutic efficacy in acute infections warrants additional exploration, despite the fact that the poly(I:C) derivate poly(ICLC) protected mice from deadly anthrax inhalation and improved survival following intranasal *F. tularensis* challenge (Pyles et al., 2010). Due to poly(I:C)'s dual function of inducing type-I IFNs and proinflammatory mediators, as well as the fact that both pathways are effective against the majority of Gram-negative bacteria, including *Legionella pneumophila* and *P. aeruginosa,* poly(I:C) derivates are excellent options for the treatment of Gram-negative infections (Carrigan et al., 2010). However, application of poly(I:C) has been demonstrated to produce worsening of *Mycobacterium tuberculosis* infection and higher susceptibility to pulmonary secondary bacterial infections, which are not found with the TLR-7 agonist imiquimod (Tian et al., 2012). Various Adjuvants with their functions are shown in Table 6.3.

TABLE 6.3 Overview of adjuvants with their functions.

S. No	Adjuvant	Source	Clinical application
1.	Flagellin	Bacterial protein, ligand for TLR-5 and NAIP5/NLRC4	• In *S. pneumoniae*-infected mice, there was an improvement in survival, a decrease in bacterial load, and an elevation in pulmonary cytokine release and neutrophil recruitment. • In *P. aeruginosa*-infected mice, there was an improvement in survival and bacterial elimination, a reduction in dispersion, and an increase in the synthesis of antimicrobial peptides. • Antibiotic-pretreated mice infected with *Clostridium difficile* had higher survival and intestinal integrity protection, as well as a lower bacterial load. • *S. typhimurium* infection is delayed while survival is improved. • Antibiotic-treated VRE-infected mice had less colonization and more RegIIIɣ expression.
2.	MPLA	Synthetic TRIF-biased TLR-4 agonist	• In *Haemophilus influenzae* or *Moraxella catarrhalis*-infected mice, bacterial load in the nasopharynx is reduced, and neutrophils are recruited.
3.	Nigericin	NLRP3 agonist	• Augmented IL-1b production and decreased morbidity and mortality in elderly mice infected with influenza.
4.	Poly(I:C)	Synthetic ligand of RIG-I, TLR-3, MDA5 and DDX1, DDX21 and DHX36	• *E. coli* phagocytosis and intracellular death has increased. • In *Francisella tularensis*-infected mice, there was a temporary increase in cytokine release and pulmonary neutrophil recruitment, as well as longer survival. • Human macrophages produce more cytokines and fight germs. • Pretreatment increases *S. pneumoniae*-induced mortality.

TABLE 6.3 Overview of adjuvants with their functions.—cont'd

S. No	Adjuvant	Source	Clinical application
5.	mesoDAP containing PGN	NOD1 agonist	• *S. pneumoniae* killing was increased, and neutrophil functions were improved.
6.	CRX-524, CRX-527, RC-529	Synthetic TRIF-biased TLR-4 agonist	• *Listeria monocytogenes*-infected mice had a higher survival rate. • In *Yersinia pestis*-infected mice, there was a decrease in bacterial load in the lung as well as an elevation in IFN-g and IL-12p70 production. • Increased survival when combined with gentamicin. • Enhanced cytokine production in vitro and in vivo, as well as improved survival and bacterial load decrease in the liver, lung, and spleen of *F. tularensis*-infected mice.
7.	Whole bacteria and bacterial lysates	Agonists of multiple PRRs	• Lysed, aerosolized *H. influenzae* improves lifespan in infected mice with *S. pneumoniae, S. aureus, P. aeruginosa, Bacillus anthracis, K. pneumoniae, Y. pestis,* or *F. tularensis.* • In *S. pneumonia*-infected mice, local protection and AMP production were observed. • Recurrent respiratory tract illnesses in children have been reduced. • Improved symptoms in chronic bronchitis and COPD, but it are uncertain whether Luivac can reduce exacerbations.
8.	AHLs	Bacterial QSMs	• In a deadly *Aeromonas hydrophila* illness in mice, improved survival, enhanced neutrophil recruitment, and decreased bacterial load in the blood, lung, and spleen. • Increased phagocytosis by mouse macrophages.
9.	CNF1	*E. coli*-derived effector, modifies host Rho GTPases	• Enhanced antimicrobial peptide survival and transcription in CNF1-expressing Drosophila infected with severely pathogenic *P. aeruginosa.*
10.	c-di-GMP	Bacterial secondary messenger, STING/DDX41 ligand, induces type-I IF	• In a *S. aureus* mouse mastitis model, enhanced leukocyte recruitment, and decreased bacterial load. • In vitro maturation of murine and human DCs. • In a deadly *S. pneumoniae* infection, mice had improved survival and a lower bacterial load in the blood and lung. • In *Klebsiella pneumoniae*-infected mice, there was improved survival, higher chemokine/cytokine release, and decreased bacterial load in the lung and blood.

By and large, bacterial pathogens contain several PAMPs and trigger multiple signaling pathways that originate from different cellular compartments, resulting in distinct innate immune responses. Thus, rather than targeting individual PRRs, adjuvant immunotherapy for bacterial infections may benefit from a simultaneous stimulation of many receptor modules.

6.6 Development of therapeutic antibodies

Antibody engineering has advanced substantially in the three decades since the US Food and Drug Administration (US FDA) approved the first monoclonal antibody in 1986. Due to their great specificity, current antibody drugs have a rising number of side effects. As a result, therapeutic antibodies have evolved to become the most frequently developed class of innovative drugs in recent years. Immunotherapies have superseded pharmaceuticals as the most popular treatments over the last 5 years, with biologics accounting for eight of the top 10 most popular medications worldwide in 2018. Since monoclonal antibodies have become the principal treatment mechanism for a wide range of ailments during the last 25 years, it is clear that therapeutic monoclonal antibodies are becoming increasingly important in healthcare. Since that time, substantial technological breakthroughs have helped to accelerate and streamline the discovery and development of monoclonal antibody (mAb) therapeutics. According to the US FDA, 48 new monoclonal antibodies have been licensed since 2008, resulting in a global market of 61 mAbs in clinical use at the end of 2017, representing a 12% increase from 2008. Surprisingly, the US FDA approved a total of 18 new antibodies between 2018 and 2019, and this figure was compiled using data from various websites, including the antibody society (Mehraj et al., 2020)Kaplon and Reichert, 2019), the database of therapeutic antibodies (Lefranc, 2011), and company pipelines and press releases.

Muromonab-CD3 (Orthoclone OKT3) was the first therapeutic mAb licensed by the US FDA in 1986 (Ecker et al., 2015). It consists of a murine mAb directed against T-cell-expressed CD3, which acts as an immunosuppressant in the treatment of acute transplant rejection. Scientists have discovered strategies for transforming rodent antibodies into structures that are more similar to human antibodies without sacrificing their binding properties, which can be used to address the issues of decreased immunogenicity and efficacy while allowing for prolonged therapeutic use of antibodies in patients. It was in 1994 that the US FDA approved the first chimeric antibody, anti-GPIIb/IIIa antigen-binding fragment (Fab) (abciximab). This antibody was used to treat atherosclerosis in patients. The medicine was produced by fusing murine variable domain with human constant region domain sequences (Morrison et al., 1984; Lu et al., 2020). Then came the first monoclonal antibody with an oncologic indication, rituximab, a chimeric anti-CD20 IgG1 licensed by the US FDA in 1997 for non-Hodgkin lymphoma (Maloney et al., 1997).

Sir Gregory P. Winter discovered a significant discovery approach for obtaining fully human mAbs in 1990, based on the clinical success of humanized mAbs (Tsurushita et al., 2005). This technology is based on phage display, in which a library of different foreign genes is integrated into filamentous bacteriophages. The library proteins are subsequently covalently linked to a phage coat protein and shown on the phage surface, allowing for the selection of specific binders and affinity characteristics. George P. Smith (1985) pioneered the phage display technique. It is a highly effective method for rapidly identifying peptides or antibody fragments, such as single chain fragment variable (scFv) or Fab, that bind a variety of target molecules (cell-surface glycans, proteins, and receptors) (Wu et al., 2016). In addition, phage display technology has been used to develop antibodies by site-directed modification of the CDR and affinity selection, both of which have been demonstrated. Adalimumab (Humira) is first fully human therapeutic antibody, which is an antitumor necrosis factor (TNF) human antibody and was approved by the US FDA in 2002 for the treatment of rheumatoid arthritis, based on these processes (Nixon et al., 2014). Until now, the US FDA has approved nine human antibody medicines created by phage display.

Using transgenic animals is yet another method of producing fully human mAbs. In 1994, two transgenic mouse lines, the XenoMouse and the HuMabMouse were used to develop mAbs (Lu et al., 2020). Human immunoglobulin (Ig) genes were introduced into the lines' genomes, substituting native Ig genes and enabling these animals to synthesize entirely human antibodies following inoculation (Green et al., 1994; Lu et al., 2020). Panitumumab, the first human antibody produced in a transgenic mouse against the antiepidermal growth factor receptor (EGFR), was approved by the US FDA in 2006 (Moroni et al., 2005; Gibson et al., 2006). The number of totally human antibodies derived from transgenic mice has expanded fast, to the point that there are currently 19 licensed medications. The acquisition of high-affinity human antibodies can be accomplished through additional selection of hybridoma clones formed from inoculated transgenic mice, depending on the vaccination approach employed. The creation of neutralizing human antibodies from human B cells has also showed promise for the treatment of infectious diseases, despite the fact that the process is theoretically similar.

Recent advances in the creation of bispecific antibodies have opened up promising new avenues for the development of new protein therapeutics. Bispecific antibody can be created by protein engineering techniques in which two antigen binding domains are joined (such as scFvsorFabs), which allows a single antibody to bind multiple antigens at once. As a result, bispecific antibodies can be created to fulfill additional functions that aren't possible with the two parental antibodies alone. The majority of bispecific antibodies are designed to recruit immune system cytotoxic effector cells to target diseased cells (Labrijn et al., 2019).

Antibodies employed in treatments fall into two broad categories. The naked antibody is employed directly for cancer therapy in the first category. Cancer medicines in this category work in a variety of ways, including through mediated pathways (e.g., ADCC/CDC), by directly targeting cancer cells to cause apoptosis, by combating the tumor microenvironment, and by attacking immunological checkpoints. Natural killer cells or other immune cells are recruited to aid the antibody in killing cancer cells via a variety of mechanisms. The second class of antibody medications involves further modifying the antibody in order to increase its therapeutic value. Chimeric antigen receptor T-cell (CAR-T) therapy, antibody−drug conjugates, Immunocytokines, antibody-radionuclide conjugates, immunoliposomes, and bispecific antibodies are only a few of the more extensive approaches. An immunocytokine is created by fusing a cytokine to an antibody to boost the specificity of the delivery (Neri, 2019). Antibody drug conjugates are composed of an antibody directed against a cancer-specific marker and a small molecule drug; the antibody enhances delivery to the tumor site, enhancing the efficacy of the small molecule while minimizing nontarget toxicity and side effects (Beck et al., 2017). Additionally, the antibody may be linked to a radionuclide to focus radiation more precisely on the tumor spot (Larson et al., 2015). Bispecific antibodies are designed to target two distinct receptors in order to increase therapeutic benefits (Labrijn et al., 2019). Effector cell actions induced by antibodies may boost the therapeutic efficiency of bispecific antibodies. In the case of immunoliposomes, the antibody's binding site (Fab or scFv) is cleaved from the constant region and then attached to various nanodrug delivery methods, such as liposomal medicines, to enable more precise targeting (Ohradanova-Repic et al., 2018). Finally, CAR-T entails the introduction of a gene encoding a chimeric T-cell receptor-antibody directed against a specific cancer marker into T cells, enabling the transformed cells to precisely target and destroy cancer cells (June, O'Connor et al., 2018) (Fig. 6.5).

Because of its rapid and cost-effective in vitro selection procedure, phage display remains a useful platform for the development of human antibody drugs, even if novel technologies such as human antibody transgenic mice and human single B cell antibody techniques have been well established. High-throughput robotic screening has recently emerged as a sophisticated approach (Chen et al., 2017). Next generation sequencing and single cell sequencing have been applied to antibody discovery (Lu et al., 2020). There is a good chance that these methods will speed up our ability to find new phage-binding proteins, which will help us to develop new monoclonal antibodies for use in pharmaceuticals, human disease research, and clinical diagnostics. One can easily appreciate how sophisticated formats were designed in response to therapeutic indication-specific constraints by analyzing recently approved monoclonal antibodies. These mAb engineering alternatives include bispecific mAbs, glycoengineered mAbs, antibody-drug conjugates, CAR-T cells, and immunomodulators.

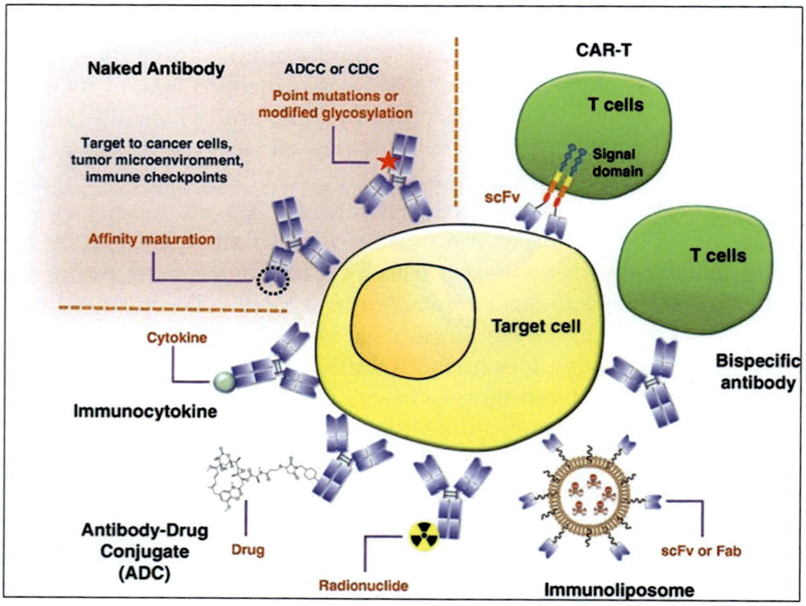

FIGURE 6.5 Schematic overview showing the development of antibody-based therapeutics for the treatment of various infections.

6.7 Future perspectives

Immunotherapies show tremendous potential for the treatment of autoimmunity, infectious diseases, and cancer in the future. Classical vaccination is the most effective and oldest immunotherapy probably; it is also the most cost-effective public health strategy against infectious diseases. Antibiotic resistance is speedily increasing, particularly among Gram-negative bacteria, and poses a massive medical and socioeconomic problem in the near future. The progressive demise of efficient antibiotics poses a serious danger to modern medicine as we know it. As a result, other solutions are critical. As a result, we should begin utilizing our growing understanding of the molecular mechanisms underlying innate immune responses to develop innovative antimicrobial immunotherapies. Identifying molecular switches that can be used to control the immune system's risk assessment machinery may be critical for developing effective adjuvant immunotherapies for infectious diseases. For the appropriate treatment of the growing number of infectious diseases, there is a need for the establishment of various molecular, genetic, and immunological procedures (Mir and Albaradie, 2014; Mir, 2015; Mir et al., 2021).

Traditional chemotherapeutics has several drawbacks, including ineffectiveness, harsh side effects, and the growing threat of drug resistance. Immune-based methods try to address these issues. As is the case with the majority of medicines for any disease, it has become increasingly clear that treating numerous infectious diseases requires a multifaceted strategy. Future preclinical and clinical trials must employ the most effective combinatorial strategies possible to produce the best patient results. This could include a conjunction of immunotherapeutic and traditional therapy methods to eliminate a disease, such as tuberculosis, which is characterized by mycobacteria at various stages of replication, or malaria, which necessitates many levels of control to prevent infection and spread. The advantage of this technique is that it promotes the development of therapies with a high degree of specificity and selectivity, bringing in an era of precision medicine. Novel vaccination platforms, such as those based on DNA, mRNA, and viral vectors, may enable more quick and cost-effective vaccine development pipelines, overcoming past restrictions associated with peptide-based vaccines. Simultaneous developments in targeted delivery and vehicle technology offer the possibility to avoid drug safety problems associated with systemic immunotherapy administration, hence expanding experimental intervention possibilities. Thus, immunotherapeutic advancements are becoming more appealing as a therapy option for infectious disorders.

The development of therapeutic antibodies has accelerated recently, making it a major player in the therapeutics industry. But the therapeutic antibody sector has a lot of capacity to develop. Traditionally, antibodies were primarily used in the treatment of cancer as well as autoimmune and viral diseases. Antibodies may be a potential therapy option if the molecular pathways of a disease are completely known and the individual proteins or molecules involved in pathogenesis are identified.

References

Airhart, C.L., Rohde, H.N., Bohach, G.A., Hovde, C.J., Deobald, C.F., Lee, S.S., Minnich, S.A., 2008. Induction of innate immunity by lipid A mimetics increases survival from pneumonic plague. Microbiology 154 (7), 2131–2138.

Ali, S.O., Yu, X.Q., Robbie, G.J., Wu, Y., Shoemaker, K., Yu, L., DiGiandomenico, A., Keller, A.E., Anude, C., Hernandez-Illas, M., 2019. Phase 1 study of MEDI3902, an investigational anti–*Pseudomonas aeruginosa* PcrV and Psl bispecific human monoclonal antibody, in healthy adults. Clin. Microbiol. Infect. 25 (5), 629.e621–629.e626.

Anand, K., Sahu, G., Burns, E., Ensor, A., Ensor, J., Pingali, S.R., Subbiah, V., Iyer, S.P., 2020. Mycobacterial infections due to PD-1 and PD-L1 checkpoint inhibitors. ESMO Open 5 (4), e000866.

Ardolino, M., Raulet, D.H., 2016. Cytokine therapy restores antitumor responses of NK cells rendered anergic in MHC I-deficient tumors. OncoImmunology 5 (1), e1002725.

Arias, C.A., Murray, B.E., 2009. Antibiotic-resistant bugs in the 21st century—a clinical super-challenge. N. Engl. J. Med. 360 (5), 439–443.

Arpaia, N., Godec, J., Lau, L., Sivick, K.E., McLaughlin, L.M., Jones, M.B., Dracheva, T., Peterson, S.N., Monack, D.M., Barton, G.M., 2011. TLR signaling is required for *Salmonella typhimurium* virulence. Cell 144 (5), 675–688.

Baer, M., Sawa, T., Flynn, P., Luehrsen, K., Martinez, D., Wiener-Kronish, J.P., Yarranton, G., Bebbington, C., 2009. An engineered human antibody fab fragment specific for *Pseudomonas aeruginosa* PcrV antigen has potent antibacterial activity. Infect. Immun. 77 (3), 1083–1090.

Barber, D.L., Sakai, S., Kudchadkar, R.R., Fling, S.P., Day, T.A., Vergara, J.A., Ashkin, D., Cheng, J.H., Lundgren, L.M., Raabe, V.N., 2019. Tuberculosis following PD-1 blockade for cancer immunotherapy. Sci. Transl. Med. 11 (475).

Batlle, E., Massagué, J., 2019. Transforming growth factor-β signaling in immunity and cancer. Immunity 50 (4), 924–940.

Beck, A., Goetsch, L., Dumontet, C., Corvaïa, N., 2017. Strategies and challenges for the next generation of antibody–drug conjugates. Nat. Rev. Drug Discov. 16 (5), 315–337.

Bekeredjian-Ding, I., 2020. Challenges for clinical development of vaccines for prevention of hospital-acquired bacterial infections. Front. Immunol. 11.

Blander, J.M., Sander, L.E., 2012. Beyond pattern recognition: five immune checkpoints for scaling the microbial threat. Nat. Rev. Immunol. 12 (3), 215–225.

Brötz-Oesterhelt, H., Sass, P., 2010. Postgenomic strategies in antibacterial drug discovery. Fut. Microbiol. 5 (10), 1553–1579.

Calfee, D.P., 2012. Methicillin-resistant *Staphylococcus aureus* and vancomycin-resistant enterococci, and other Gram-positives in healthcare. Curr. Opin. Infect. Dis. 25 (4), 385–394.

Carrigan, S.O., Junkins, R., Yang, Y.J., MacNeil, A., Richardson, C., Johnston, B., Lin, T.-J., 2010. IFN regulatory factor 3 contributes to the host response during *Pseudomonas aeruginosa* lung infection in mice. J. Immunol. 185 (6), 3602–3609.

Chan, M., 2012. Antimicrobial Resistance in the European: Union: and the World. WHO.

Chen, C., Chiu, Y.-K., Yu, C.-M., Lee, C.-C., Tung, C.-P., Tsou, Y.-L., Huang, Y.-J., Lin, C.-L., Chen, H.-S., Wang, A.H.-J., 2017. High throughput discovery of influenza virus neutralizing antibodies from phage-displayed synthetic antibody libraries. Sci. Rep. 7 (1), 1–17.

Coffman, R.L., Sher, A., Seder, R.A., 2010. Vaccine adjuvants: putting innate immunity to work. Immunity 33 (4), 492–503.

Day, C.L., Abrahams, D.A., Bunjun, R., Stone, L., De Kock, M., Walzl, G., Wilkinson, R.J., Burgers, W.A., Hanekom, W.A., 2018. PD-1 expression on Mycobacterium tuberculosis-specific CD4 T cells is associated with bacterial load in human tuberculosis. Front. Immunol. 9, 1995.

Detmer, A., Glenting, J., 2006. Live bacterial vaccines–a review and identification of potential hazards. Microb. Cell Factories 5 (1), 1–12.

Ecker, D.M., Jones, S.D., Levine, H.L., 2015. The Therapeutic Monoclonal Antibody Market. MAbs, Taylor & Francis.

Evans, S.E., Scott, B.L., Clement, C.G., Larson, D.T., Kontoyiannis, D., Lewis, R.E., LaSala, P.R., Pawlik, J., Peterson, J.W., Chopra, A.K., 2010. Stimulated innate resistance of lung epithelium protects mice broadly against bacteria and fungi. Am. J. Respir. Cell Mol. Biol. 42 (1), 40–50.

Feigman, M.S., Kim, S., Pidgeon, S.E., Yu, Y., Ongwae, G.M., Patel, D.S., Regen, S., Im, W., Pires, M.M., 2018. Synthetic immunotherapeutics against Gram-negative pathogens. Cell Chem. Biol. 25 (10), 1185.e1185–1194.e1185.

Feldmann, M., 2008. Many cytokines are very useful therapeutic targets in disease. J. Clin. Invest. 118 (11), 3533–3536.

François, B., Mercier, E., Gonzalez, C., Asehnoune, K., Nseir, S., Fiancette, M., Desachy, A., Plantefève, G., Meziani, F., De Lame, P.-A., 2018. Safety and tolerability of a single administration of AR-301, a human monoclonal antibody, in ICU patients with severe pneumonia caused by *Staphylococcus aureus*: first-in-human trial. Intensive Care Med. 44 (11), 1787–1796.

Gern, B.H., Adams, K.N., Plumlee, C.R., Stoltzfus, C.R., Shehata, L., Moguche, A.O., Busman-Sahay, K., Hansen, S.G., Axthelm, M.K., Picker, L.J., 2021. TGFβ restricts expansion, survival, and function of T cells within the tuberculous granuloma. Cell Host Microbe 29 (4), 594.e596–606.e596.

Gibson, T.B., Ranganathan, A., Grothey, A., 2006. Randomized phase III trial results of Panitumumab, a fully human anti—epidermal growth factor receptor monoclonal antibody, in metastatic colorectal cancer. Clin. Colorectal Cancer 6 (1), 29–31.

Grange, J., Brunet, L.R., Rieder, H., 2011. Immune protection against tuberculosis—When is immunotherapy preferable to vaccination? Tuberculosis 91 (2), 179–185.

Green, L.L., Hardy, M., Maynard-Currie, C., Tsuda, H., Louie, D., Mendez, M., Abderrahim, H., Noguchi, M., Smith, D., Zeng, Y., 1994. Antigen—specific human monoclonal antibodies from mice engineered with human Ig heavy and light chain YACs. Nat. Genet. 7 (1), 13–21.

Hamdani, S.S., Bhat, B.A., Tariq, L., Yaseen, S.I., Ara, I., Rafi, B., Hamdani, S.N., Hassan, T. O. J. I. J. F. R. I. A. S. Rashid and Biotechnology, 2020. Antibiotic resistance: the future disaster, 7.

Harjunpää, H., Guillerey, C., 2020. TIGIT as an emerging immune checkpoint. Clin. Exp. Immunol. 200 (2), 108–119.

Helbig, E.T., Opitz, B., Sander, L.E., 2013. Adjuvant immunotherapies as a novel approach to bacterial infections. Immunotherapy 5 (4), 365–381.

Hennessy, E.J., Parker, A.E., O'neill, L.A., 2010. Targeting Toll-like receptors: emerging therapeutics? Nat. Rev. Drug Discov. 9 (4), 293–307.

Hooks, M.A., Wade, C.S., Millikan Jr., W.J., 1991. Muromonab CD-3: a review of its pharmacology, pharmacokinetics, and clinical use in transplantation. Pharmacotherapy 11 (1), 26–37.

Horn, M.P., Zuercher, A.W., Imboden, M.A., Rudolf, M.P., Lazar, H., Wu, H., Hoiby, N., Fas, S.C., Lang, A.B., 2010. Preclinical in vitro and in vivo characterization of the fully human monoclonal IgM antibody KBPA101 specific for *Pseudomonas aeruginosa* serotype IATS-O11. Antimicrob. Agents Chemother. 54 (6), 2338–2344.

Iwasaki, A., Medzhitov, R., 2010. Regulation of adaptive immunity by the innate immune system. Science 327 (5963), 291–295.

Janeway Jr., C.A., Travers, P., Walport, M., Shlomchik, M.J., 2001. Principles of innate and adaptive immunity. In: Immunobiology: the Immune System in Health and disease, fifth ed. Garland Science.

Jayaraman, P., Jacques, M.K., Zhu, C., Steblenko, K.M., Stowell, B.L., Madi, A., Anderson, A.C., Kuchroo, V.K., Behar, S.M., 2016. TIM3 mediates T cell exhaustion during *Mycobacterium tuberculosis* infection. PLoS Pathog. 12 (3), e1005490.

Jean, S.-S., Hsueh, P.-R., 2011. High burden of antimicrobial resistance in Asia. Int. J. Antimicrob. Agents 37 (4), 291–295.

Juárez, E., Carranza, C., Hernández-Sánchez, F., León-Contreras, J.C., Hernández-Pando, R., Escobedo, D., Torres, M., Sada, E., 2012. NOD 2 enhances the innate response of alveolar macrophages to *Mycobacterium tuberculosis* in humans. Eur. J. Immunol. 42 (4), 880–889.

June, C.H., O'Connor, R.S., Kawalekar, O.U., Ghassemi, S., Milone, M.C., 2018. CAR T cell immunotherapy for human cancer. Science 359 (6382), 1361–1365.

Kak, V., Sundareshan, V., Modi, J., Khardori, N.M., 2012. Immunotherapies in infectious diseases. Med. Clin. 96 (3), 455–474.

Kalinina, A.A., Nesterenko, L.N., Bruter, A.V., Balunets, D.V., Chudakov, D.M., Izraelson, M., Britanova, O.V., Khromykh, L.M., Kazansky, D.B., 2020. Adoptive immunotherapy based on chain-centric TCRs in treatment of infectious diseases. iScience 23 (12), 101854.

Kalliolias, G.D., Ivashkiv, L.B., 2016. TNF biology, pathogenic mechanisms and emerging therapeutic strategies. Nat. Rev. Rheumatol. 12 (1), 49–62.

Kaplon, H., Reichert, J.M., 2019. Antibodies to Watch in 2019. MAbs. Taylor & Francis.

Kaufmann, S.H., Hussey, G., Lambert, P.-H., 2010. New vaccines for tuberculosis. Lancet 375 (9731), 2110–2119.

Khan, N., Vidyarthi, A., Amir, M., Mushtaq, K., Agrewala, J.N., 2017. T-cell exhaustion in tuberculosis: pitfalls and prospects. Crit. Rev. Microbiol. 43 (2), 133–141.

Köhler, G., Milstein, C., 1975. Continuous cultures of fused cells secreting antibody of predefined specificity. Nature 256 (5517), 495–497.

La Manna, M.P., Orlando, V., Tamburini, B., Badami, G.D., Dieli, F., Caccamo, N., 2020. Harnessing unconventional T cells for immunotherapy of tuberculosis. Front. Immunol. 11, 2107.

Labrijn, A.F., Janmaat, M.L., Reichert, J.M., Parren, P.W., 2019. Bispecific antibodies: a mechanistic review of the pipeline. Nat. Rev. Drug Discov. 18 (8), 585–608.

Langan, E.A., Graetz, V., Allerheiligen, J., Zillikens, D., Rupp, J., Terheyden, P., 2020. Immune checkpoint inhibitors and tuberculosis: an old disease in a new context. Lancet Oncol. 21 (1), e55–e65.

Larson, S.M., Carrasquillo, J.A., Cheung, N.-K.V., Press, O.W., 2015. Radioimmunotherapy of human tumours. Nat. Rev. Cancer 15 (6), 347–360.

Laxminarayan, R., Jamison, D.T., Krupnick, A.J., Norheim, O.F., 2014. Valuing vaccines using value of statistical life measures. Vaccine 32 (39), 5065–5070.

Lee, S.A., Wang, Y., Liu, F., Riordan, S.M., Liu, L., Zhang, L., 2020. *Escherichia coli* K12 upregulates programmed cell death ligand 1 (PD-L1) expression in gamma interferon-sensitized intestinal epithelial cells via the NF-κB pathway. Infect. Immun. 89 (1), e00618–00620.

Lefranc, M.-P., 2011. IMGT, the international ImMunoGeneTics information system. Cold Spring Harb. Protoc. 2011 (6), top115.

Lembo, A., Pelletier, M., Iyer, R., Timko, M., Dudda, J.C., West, T.E., Wilson, C.B., Hajjar, A.M., Skerrett, S.J., 2008. Administration of a synthetic TLR4 agonist protects mice from pneumonic tularemia. J. Immunol. 180 (11), 7574–7581.

Lin, C.-Y., Gobius, I., Souza-Fonseca-Guimaraes, F., 2020. Natural killer cell engineering—a new hope for cancer immunotherapy. Semin. Hematol. (Elsevier).

Livermore, D.M., 2012. Current epidemiology and growing resistance of gram-negative pathogens. Kor. J. Intern. Med. 27 (2), 128.

Lu, L.L., Chung, A.W., Rosebrock, T.R., Ghebremichael, M., Yu, W.H., Grace, P.S., Schoen, M.K., Tafesse, F., Martin, C., Leung, V., 2016. A functional role for antibodies in tuberculosis. Cell 167 (2), 433.e414–443.e414.

Lu, R.-M., Hwang, Y.-C., Liu, I.-J., Lee, C.-C., Tsai, H.-Z., Li, H.-J., Wu, H.-C., 2020. Development of therapeutic antibodies for the treatment of diseases. J. Biomed. Sci. 27 (1), 1–30.

Lysenko, E.S., Clarke, T.B., Shchepetov, M., Ratner, A.J., Roper, D.I., Dowson, C.G., Weiser, J.N., 2007. Nod1 signaling overcomes resistance of S. pneumoniae to opsonophagocytic killing. PLoS Pathog. 3 (8), e118.

Maloney, D.G., Grillo-López, A.J., White, C.A., Bodkin, D., Schilder, R.J., Neidhart, J.A., Janakiraman, N., Foon, K.A., Liles, T.-M., Dallaire, B.K., 1997. IDEC-C2B8 (Rituximab) anti-CD20 monoclonal antibody therapy in patients with relapsed low-grade non-Hodgkin's lymphoma. Blood 90 (6), 2188–2195.

McCulloch, T.R., Wells, T.J., Souza-Fonseca-Guimaraes, F., 2021. Towards efficient immunotherapy for bacterial infection. Trends Microbiol.

Medzhitov, R., 2007. Recognition of microorganisms and activation of the immune response. Nature 449 (7164), 819–826.

Mehraj, U., Nisar, S., Hamdani, S.S., Sheikh, B.A., Bhat, B.A., Qayoom, H., Mir, M.A., 2020. Cells of the immune system. In: Basics and Fundamentals of Immunology, 25.

Mir, M., Albaradeh, R., Agrewala, J., 1993. Innate—effector immune response elicitation against tuberculosis through anti-B7-1 (CD80) and anti-B7-2 (CD86) signaling IN macrophages.

Mir, M.A., 2015. Developing Costimulatory Molecules for Immunotherapy of Diseases. Academic Press.

Mir, M.A., Al-baradie, R.J.M.J.O.H.S., 2013. Tuberculosis time bomb-a global emergency: need for alternative vaccines. Majmaah J. Health Sci. 1 (1), 77–82.

Mir, M.A., Albaradie, R.S.J.A.I.N.B., 2014. Inflammatory mechanisms as potential therapeutic targets in stroke. Adv. Neuroimmune Biol. 5 (4), 199–216.

Mir, M.A., Bhat, B.A., Sheikh, B.A., Rather, G.A., Mehraj, S., Mir, W.R., 2021. Nanomedicine in human health therapeutics and drug delivery: nanobiotechnology and nanobiomedicine. Applications of nanomaterials in agriculture, Food science, and medicine. IGI Glob. 229–251.

Mir, M.A., Hamdani, S.S., Mehraj, U., 2020. Antigens and immunogens. In: Basics and Fundamentals of Immunology, 77.

Mir, M.A., Mehraj, U., Sheikh, B.A., Hamdani, S.S.J.H.A., 2020. Nanobodies: the "magic bullets" in therapeutics, drug delivery and diagnostics. Pharmaceutics 28 (1), 29–51.

Moroni, M., Veronese, S., Benvenuti, S., Marrapese, G., Sartore-Bianchi, A., Di Nicolantonio, F., Gambacorta, M., Siena, S., Bardelli, A., 2005. Gene copy number for epidermal growth factor receptor (EGFR) and clinical response to antiEGFR treatment in colorectal cancer: a cohort study. Lancet Oncol. 6 (5), 279–286.

Morrison, S.L., Johnson, M.J., Herzenberg, L.A., Oi, V.T., 1984. Chimeric human antibody molecules: mouse antigen-binding domains with human constant region domains. Proc. Natl. Acad. Sci. USA 81 (21), 6851–6855.

Munoz, N., Van Maele, L., Marqués, J.M., Rial, A., Sirard, J.-C., Chabalgoity, J.A., 2010. Mucosal administration of flagellin protects mice from *Streptococcus pneumoniae* lung infection. Infect. Immun. 78 (10), 4226–4233.

Naran, K., Nundalall, T., Chetty, S., Barth, S., 2018. Principles of immunotherapy: implications for treatment strategies in cancer and infectious diseases. Front. Microbiol. 9, 3158.

Neri, D., 2019. Antibody–cytokine fusions: versatile products for the modulation of anticancer immunity. Cancer Immunol. Res. 7 (3), 348–354.

Nixon, A.E., Sexton, D.J., Ladner, R.C., 2014. Drugs Derived from Phage Display: From Candidate Identification to Clinical Practice. MAbs, Taylor & Francis.

Ohradanova-Repic, A., Nogueira, E., Hartl, I., Gomes, A.C., Preto, A., Steinhuber, E., Mühlgrabner, V., Repic, M., Kuttke, M., Zwirzitz, A., 2018. Fab antibody fragment-functionalized liposomes for specific targeting of antigen-positive cells. Nanomed. Nanotechnol. Biol. Med. 14 (1), 123–130.

Organization, W.H., 2013. Global Tuberculosis Report 2013. World Health Organization.

Papaioannou, N.E., Beniata, O.V., Vitsos, P., Tsitsilonis, O., Samara, P., 2016. Harnessing the immune system to improve cancer therapy. Ann. Transl. Med. 4 (14).

Pardoll, D.M., 2012. The blockade of immune checkpoints in cancer immunotherapy. Nat. Rev. Cancer 12 (4), 252–264.

Pauken, K.E., Wherry, E.J., 2015. Overcoming T cell exhaustion in infection and cancer. Trends Immunol. 36 (4), 265–276.

Peck, M., Rothenberg, M.E., Deng, R., Lewin-Koh, N., She, G., Kamath, A.V., Carrasco-Triguero, M., Saad, O., Castro, A., Teufel, L., 2019. A phase 1, randomized, single-ascending-dose study to investigate the safety, tolerability, and pharmacokinetics of DSTA4637S, an anti-*Staphylococcus aureus* thiomab antibody-antibiotic conjugate, in healthy volunteers. Antimicrob. Agents Chemother. 63 (6) e02588-02518.

Phillips, B.L., Gautam, U.S., Bucsan, A.N., Foreman, T.W., Golden, N.A., Niu, T., Kaushal, D., Mehra, S., 2017. LAG-3 potentiates the survival of *Mycobacterium tuberculosis* in host phagocytes by modulating mitochondrial signaling in an in-vitro granuloma model. PLoS One 12 (9), e0180413.

Pitt, J.M., Stavropoulos, E., Redford, P.S., Beebe, A.M., Bancroft, G.J., Young, D.B., O'Garra, A., 2012. Blockade of IL-10 signaling during BCG vaccination enhances and sustains Th1, Th17, and innate lymphoid IFN-γ and IL-17 responses and increases protection to *Mycobacterium tuberculosis* infection. J. Immunol. 189 (8), 4079.

Pose, E., Coll, M., Martínez-Sánchez, C., Zeng, Z., Surewaard, B.G., Català, C., Velasco-de Andrés, M., Lozano, J.J., Ariño, S., Fuster, D., 2021. Programmed death ligand 1 is overexpressed in liver macrophages in chronic liver diseases, and its blockade improves the antibacterial activity against infections. Hepatology 74 (1), 296.

Pyles, R.B., Jezek, G.E., Eaves-Pyles, T.D., 2010. Toll-like receptor 3 agonist protection against experimental *Francisella tularensis* respiratory tract infection. Infect. Immun. 78 (4), 1700–1710.

Qadri, H., Haseeb, A., Mir, M.J.C.D.T., 2021. Novel strategies to combat the emerging drug resistance in human pathogenic microbes. Curr. Drug Targets 22, 1–13.

Qayoom, H., Mehraj, U., Aisha, S., Sofi, S., Mir, M.A., 2021. Integrating Immunotherapy With Chemotherapy: a New Approach to Drug Repurposing. Intech Open.

Que, Y.-A., Lazar, H., Wolff, M., François, B., Laterre, P.-F., Mercier, E., Garbino, J., Pagani, J.-L., Revelly, J.-P., Mus, E., 2014. Assessment of panobacumab as adjunctive immunotherapy for the treatment of nosocomial *Pseudomonas aeruginosa* pneumonia. Eur. J. Clin. Microbiol. Infect. Dis. 33 (10), 1861–1867.

Ramamurthy, D., Nundalall, T., Cingo, S., Mungra, N., Karaan, M., Naran, K., Barth, S., 2021. Recent advances in immunotherapies against infectious diseases. Immunother. Adv. 1 (1), ltaa007.

Rappuoli, R., Pizza, M., Del Giudice, G., De Gregorio, E., 2014. Vaccines, new opportunities for a new society. Proc. Natl. Acad. Sci. USA 111 (34), 12288–12293.

Ribes, S., Ebert, S., Czesnik, D., Regen, T., Zeug, A., Bukowski, S., Mildner, A., Eiffert, H., Hanisch, U.-K., Hammerschmidt, S., 2009. Toll-like receptor prestimulation increases phagocytosis of *Escherichia coli* DH5α and *Escherichia coli* K1 strains by murine microglial cells. Infect. Immun. 77 (1), 557–564.

Riley, R., June, C.H., Langer, R., Mitchell, M.J., 2019. Nat. Rev. Drug Discov. 18, 175–196.

Roca, F.J., Ramakrishnan, L., 2013. TNF dually mediates resistance and susceptibility to mycobacteria via mitochondrial reactive oxygen species. Cell 153 (3), 521–534.

Ruzin, A., Wu, Y., Yu, L., Yu, X.Q., Tabor, D.E., Mok, H., Tkaczyk, C., Jensen, K., Bellamy, T., Roskos, L., 2018. Characterisation of anti-alpha toxin antibody levels and colonisation status after administration of an investigational human monoclonal antibody, MEDI4893, against *Staphylococcus aureus* alpha toxin. Clin. Trans. Immunol. 7 (1), e1009.

Sable, S.B., Posey, J.E., Scriba, T.J., 2019. Tuberculosis vaccine development: progress in clinical evaluation. Clin. Microbiol. Rev. 33 (1), e00100–00119.

Said, E.A., Dupuy, F.P., Trautmann, L., Zhang, Y., Shi, Y., El-Far, M., Hill, B.J., Noto, A., Ancuta, P., Peretz, Y., 2010. Programmed death-1–induced interleukin-10 production by monocytes impairs $CD4^+$ T cell activation during HIV infection. Nat. Med. 16 (4), 452–459.

Schietinger, A., Greenberg, P.D., 2014. Tolerance and exhaustion: defining mechanisms of T cell dysfunction. Trends Immunol. 35 (2), 51–60.

See, J.-X., Samudi, C., Saeidi, A., Menon, N., Choh, L.-C., Vadivelu, J., Shankar, E.M., 2016. Experimental persistent infection of BALB/c mice with small-colony variants of *Burkholderia pseudomallei* leads to concurrent upregulation of PD-1 on T cells and skewed Th1 and Th17 responses. PLoS Neglected Tropical Diseases 10 (3), e0004503.

Sharpe, A.H., Wherry, E.J., Ahmed, R., Freeman, G.J., 2007. The function of programmed cell death 1 and its ligands in regulating autoimmunity and infection. Nat. Immunol. 8 (3), 239–245.

Sheikh, B.A., Bhat, B.A., Mehraj, U., Mir, W., Hamadani, S., Mir, M.A., 2021. Development of new therapeutics to meet the current challenge of drug resistant tuberculosis. Curr. Pharmaceut. Biotechnol. 22 (4), 480–500.

Shen, B., Qian, A., Lao, W., Li, W., Chen, X., Zhang, B., Wang, H., Yuan, F., Sun, Y., 2019. Relationship between *Helicobacter pylori* and expression of programmed death-1 and its ligand in gastric intraepithelial neoplasia and early-stage gastric cancer. Cancer Manag. Res. 11, 3909.

Smith, G.P., 1985. Filamentous fusion phage: novel expression vectors that display cloned antigens on the virion surface. Science 228 (4705), 1315–1317.

Springer, T.A., 2002. César Milstein, the father of modern immunology. Nature Immunology 3 (6), 501–503.

Suurs, F.V., Lub-de Hooge, M.N., de Vries, E.G., de Groot, D.J.A., 2019. A review of bispecific antibodies and antibody constructs in oncology and clinical challenges. Pharmacol. Therapeut. 201, 103–119.

Tacke, R., Sun, J., Uchiyama, S., Polovina, A., Nguyen, D.G., Nizet, V., 2019. Protection against lethal multidrug-resistant bacterial infections using macrophage cell therapy. Infect. Microb. Dis. 1 (2), 61–69.

Tait, D., Hatherill, M., Van Der Meeren, O., Ginsberg, A., Van Brakel, E., Salaun, B., Scriba, T., Akite, E., Ayles, H., Bollaerts, A., Mesia Vela, D., Muyoyeta, M., Nduba, V., Pascal, T.G., Tameris, M., Thienemann, F., Wilkinson, R.J., Roman, F., 2019. Final analysis of a trial of M72/AS01 E vaccine to prevent tuberculosis. N. Engl. J. Med. 381, 2429–2439.

Tameris, M.D., Hatherill, M., Landry, B.S., Scriba, T.J., Snowden, M.A., Lockhart, S., Shea, J.E., McClain, J.B., Hussey, G.D., Hanekom, W.A., 2013. Safety and efficacy of MVA85A, a new tuberculosis vaccine, in infants previously vaccinated with BCG: a randomised, placebo-controlled phase 2b trial. Lancet 381 (9871), 1021–1028.

Tezera, L.B., Bielecka, M.K., Ogongo, P., Walker, N.F., Ellis, M., Garay-Baquero, D.J., Thomas, K., Reichmann, M.T., Johnston, D.A., Wilkinson, K.A., 2020. Anti-PD-1 immunotherapy leads to tuberculosis reactivation via dysregulation of TNF-α. Elife 9, e52668.

Tian, X., Xu, F., Lung, W.Y., Meyerson, C., Ghaffari, A.A., Cheng, G., Deng, J.C., 2012. Poly I: C enhances susceptibility to secondary pulmonary infections by gram-positive bacteria. PLoS One.

Tsurushita, N., Hinton, P.R., Kumar, S., 2005. Design of humanized antibodies: from anti-Tac to Zenapax. Methods 36 (1), 69–83.

Tur, M., Barth, S., 2014. Immunotherapy. In: Schwab, M. (Ed.), Encyclopedia of Cancer. Springer, Berlin.

Waldman, A.D., Fritz, J.M., Lenardo, M.J., 2020. A guide to cancer immunotherapy: from T cell basic science to clinical practice. Nat. Rev. Immunol. 20 (11), 651–668.

Waldmann, T.A., 2003. Immunotherapy: past, present and future. Nat. Med. 9 (3), 269–277.

Wang, F., Hou, H., Wu, S., Tang, Q., Huang, M., Yin, B., Huang, J., Liu, W., Mao, L., Lu, Y., 2015. Tim-3 pathway affects NK cell impairment in patients with active tuberculosis. Cytokine 76 (2), 270–279.

Warsinske, H.C., Pienaar, E., Linderman, J.J., Mattila, J.T., Kirschner, D.E., 2017. Deletion of TGF-β1 increases bacterial clearance by cytotoxic t cells in a tuberculosis granuloma model. Front. Immunol. 8, 1843.

Washburn, M.L., Wang, Z., Walton, A.H., Goedegebuure, S.P., Figueroa, D.J., Van Horn, S., Grossman, J., Remlinger, K., Madsen, H., Brown, J., 2019. T cell—and monocyte-specific RNA-sequencing analysis in septic and nonseptic critically ill patients and in patients with cancer. J. Immunol. 203 (7), 1897—1908.
Wherry, E.J., 2011. T cell exhaustion. Nat. Immunol. 12 (6), 492—499.
Wherry, E.J., Kurachi, M., 2015. Molecular and cellular insights into T cell exhaustion. Nat. Rev. Immunol. 15 (8), 486—499.
Wu, C.-H., Liu, I.-J., Lu, R.-M., Wu, H.-C., 2016. Advancement and applications of peptide phage display technology in biomedical science. J. Biomed. Sci. 23 (1), 1—14.
Wykes, M.N., Lewin, S.R., 2018. Immune checkpoint blockade in infectious diseases. Nat. Rev. Immunol. 18 (2), 91—104.
Ye, Z., Ting, J.P.-Y., 2008. NLR, the nucleotide-binding domain leucine-rich repeat containing gene family. Curr. Opin. Immunol. 20 (1), 3—9.

Further readings

Mehraj, U., Nisar, S., Qayoom, H., Mir, M.A., 2020. Monoclonal antibodies in therapeutics. Immunoglobulins, Magic Bullets and Therapeutic Antibodies 155.
Mehraj, U., Nisar, S., Qayoom, H., Mir, M.A., 2020. Hybridoma technology. In: Immunoglobulins, magic bullets and therapeutic antibodies, 209.

CHAPTER

7

Significance of immunotherapy for human fungal diseases and antifungal drug discovery

Manzoor Ahmad Mir, Ulfat Jan and Hafsa Qadri

Department of Bioresources, School of Biological Sciences, University of Kashmir, Srinagar, Jammu and Kashmir, India

7.1 Introduction

There are a wide variety of microbes that are pathogenic to humans as well as plants. Common pathogenic microorganisms include bacteria, protozoa, viruses, fungi, etc.; among these pathogens, fungi have recently emerged as a global pathogen causing disease in both plants and animals (Rhodes, 2019). There are various fungi species but most common infectious fungi belong to genera *Aspergillus, Candida, Pneumocystis,* and *Cryptococcus* leading toward death of 1/2 million persons every year (Brown et al., 2012). Since the past few years, the mortality rate caused by these fungal pathogens is adequately increasing at an alarming rate globally posing a threat to human life (Ji et al., 2019). The death rate is increasing despite the presence of various antifungal drugs like azoles, echinocandins, polyenes, and flucytosine due to drug resistance especially in *Candida* and *Aspergillus* species (Van Daele et al., 2019). Globally, the fungal drug resistance is posing an earnest risk to health and healthcare systems, as these resistant fungi are not susceptible to typical antifungals, although they provide good results but death rate is still increasing especially in immunocompromised individuals, so there is vital importance of the new antifungal agents possessing an extended-spectrum antimicrobial activity and can easily target resistant fungal species (Marquez and Quave, 2020).

Among the millions of fungal species, there are certain species that are infecting humans such as yeasts, including *C. albicans, C. neoformans, P. jirovecii;* molds, including *A. fumigatus, Mucor circinelloides;* and dimorphic fungal pathogens including, *Histoplasma capsulatum* and *C. immitis.* These fungal pathogens attack various body parts such as, lungs, skin, and brain, causing allergic, persistent, or invasive infections, especially more dangerous infections in

Human Pathogenic Microbes
https://doi.org/10.1016/B978-0-323-96127-1.00001-2

immunocompromised individuals (Bruch et al., 2021). Fungal infection accounts for 10% of all healthcare-associated infections (Khan et al., 2017). Fungal pathogens cause both superficial and invasive infections; but invasive infections are more life threatening, that is, they are concerned with more mortality rates. Fungal infections are mainly associated with opportunistic fungi that become pathogenic in immunocompromised people and lead to various fungal infections (Goyal et al., 2018). The most frequent infectious diseases caused by *Aspergillus*, *Cryptococcus*, and *Candida*, are *Aspergillosis*, *Cryptococcosis*, and *Candidiasis*, respectively, accounting for about more than 80% of total 1.5 million invasive fungal deaths each year (Pathakumari et al., 2020). *C. albicans* is the main causative agent for invasive *candidiasis* that mainly colonizes GIT, skin, and other mucosal surfaces and causes thrush, vaginal and invasive *candidiasis*. Invasive or deep tissue *candidiasis* can affect lungs, blood, heart, brain, eyes, bones, and various other parts of the body accounting for about 46%–75% of all invasive fungal infections, leading to high death rates (40%) in immunosuppressed individuals (Brown et al., 2012; Pathakumari et al., 2020). C. *albicans*, a universal opportunistic organism that mainly occupies oral mucosa, gastrointestinal tract, and reproductive tract can be easily identified in healthy people as a harmless entity. Any changes in host immune system that favor its multiplication help this pathogen to invade deep tissues leading to both superficial and deep tissue infections involving multiple organs (Vila et al., 2020). Out of the total *candida* species, the majority of fungal diseases are caused by *C. albicans* (37%), followed by *C. glabrata* (27%), *C. parapsilosis* (14%), *C. krusei* (2%), *Candida tropicalis* (8%), *Candida dubliniensis* (2%) (Bhattacharya et al., 2020), and recently *C. auris*, an emerging multidrug-resistant nosocomial fungal pathogen causing severe deep tissue infections accompanied by a crude mortality extending from 30% to 72% (Lone and Ahmad, 2019) [https://www.cdc.gov/drugresistance/pdf/threats-report/candida-508.pdf] After *candidiasis*, another crucial heath issue is *aspergillosis* caused by *Aspergillus* species consisting of a rapidly developing epidemiology and mainly infecting immunocompromised patients (Latgé and Chamilos, 2019). There are more than 200 species of the genus, out of which 30 species are believed to cause infections and out of these 30 species *A. fumigatus* primarily causes severe infections followed by *A. niger*, *A. flavus*, *A. terreus* (Pathakumari et al., 2020). *Aspergillus* infections in severely immune compromised individuals with acute leukemia and hematopoietic stem cell transplantation recipients (HSCTs) are connected with significant death rates up to 40%–50% with a three to four fold increase in infectious frequency over the past two decades, majorly at cancer centers (Latgé and Chamilos, 2019). Another globally extended infective genus is *Cryptococcus*, encapsulated yeasts associated with phylum Basidiomycota and is commonly found in lignaceous environments like flowering plants, bird feces, cold climates, and under severe pH. Genus Cryptococcus comprises of more than 70 species, out of which *C. neoformans* as well as *C. gatti* are highly pathogenic for both animals and humans and commonly cause meningitis and airways infections (Setianingrum et al., 2019). These yeasts not only infect immunocompromised patients but have also been emerging as a disease causing agent in immunocompetent individuals (Fisher et al., 2016; Rigby and Glanville, 2012). The disease caused by this obscure fungus in humans is known as *cryptococcosis* and its global prevalence is one million HIV infected persons per year, and of this, 62,000 deaths have also been reported. On the basis of infection site, cryptococcosis has been divided into four types—cerebral cryptococcosis, cutaneous cryptococcosis, pulmonary cryptococcosis, and disseminated cryptococcosis; out of all these, invasive cerebral cryptococcosis is commonly found in HIV infected persons (Pathakumari et al., 2020).

TABLE 7.1 Various antifungal drugs and their mode of action (DiDomenico, 1999).

S.No	Antifungal drug	Mechanism of action
1.	FLC	Specific repression of the enzyme lanosterol-14-α-demethylase
2.	Caspofungin	Obstruction of β-1,3-glucan synthase
3.	AmpB	Particularly binds to ergosterol and disrupts cell membrane
4.	Nystatin	Particularly binds to ergosterol and disrupts cell membrane

Nowadays, five main categories of antifungal drugs are utilized, especially for the treatment of fungal diseases, including azoles, polyenes, echinocandins, allylamines, and antimetabolites (Table 7.1).

Azoles are the first-line of treatment against various invasive fungal infections (IFIs), carrying a strong fungicidal action and a broadspectrum activity. These drugs work by inhibiting the cytochrome P450-dependent enzyme: 14α-lanosterol demethylase (Gintjee et al., 2020) encoded by ERG11 gene that in turn inhibits ergosterol biosynthesis and thus blocks fungal membrane formation, attacking cell division and cell expansion under drastic ergosterol exhaustion. When 14α-lanosterol demethylase is inhibited, toxic methylated sterols agglomerate in the cell membrane and lead toward membrane stress. When azoles like fluconazole are given against *C. neoformans,* there is accumulation of obtusifolione mostly due to ERG 27 inhibition and leads to membrane stress. There are two subclasses of azoles including imidazoles and triazoles. Imidazoles consist of clotrimazole, tioconazole, ketoconazole, oxiconazole, miconazole, kecoconazole; these imidazoles consist of two atoms of nitrogen in an azole ring. The second subclass triazoles consist of fluconazole, terconazole, voriconazole, itraconazole, and posaconazole; all these triazoles are having three nitrogen atoms in an azole ring. Imidazoles are given against mucosal infections; however, triazoles are preferred for both mucosal as well pervasive fungal infections. Although having such a broadspectrum fungistatic activity, triazoles have certain limitations like drug resistance, lack of fungicidal properties, etc. (Prasad et al., 2016; Gintjee et al., 2020). Polyenes are biological cyclic macrolides comprising of 20–40 carbon macrolactone rings conjoined with D-mycosamine. There are more than 200 molecules of polyenes primarily obtained from *Streptomyces* (Vandeputte et al., 2012). Polyenes are fungicidal antifungals that directly bind with ergosterol, resulting in formation of pores in the cell membrane that causes leakage of intracellular components, loss of membrane integrity, and finally cell death (Sanglard et al., 2003). One of the most important clinically used polyene for IFIs is amphotericin B (AmpB), possessing a broadspectrum fungicidal action. Amphotericin B is usually effective against various IFIs generally caused by *Cryptococcus, Candida* and *Aspergillus*. It can only be used as intravenous agent, i.e., oral formulations are absent. There are various limitations of AmpB—insertion reactions, nephrotoxicity (Gintjee et al., 2020; Sanglard et al., 2003). Another class of antifungal drug is echinocandins. These drugs target 1,3-β-D-glucan, main element of fungal cell wall and hinder its synthesis resulting in cell wall disintegration and finally death of fungi. Echinocandins are used against molds and yeasts, but it is a choice of drug for invasive candidiasis (Gintjee et al., 2020). Allylamines are another drug category that inhibits squalene epoxidase (ERG1) that in turn obstructs ergosterol biosynthesis. Terbinafine, an allylamine acts as a tropical agent used against dermatophytes and molds.

Taking into account that the availability of effective antifungal drugs is very low, there is a constant search for novel antifungal therapies to tackle such a life threatening fungal infections (Prasad et al., 2016). Apart from lesser availability of antifungals, there are various other reasons for increase in death rates and IFIs such as drug resistance, immunosuppression, organ transplantations, cancers, etc. (Loeffler and Stevens, 2003). There are various mechanisms through which these fungal pathogens impose drug resistance. Example, mitochondrial mutation leads toward the abnormal functioning of the mitochondria that in turn leads toward azole resistance through the activation of the pleiotropic drug resistance (pdr) pathway, this pathway regulates the expression of drug efflux pumps (MDR1, CDR1, and CDR2). Calcium signaling also takes part in azole resistance in various fungal pathogens via the inadequacy in calcium homeostasis that leads toward the preservation of cytoplasmic calcium, following activation of calcineurin phosphatase, that in turn dephosphorylates Crz1 (zinc finger transcription factor) and further nuclear translocation and initiation of genes engaged in azole resistance in *C. albicans* (Li et al., 2019; Xu et al., 2020). Due to the continuous usage of classical antifungal, immunosuppressive drugs, drug-resistant fungal strains have developed; death rate is very high relatively. The number of cancer patients is increasing; autoimmune diseases and HIV patients are also increasing day by day. This means immunocompromised patients are increasing every day, and immunotherapy is the only hope to combat such diseases. These immune therapies host a robust immune system against such infections. Immunotherapies include various treatments such as cytokine therapy, clonal antibodies, vaccines, etc. (Williams et al., 2020). These therapies regulate inborn as well as specific immune responses of the host, and this may be a successful treatment against various fungal infections (Posch et al., 2017).

7.2 Human fungal infections

Kingdom fungi is universally dispersed in the nature, consisting of 100,000 species out of which 300 species are thought to be infective in both animals as well as human beings (Gupta et al., 2017). Out of the total world population, 25% people suffer from superficial fungal infections, while as invasive fungal infections [IFI] though possessing lower incidences these IFIs are associated with higher mortalities than malaria and tuberculosis. At least one million people every year die because of these IFIs mainly caused by *Aspergillus, Pneumocystis, Candida* and *Cryptococcus* species (Gonzalez-Lara et al., 2017). Normally, under ideal conditions fungi colonize epithelial surfaces as harmless entities unless and until they come across an immunocompromised patient, having a weak immune system that can easily allow these pathogens to invade the body and cause life threatening invasive fungal infections (de Pauw, 2011). It has been seen that fungal infections (Fig. 7.1) in human beings range from simple to life-threatening diseases, including skin, hair, nails, mucosa, blood, etc. (Brown et al., 2012).

Globally it has been estimated that IFIs exceed to two million infections every year including candidiasis [>400,000], cryptococcalmeningitis [>1,000,000], pneumocystosis [>400,000], aspergillosis [>200,000] (Table 7.2) (Shaw and Ibrahim, 2020). Due to the increase in use of antibiotics that disrupt normal biota of mucosal membranes, and immunosuppressive drugs, the frequency of these infections is increasing (Perlin et al., 2017).

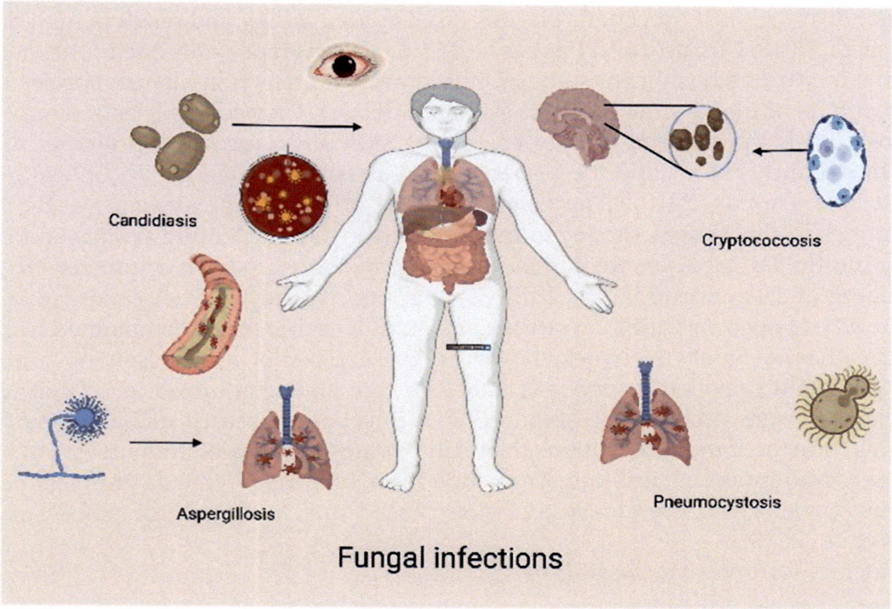

FIGURE 7.1 Various types of human fungal infections viz. candidiasis, aspergillosis, cryptococcosis, and pneumocytosis.

TABLE 7.2 Human fungal infections (Shaw and Ibrahim, 2020).

Disease	Causative agent	Type of fungi	Infections/yr
Candidiasis	Genus *Candida*	Yeasts	>400,000
Aspergillosis	*Aspergillus fumigatus, A. flavus,* etc.	Molds	>200,000
Cryptococcosis	*C. Neoformans* and *C. gatti*	Yeasts	>1,000,000
Pneumocystosis	Genus *Pneumocystis*	Yeasts	>400,000

7.2.1 Candidiasis

Candidiasis refers to the infections caused by genus *candida;* these clinical manifestations range from superficial infections to mucosal and deep-seated organs. Candida species cause both superficial and deep tissue invasive candidiasis. Invasive candidiasis is an emerging IFI, which is the leading cause of mortality and morbidity in the hospitals. There are at least 200 species of genus *Candida,* out of which about 15 different species can cause disease in humans, but *C. albicans, C. glabrata, C. tropicalis, C. parapsilosis,* and *C. krusei* represent the most frequent *Candida* species associated with the disease candidiasis. *C. auris,* a newly emerged major fungal pathogen in some parts of the world, has been described primarily as a major nosocomial pathogen, first time described in 2009 in Japan and is less susceptible to the main antifungal drugs. *C. auris* infections possess rapid epidemiology and result in high death rates (Pappas et al., 2018; McCarty and Pappas, 2016). *C. albicans* represents one of the major fungal pathogens in various healthcare settings, but in certain areas non-albican species constitute more than 50% of blood stream infections (candidemia). Candida species are opportunistic, commensal fungal pathogens usually colonizing skin and mucosal membranes along with normal human microbiota, and these species can be easily detected in 60% of healthy persons (McCarty and Pappas, 2016). Any changes in the host immune system are the favorable opportunities to this fungal species to divide rapidly and enter body tissues and cause IFIs affecting multiple organs, i.e., the fungi changes its relationship from commensal to an amensal (Kullberg and Arendrup, 2015). It has be reported that in the USA candidemia accounts for about 22% of nosocomial-blood stream infections based on the data obtained from the National Health and Safety Network; however, these patients were receiving antimicrobial agents. In a point prevalence survey of intensive care units conducted in 2009, it was found that *Candida* species account for about 18% of infectious diseases globally (Pappas et al., 2018). There are three main conditions that lead toward deep seated human fungal infections, including consumption of antibiotics for a long term, breakage in the gastrointestinal and the cutaneous barriers, and immunosuppression caused due to therapy or physician activities (Pappas et al., 2018).

Risk factors for invasive candidiasis:

1. Persons who are critically ill and stay in intensive care units for longer durations.
2. Patients having repeated laparotomy.
3. Blood cancer, multiple myeloma, and lymphoma.
4. Patients with acute necrotizing pancreatitis.
5. Patients receiving solid organ transplantations.
6. Having tumors
7. Newborn infants with low birth weight, premature babies
8. Persons taking wide-ranging antibiotics
9. Use of steroids and chemotherapies
10 Multifactorial *Candida* settlements
11. Dialysis
12. Very low neutrophil count <cells per cubic millimeter.

(Arendrup et al., 2011; Kullberg and Arendrup, 2015; Pappas et al., 2018).

7.2.1.1 Diagnosis and prevention of invasive candidiasis

7.2.1.1.1 Diagnosis

There are no particular clinical manifestations for invasive candidiasis. Consequently, these diseases have to be suspected in individuals having familiar risk factors, having inexplicable fever showing no response to antibacterial drugs. Early diagnosis is favorable for treatment of this invasive infection. It has been seen that there is doubling of deaths if an efficacious antifungal therapy is delayed by 1 or 2 days (Morrell et al., 2005; Garey et al., 2006). There are various amalgamated diagnostic methods that enable earlier and more perceptive diagnosis/detection of invasive candidiasis including histopathology, mannan antigen and anti-mannan antibody detection, β--D-glucan detection, polymerase chain reaction, microscopy, and fungal culture.

7.2.1.2 Prevention of invasive candidiasis

7.2.1.2.1 Antifungal prophylaxis

A report suggests that when fluconazole prophylaxis was given to patients in intensive care units having invasive candidiasis, it lowers the frequency of this infection; however, mortality rates remain the same (Shorr et al., 2005). It is a fact that antifungal prophylaxis in spite of showing promising results can favor the resistance to such treatment and the results of the clinical trials were unconvincing (Ostrosky-Zeichner et al., 2014).

7.2.2 Aspergillosis

Aspergillus is a ubiquitous filamentous fungus that causes aspergillosis, exhibiting various clinical manifestations. In spite of the today's robust medical technologies, aspergillosis still remains a crucial fungal infection, with a rapidly developing epidemiology and infection rates in immunocompromised patients. *Aspergilli* can cause ABPA (allergic bronchopulmonary aspergillosis) (Fig. 7.2). In patients having problems in functioning of lungs like cystic fibrosis and asthma, ABPA is a hypersensitive immune response to different fungal components. There are many species of the genus Aspergillus, but invasive aspergillosis and other pulmonary infections are caused by *A. fumigatus* followed by *A. flavus, A. terreus, A. niger,* and *A. nidulans. A. fumigatus* is omnipresent, lysotrophic airborne filamentous fungal pathogen that naturally lives in the soil and organic matter, heat and wetness promote its growth. Human beings continuously inhale plentiful conidial spores that are commonly removed from the human body by innate immune responses usually in immunocompetent individuals in absence of primary lung diseases (Chabi et al., 2015; Dagenais and Keller, 2009; Latgé and Chamilos, 2019).

The conidia spores of *A. fumigatus* are found everywhere, and their concentrations range between 1 and 100 spores per cubic millimeter but sometimes concentrations can reach up to 10^8 spores per cubic millimeter. Therefore, it is common that isolates of *Aspergillus* species

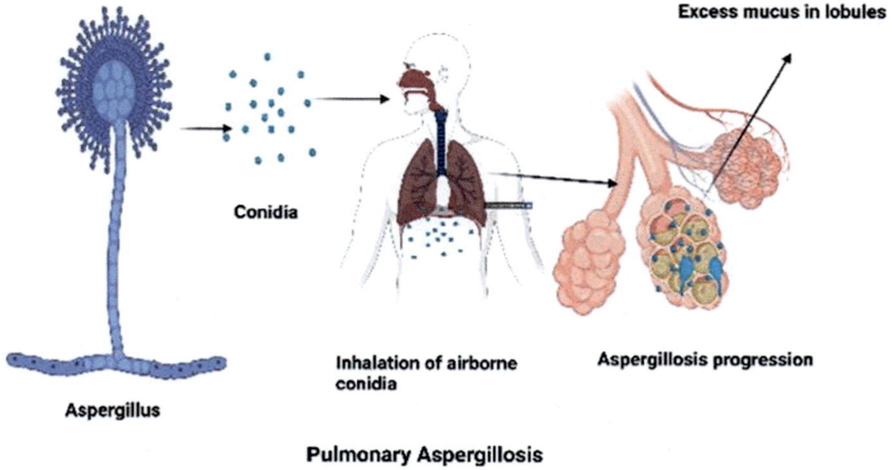

FIGURE 7.2 Stage-wise schematic representation of pulmonary aspergillosis.

can be easily detected from the cultures of the respiratory tracts of healthy individuals and the DNA of *Aspergillus* was identified in 37% lung biopsy specimens of healthy individuals. There are only three antifungal drug classes that are employed to treat IA including two fungicidal, triazoles and polyenes and other fungistatic echinocandins, but due to azole resistance and toxicity with amphotericin B and azoles, echinocandins are a drug of choice for the treatment of IA. Echinocandins are obtained from secondary metabolites of fungi and are lipopeptides in nature. There are three marketably accessible echinocandins, that is, micafungin, caspofungin, and anidulafungin, and rezafungin (CD101), a novel echinocandin nowadays in phase III clinical trials. All the three available drugs are administered intravenously and these drugs target β-1.3-glucan synthase, a protease enzyme encoded by the gene fsk 1 noncompetitively inhibiting the biosynthesis of β-1,3-glucan(major fungal cell wall polysaccharide). This cell wall polysaccharide plays an important role in protection, development of hyphae, invasion into the host tissues, and activates host innate immune responses via Dectin-1 receptor present on host immune cell surfaces. It was proven by culturing method that *Aspergillus* colonization reaches up to 30% of COPD patients (chronic obstructive pulmonary disease). Recently it was established that there is a close genetic association with impaired expression of ZNF77 (transcriptional factor) in bronchial epithelia that results in conidial adhesion and finally invasive pulmonary aspergillosis (IPA) usually in immunocompromised patients. The epidemiology of aspergillosis can be nosocomial or through environmental exposures (Latgé and Chamilos, 2019; Aruanno et al., 2019). IPA carries a poor prognosis and it is very difficult to identify and treat such disease. There are various risk factors that lead to IPA including hematopoietic stem cell transplantations, neutropenia, and excessive intake of steroids, immunosuppressive medications, immunodeficiency diseases and various critically ill patients. The two main characteristics of IPA that can be identified though CT scans are angioinvasive and airway invasive aspergillosis. But CT scans are not accurate tests to diagnose such disease at an early stage (Ledoux et al., 2020). The death rates of *Aspergilli*-infected people range from 40% to 90%, particularly in at-risk people, and these death rates depend on various factors including hosts' prior immunity, infection site, and the plan of applied treatment (Lin et al., 2001).

7.2.3 Cryptococcosis

Cryptococcosis is a fungal infection caused by the encapsulated yeasts belonging to the genus *Cryptococcus*, having a global dispense and vast clinical manifestations. Before 1970s cryptococcosis was not a common disease but during the AIDS pandemic in 1970s this mycotic infection significantly increased. The significant rise in infection was mainly associated with immunosuppressive treatments, malignancies, and solid organ transplantations with more than 80% of people associated with HIV or AIDS. The genus *Cryptococcus* consists of more than 30 spp. distributed all-over the environment. There are two main species of the genus *Cryptococcus* that usually cause cryptococcal meningitis in humans—*C. neoformans* and *C. gattii*. There are approximately 625,000 deaths yearly and globally one million persons are infected per year (Maziarz and Perfect, 2016; Kwon-Chung et al., 2017). *C. neoformans* possess a well-defined epidemiology, and it infects both immune compromised and immunocompetent patients and contrarily *C. gattii* classically infects immunocompetent hosts, although there are various other risk factors including hosts prior diseases, immune defects like anti-GMCSF autoantibodies, etc. (MacDougall et al., 2011; Rosen et al., 2013). The genus

Cryptococci occupies various ecological niches and majorly hollows of dead and decayed trees. *C. neoformans* is ubiquitous while *C. gattii* is a tropical or a subtropical fungal species, but nowadays *C. gattii* infections are widely distributed. Cryptococcus neoformans is especially found in bird droppings especially associated with street pigeons and it could become a main source of cryptococcal infection in overpopulated municipal areas. These two *Cryptococcus* spp. have the ability to live and divide in free-living amoebae and various round worms. The cause for global distribution and virulence of these fungal pathogens may be optional hosts (May et al., 2016). Cryptococcal meningitis usually infects those persons that are having impaired cell-mediated immunity particularly in AIDS patients, whose $CD4^+$ cell number falls up to 100 cells/μL. In developed countries the prevalence of this disease decreases due to the application of antiretroviral therapy (ART), although the prevalence of this disease is high in others not having ART and healthcare resources. In sub-Saharan Africa there were 15% of deaths due to AIDS and 75% deaths due to Cryptococcal meningitis. Due to abnormal functioning of the $CD4^+$ T cells there is downregulation of helper T cells accompanied by the elevated expression of the Th t2 phenotypes than Th1 that in turn contributes to the pathogenesis of cryptococcal infections in murine models but in human beings effector T-cell reactions against *cryptococcus* are found that are different from murine models (Maziarz and Perfect, 2016; Tugume et al., 2019). There are three stages for the management of cryptococcal meningitis including induction, consolidation, and maintenance therapy. In induction therapy CSF is purified quantitatively at the rate of yeast clearance CSF/mL/day using different antifungals called as early fungicidal activity (EFA) (Bicanic et al., 2009). After 2 weeks of induction therapy, consolidation therapy is given and nowadays it comprises of fluconazole 400–800 mg/day for about 8 weeks. An experiment was done in Uganda, in which 56% of patients possess positive cultures after 2 weeks, as fluconazole is behaving as a fungistatic drug at a concentration of 400 mg per day. After following prosperous induction and consolidation treatments up to 10 weeks, patients showing negative cultures are given secondary prophylaxis, i.e., fluconazole 200 mg per day as a protection therapy to avoid relapse of this infection. The patients taking ART and having unnoticeable HIV RNA levels >3 months, and having $CD4^+$ cell number greater than or equal to 100 cells per microliter can carefully stop secondary prophylactic treatment (Abassi et al., 2015).

7.3 Potential immunotherapies for fungal diseases

There side effects of the conventional antifungal drugs may be one or more of the following: liver toxicity, renal toxicity, treatment spectrum, resistance against classical antifungals, multidrug interactions, and rise in mycotic death rates, e.g., a conventional antifungal drug amphotericin B shows significant nephrotoxicity and it can be only given intravenously (IV) (Laniado-Laborín and Cabrales-Vargas, 2009). The mortality rate is still high in case of IFIs in spite of the conventional antifungals having high efficacy; this makes such drug therapy inefficient. So, antifungal immunotherapies provide an additional promising strategy against such resistant fungal pathogens. Immunotherapy is a therapeutic approach by the help of which we can modulate the immune response against disease causing pathogen. Immunotherapies modulate both innate as well as adaptive immunity of the host in order to eliminate such kind of infections (Valedkarimi et al., 2017). There may be two scenarios of immunotherapies against various microbes, one scenario is the prevention and treatment at host level and another scenario is the prevention and treatment at the pathogen level. At the host level following mechanisms can be used—vaccines that can influence hosts immunity, activation of antibodies, phagocytic cells, and T-cells, and developing chemokines, cytokines, and various antimicrobial peptides. At the pathogen level, there are also similar methods but they are not dependent on hosts' prior immunity; these methods include mAbs, synthetic AMPs, inhibitors of various signal transduction pathways, organic molecules that can affect the pathogen directly or through complement system that stops their growth and functions (Gellin et al., 2001). Immunotherapies notably strengthen the host immune system against these fungal pathogens or provide artificial immunity against pathogens. There are various approaches by which host immune system can be modified to combat against fungal pathogens including implementation of immune cells mainly effector and regulatory cells (e.g., antigen-specific T cells, granulocytes), recombinant cytokines, antimicrobial peptides (AMPs) and growth factors (e.g., interferon-γ (IFN-γ), granulocyte-colony stimulating factor (G-CSF), granulocyte-macrophage colony stimulating factor (GM-CSF), TNF-α, IL-15) and vaccinations (Fig. 7.3) (Posch et al., 2017; Qadri et al., 2021a).

The results of antifungal immunotherapy like cytokines, vaccines, GM-CFS, improve the effect of antifungals used as well as improve host immune system; however, there are some limitations including potential toxicity, resistance, hyper-inflammatory responses, and strain substitutes. Example, while using a mouse anticryptococcal GXM mAb bound with 18 B_7 mAb, unfavorable conditions were observed during phase-I examination (Datta and Hamad, 2015).

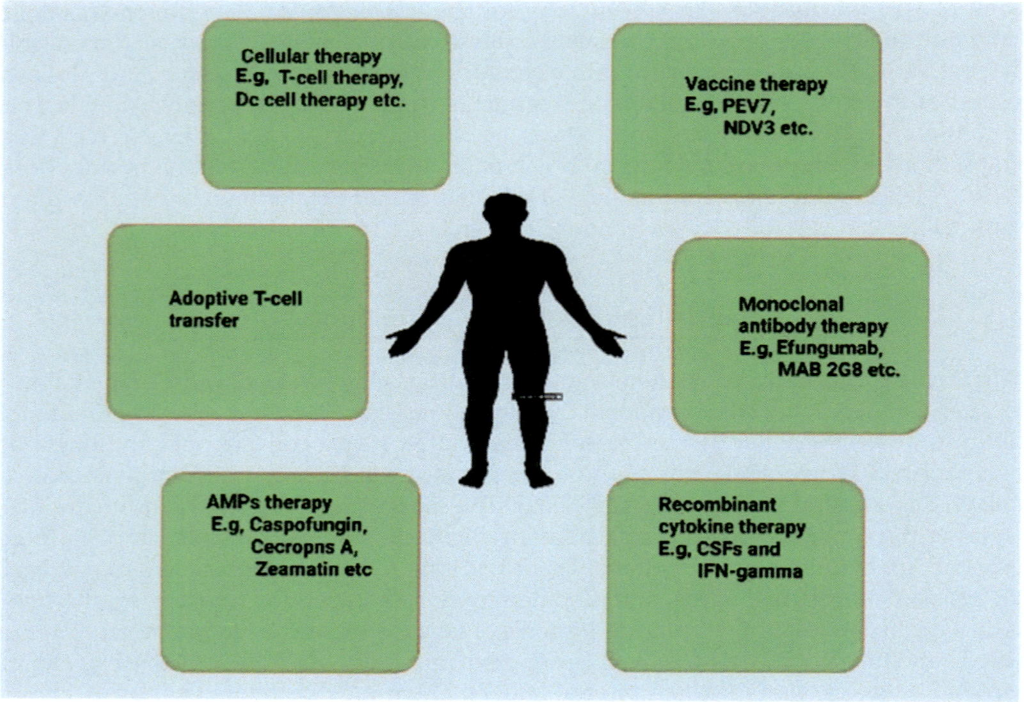

FIGURE 7.3 Various kinds of immunotherapeutic approaches against pathogenic fungi.

7.4 Different immunotherapeutic approaches against human fungal infections

Invasive fungal infection (IFI) rates are increasing day by day because of various reasons including increase in immunocompromised and immunosuppressed patients, climate alterations, and due to the common use of invasive devices. To combat such infection rates medical mycology community finds out some new immunotherapeutic approaches including vaccines, monoclonal antibodies (mAbs), immune cell activation and enhancement, recombinant cytokines, etc. (Cassone, 2008; Clark and Drummond, 2019; Medici and Del Poeta, 2015; Mir et al., 2020).

7.4.1 Recombinant cytokine therapy

Cytokines are hypomorphic glycoproteins having molecular weight lower than 30 kDa. Various immune cells produce cytokines including leukocytes, regulating immunity, inflammation, and hematopoiesis (Deverman and Patterson, 2009). Cytokines control the development and immune responses arbitrated by various immune cells. On the basis of function, cytokines are proinflammatory or noninflammatory hematopoietic-CSFs and control immune homeostasis and cell–cell communications (Chauhan et al., 2021; Berraondo et al., 2019). IL-2 and IFN-α have been approved by FDA as a treatment option for various cancerous diseases, IL-2 for progressed renal cell carcinoma, and interferon-alpha for hairy cell leukemia, melanoma, etc. (Berraondo et al., 2019). The clinical use of these recombinant cytokines in cancer diseases has facilitated the use of such foremost immunotherapy to treat fungal diseases, particularly in patients having T-cell (T-helper cell) and neutrophil immunodeficiencies. G-CSF and GM-CSF can augment such immunodeficiencies by myeloid cell reclamation and stimulation. It has been revealed in many cases that GM-CSF and G-CSF therapies seem to be successful immunotherapies in treating the patients with CNS candidiasis having CARD9 deficiencies. However, for cryptococcosis such evaluations as GM-CSF can be used as a treatment therapy are not done (Williams et al., 2020). In invasive aspergillosis, a combined therapy of GM-CSF and IFN-γ has shown to be an effective treatment for such IFIs (Bandera et al., 2008).

7.4.1.1 Colony stimulating factors in activation and enrichment of neutrophils

One of the favorable factors for opportunistic fungal pathogens is the defect in neutrophils, so neutrophil activation and enrichment becomes a promising factor to tackle fungal pathogens. Generally there is a need for the activation of nonfungus-specific innate immune cells, mainly neutrophils (Puccetti et al., 1995). Mainly for IFIs, one of the best options for neutrophil activation is granulocyte transmission (Nami et al., 2019). Granulocyte macrophage colony stimulating factor (GM-CSF), a cytokine, promotes production of neutrophils via stimulating bone marrow and inhibition of neutrophil apoptosis (Alspaugh, 2019). Along with increase in neutrophil number, GM-CSF increases their functions in the form of oxidative burst stimulation, phagocytosis, and upregulation of receptors for complement proteins (Al-Shami et al., 1998). Granulocyte colony stimulating factor (G-CSF) improves functions of neutrophils, inhibits neutrophil apoptosis, differentiation of neutrophils within the lineage, and triggering functions of monocytes and macrophages. It was seen in an adult cancer patients that prophylactic G-CSF reduces the neutropenic fever and death (Kuderer et al., 2007). It was experimentally shown that filamentous fungal infections are closely linked with activation of neutrophils that jointly under oxidative and nonoxidative paths damage the hypha of various fungi, e.g., *Aspergillus fumigatus* and *Rhizopus oryzae* and effectively controls fungal invasion and distribution (Ellett et al., 2017). Extensive neutrophil activation induces a hyperinflammatory response that can be dangerous for host body, though it has to be taken into account before application purposes. Due to less available clinical data, CSFs cannot be recommended to patients with IFIs (Patterson et al., 2016).

7.4.1.2 Recombinant human IFN-γ as an immunotherapy

In animal models, it is proved that the sensitivity toward IFIs is directly linked with diminished IFN responses in the lungs that in turn trigger inflammatory pathogenesis by Th_{17} reaction, although Th_1 reactions are poor (Roilides et al., 2002). In addition to this it has been revealed that IFN-γ is required for immunity against aspergillosis, and IFN-γ knockout rats are more susceptible to aspergillosis (Ma et al., 2002). From human genetic studies we come to know about that there is a connecting link between IFNG gene polymorphism and aspergillosis in patients with SCT (stem cell transplantation). From these studies, human recombinant IFN-γ (rINF-γ) can be used in the treatment of IFIs. As far as treatment with recombinantIFN-γ is concerned, there are chances of transplant rejection in transplant patients, though this immunotherapy may be beneficial in chronic allograft (Nami et al., 2019). Furthermore, studies have shown that recombinant IFN-γ is liable for IL-1β and TNFα secretions and also upregulates the MHC class II receptors present on leukocytes that consecutively produce elevated T-cell responses against invasive aspergillosis and candidemia. In addition to this, rINF-γ therapies magnify IL-17 and IL-22 by APCs that in turn stimulates TH17 cells that gave defensive immune responses in patients against candidemia and invasive aspergillosis. Recombinant IFN-γ is also combined with classical antifungal drugs to decrease their load and is so far used as a precautionary vehicle for various diseases (Posch et al., 2017).

7.4.2 Antibodies as immunotherapy

The discovery of hybridoma technology around 40 years ago for the generation of mAbs illuminates the role of antibody-mediated immunity in fungal infections in a highly specific and versatile manner. mAbs target single epitope of the pathogen by activating complement-mediated lysis, opsonization followed by phagocytosis, production of antimicrobial agents, and Fc-mediated release of cytokines. These antibodies further change the vital functions of fungi, involving the production of extracellular virulence factors in vesicles (Boniche et al., 2020; Mehraj et al.). The first monoclonal antibody used as an immunotherapy against fungal diseases is the 18B7 mAb attached to the polysaccharide capsule of *C. neoformans*, and its outcomes were estimated in the phase II trial. Nowadays, the group investigates radiolabeled antibodies, recognizing antifungal substitutes and fungal antigens (Nami et al., 2019). According to several studies whenever a person gets infected with fungal pathogen, protective antibodies are produced. This concept of antibody production can act as a vaccine against systemic fungal infections in immunocompromised persons, who are themselves unable to produce such antibody-mediated immune responses. Thus passive transfer of antibodies permits for the regulation of the protective monoclonal antibodies against specific pathogens and thus can provide protection to host if the host themselves is unable to form such antibodies (Boniche et al., 2020). The rich diversity of antigenic determinants in humans and fungi favors the mAbs-based immunotherapy. Fungal cell wall glycoproteins, heat shock proteins (Hsp) are favorable positions for antibody-based fungal vaccines (Boniche et al., 2020). A human recombinant antibody fragment efungumab, linked with fungal Hsp90, has been formerly incorporated with amphotericin B in *Candida albicans* and its overall clinical response was enhanced in phase III study of CI patients (Nami et al., 2019). It has been suggested that while using analogs of geldanamycin, an inhibitor of Hsp90, a chaperone molecule mainly engaged in antifungal resistance, combined with caspofungin and fluconazole (FLC), showed favorable outcomes in *Aspergillus* and *Candida* infections. Efungumab was not able to obtain a market license because of incompatibility of original efungumab formulations, so a C28Y form of efungumab was formed to reduce such incompatibility, and this form showed positive effects when fused with amphotericin B and exhibited inherent antifungal activity. But the target of using Hsp90 has not been achieved completely (Pachl et al., 2006; Louie et al., 2011). Another monoclonal antibody produced was MAb 2G8; it binds with laminarin and inhibits growth of *C. neoformans*, reduces the capsule thickness, and a single dose of this monoclonal antibody reduces fungal load in liver and brain of infected mice (Rachini et al., 2007).

7.4.3 Vaccines as immunotherapeutic approach

A vaccine is a pharmaceutical substance composed of antigen preparations, that when delivered into the body evokes immune responses and builds up life-long immunological memory, i.e., it produces strong immune response on the second encounter in the form of antibodies against these antigens. Prophylactic vaccine therapy had effectively eliminated the recurrence of contagious diseases such as small pox, measles, diphtheria, polio, mumps, rubella, typhoid, etc. These successful vaccines build up the importance of an efficient vaccine. But, effective vaccines are unavailable for a number of principal diseases, including malaria, tuberculosis, HIV, AIDS, etc. (Irvine et al., 2015; Mir and Al-baradie, 2013; Mir et al., 2021). There are various infectious fungi including *C. albicans*, *A. fumigatus*, *C. neoformans*, etc., against which vaccine development is the main priority as these are the leading causes of mycotic deaths. There are many challenges ahead of vaccine development against fungal pathogens including different modes of fungal pathogenesis, various host risk factors. Specific tailored vaccines are needed against leading fungal pathogens as different pathogens use different antigens or epitopes (Segal et al., 2006). One of the effective methods of preventing IFIs mainly invasive candidiasis (IC) is to generate an immune response that mimics innate immune response to Candida, and this method might be a successful method to prevent such invasive infections. Based on this method, PEV7 and NDV3 vaccines completed phase I testing and are now tested for toxicity and immunogenicity. PEV7 vaccine is subjected to provide protection against chronic vulvovaginal candidiasis (VVC). PEV7 comprises of shortened recombinant secreted aspartic protease 2 (Sap2), presently going through clinical testing by PevionBiotechAG. NDV3 attacks N-terminal region of the recombinant agglutinin-like sequence-3 protein (rAls3p-N) developed by NovaDigm therapeutics (Ibrahim et al., 2013). An experiment was conducted in which mice having both systemic and mucosal candidiasis were immunized with diphtheria toxoid CRM197 fused with the algal antigen laminarin (Lam), having immunogenic properties; this shows protective results against candidiasis (Torosantucci et al., 2005). Efungumab, monoclonal antibody-based vaccine targeting fungal Hsp90 progressed up to phase III trial but it was rejected due to incompatibility concerns. Eventually, MAb B6.1, a therapeutic and prophylactic IgM-monoclonal antibody formed by LigoCyte Pharmaceuticals targetting (1 → 2)-β-mannotriose was also rejected throughout its development. So, finally we can say that no vaccine is approved by FDA for human use (Ibrahim et al., 2013; Tarang et al., 2020).

7.4.4 Dendritic cells as immunotherapeutic approach

Another immunotherapeutic approach to tackle human fungal infections is the induction of T-cell responses via dendritic cells (Datta and Hamad, 2015; Mehraj et al.). DCs are APCs and act as a bridge between inborn and specific immunity via recognizing the fungi through PRRs, phagocytosing it, then processing and production of chemokines and cytokines into the surroundings and then presenting fungal antigens to helper T-cells for the induction of adaptive immune responses. This robust functional plasticity of dendritic cells has been investigated by the scientists for successful vaccine development or other kind of immunotherapy. However, when DCs were transfected with fungal yeast cells, Yeast RNA, or conidia, Th1 response was produced (Loreto et al., 2017). It has been shown that when DCs are induced with FLT3 ligands similar to GM-CSF/IL-4, the mice are not infected with fungi and also prevents transplantation diseases over the host. This activates Th1 cells, producing IFN-γ and the amplification of $FoxP3^{+}$ $IL\text{-}10^{+}CD25^{+}$Treg cells. In addition to this, myeloid DCs intended to be immunotoxic (Nami et al., 2019). It was seen that when DCs were pulsed with fructose-biphosphatealdolase-derived cell wall proteins, it induced a powerful antibody-dependent immune responses against *C. albicans* (Xu et al., 2020). Another experiment was conducted in which bone marrow–derived DCs were fused with the acapsular cells, and it was seen that these fused cells provide protection against *C. gattii* (Ueno et al., 2015).

7.4.5 Adoptive T-cell transfer

It is well known that $CD4^{+}$ T-helper cells play a crucial role against mycotic infections. It was experimentally seen that the exhaustion of $CD4^{+}$ T-cells in rodents results in increase in sensitivity toward *pneumocystis* pneumonia and various other fungal infections (Harmsen and Stankiewicz, 1990; Thullen et al., 2003). One of the determining factors in humans to resist fungal infections is Th to Th1 differentiation (Matthews and Burnie, 2004). Th1 mediated immune responses provide remarkable defense against different fungal infections by secreting IFN-γ, IL-6, IL-10, and IL-12 (Brandt, 2002; Hebart et al., 2002). An experiment was conducted in which mice were stimulated with *Aspergillus fumigatus* and then $CD4^{+}$ splenocytes were taken from these stimulated mice and then transferred into naive mice; it was seen that naive mice show prolonged survival ensuring intravenous challenge with viable *A. fumigatus* conidia (Cenci et al., 2000). The results of such kinds of experiments act as a base for antigen-specific T-cell immunotherapy mainly in immunocompromised individuals, as they can modulate immune system and provide antifungal activity. On combining IL-12 and anti-CD40, which averts CD40 and CD154 binding, it increases the survival of the *C. neoformans* infected mice. This protection in mice was positively corresponded with reduced fungal load in brain and kidney tissues and elevated serum concentration of TNF-α and interferon gamma. However, protection depends on IFN-gamma was proved by the lack of protection in IFN-gamma knockout mice. The results of such experiments reveal that adoptive T-cell transfer is worthy for treating fungal infection (Hamad, 2008). Along with $CD4^+$ T-cells, fungi-specific effector $CD8^+$ cells are efficient in generating defensive IFN-γ levels against pulmonary cryptococcosis in $CD4^-$ mice. It indicates that $CD8^+$ subset are also involved in protection against mycotic infections (Zhou et al., 2006).

7.4.6 B-cell and natural killer cell treatments

B cells have also been examined for antimicrobial activity via adoptive cell transfer techniques. Hoyt et al. revealed that when activated B cells were transferred into the mice having impaired lymphocytes, impaired type 1-IFN receptors, and also lung pneumocystosis, they show hematopoietic progenitor activity against the infection, thus these B cells renew the depleted bone marrow cells in IL-10 and IL-27-dependent techniques maybe by macrophage or DCs responses (Hoyt et al., 2015; Loreto et al., 2017).

Natural killer cells (NK) are natural lymphocytes possessing characteristics of both inborn as well as specific immunity and these cells can be stimulated when there are transformed cells, infected cells, so that these NK cells recognize and organize rapid effector immune responses in the form of cytokines, antibodies, cytolytic activities, and immunological memory (Bezman et al., 2012; Cerwenka and Lanier, 2016). Park et al. reported that when activated NK cells were transferred into a neutropenic mice having invasive aspergillosis, protective response in mice was seen in the form of IFN-γ secreted by these activated NK cells during primary stages of mycosis. This experiment suggested that the transferred NK cells provide protection to both wild type host and deficient hosts; this resulted in more rapid clearance of *A. fumigatus* from the lungs of mice (Park et al., 2009). It has been observed that NK cells precisely or through IFN-γ possess antifungal activity against hyphae of *A. fumigatus* but show no activity against infectious conidia. On the basis of all these experiments, NK cell adoptive treatments constitute one of the possible immunotherapy to be used separately or in combination with other antifungals; however, they possess limited role in prophylaxis against *Aspergillus* infections (Loreto et al., 2017).

7.4.7 Antimicrobial peptides as immunotherapy against fungal diseases

There are more than 150 antimicrobial peptides (AMPs) exhibiting antifungal action such as caspofungin, anidulafungin, aureobasidins, sordarins, micafungin, nikkomycins, bacillomycin, cecropins A and B, zeamatin, etc. These AMPs are obtained from fungi, bacteria, insects, and plants. In addition to this a number of AMPs or AFPs are also produced by mammalian inborn responses such as alpha and beta-defensins, nonhuman primate peptides, cysteine-rich antimicrobial peptides like gallinacin, cathelicidins, etc. IL-6 and TNF-α also possess the ability to induce articulation of antimicrobial peptides (De Lucca and Walsh, 2000; Hamad, 2008). It has been seen that cathelicidins and other small host defense peptides regulate the cytokine production, reduce inflammation, and modify host gene expression (Bowdish et al., 2005). Most of the AMPs or AFPs usually target fungal plasma membrane based on their structure and hydrophobic nature. Consequently, they target fungi by disrupting plasma membrane that leads toward cytoplasmic leakage and cell lysis. Others interact with cytoplasmic or nuclear molecules and interrupt with signaling cascades thus inhibiting growth of fungal pathogen (De Lucca and Walsh, 2000). It was observed that a recombinant defensin Tfgd1, amalgamated from complementary DNA cloned from *Trigonella foenum-graecum* shows wide range of antifungal activity (Olli and Kirti, 2006). Another synthetic AMP D4E1 can be used against *Aspergillus* possessing a 50% lethal dose of 2.1−16.8 μg per milliliter (De Lucca and Walsh, 2000). A synthetic AMP Lfpep, synthesized from human lactoferrin gene, including 18 to 40 human lactroferrin gene sequence, is fungicidal in case of fluconazole and amphotericin B-resistant *Candida albicans* strains, confirmed by the capacity of propidium iodide to permeate Lfpep-treated *Candida* strains (Viejo-Díaz et al., 2005).

7.5 Advantages of immunotherapies for fungal diseases

No matter how medicinal interventions have increased the life expectancy of diseased persons by using broadspectrum antibiotics, immunosuppressive drugs, and various chemotherapies for cancers, but these therapeutic changes can also pose various ill effects that can lead toward vulnerability to various diseases such as bacterial, parasitic, fungal, and viral pathogens. Invasive mycotic infections are increasing worldwide as a consequence of HIV outbreak, restorative drugs, surgeries (Boniche et al., 2020; Clark and Drummond, 2019), drug toxicity, restricted activity of available drugs, drug resistance, etc. So, there is requirement of new therapies that can deal with these concerns, specifically for immunocompromised patients against various pathogens (Qadri et al., 2021a). However, through immunotherapy we can eliminate the infective agents by modulating the immune responses of the host (Nami et al., 2019). The main advantage of immunotherapies (Figs. 7.4 and 7.5) is that it fortifies the hosts' immune system against various infections, mainly tumor cells, by modulating the inherent immunity accordingly through various immune cells (Tan et al., 2020).

7.5.1 Advantages of T-cell immunotherapies

During the past few decades, it has been revealed that T-cell reactions are crucial for specific immunity against various types of fungal diseases, especially Th1 cells. This fact leads the way for transfusion of T-cells or other immune stimulating factors that can produce interferon-γ against *Aspergillus* infections. Recently it has been demonstrated that glucanase CrF1, a cell wall protein of *A. fumigatus* can develop $CD4^+$ memory responses that can be used against *C. albicans* (Stuehler et al., 2011). From all these examinations there is possibility that fungal vaccine immunotherapy will be available in near future (Goodman et al., 1996). In an experiment done by Kumaresan et al., it was seen that by using recombinant T cells through the utilization of C-type lectin receptor (CLR) function, specific for fungal pathogens

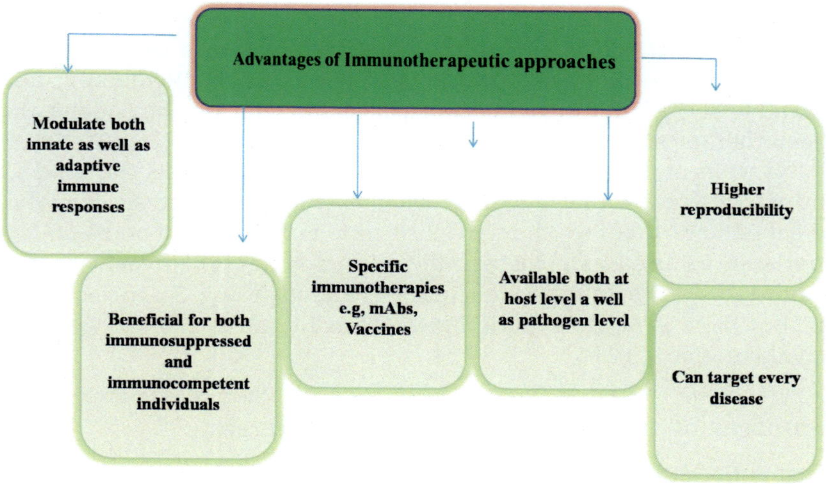

FIGURE 7.4 Advantages of immunotherapeutic approaches.

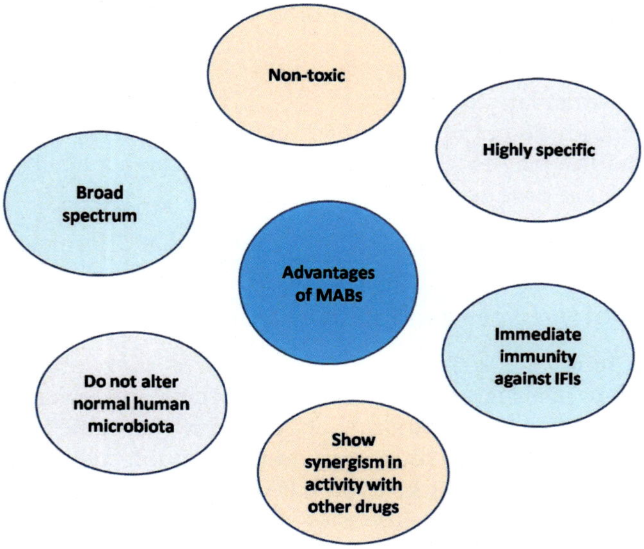

FIGURE 7.5 Advantages of monoclonal antibodies used against various microbial diseases.

can be an advantageous immunotherapy against such pathogens (Kumaresan et al., 2014; Qadri et al., 2021b). The advantage of recombinant cytokines upon the conventional antifungal drugs is that they possess possible roles in managing the equilibrium of the host immune system against pathogenic fungi. In two patients with *Candida* infection and relapsing *Candida meningoencephalitis* having deficiency of protein CARD 9, it was found that G-CSF or G-CSF + antifungal agents provide effective results (Karavalakis et al., 2021). EMA and US-FDA approved the granulocyte-CSF that usually stimulates the extension and differentiation of neutrophils and enhances the functions of neutrophil antifungal action in both artificial environments as well as in the experimental mouse models. When G-CSF was used along with classical and new antifungals, just as adjuvant therapy, fast neutrophil reclamation and lesser risk of fungal infections occurs in neutropenic patients that are taking rigorous chemotherapies for various hematological malignancies (Hazel et al., 1999; Farkas et al., 2019). NK cells and T cells chiefly produce IFN-γ that assists the production of ROS and RNS and activates macrophage that in turn intensifies phagocytosis. It has been observed in many fungal patients that the adjunctive utilization of IFN-γ can boost up hosts' immune responses and can be used as a therapeutic approach against invasive fungal infections mainly in inborn or specific immunodeficiencies patients, STAT-3 deficiency patients, AIDS patients, etc. But there is need of randomized clinical trials to confirm the useful effects of IFN-γ (Karavalakis et al., 2021).

7.5.2 Advantages of antibodies used as immunotherapy

- Monoclonal antibodies are highly specific against specific fungal strain.
- Monoclonal antibodies furnish rapid immune responses against pervasive fungal infections.
- Monoclonal antibodies along with other antifungal agents decrease the treatment time period synergistically.
- Monoclonal antibodies avert noxious risks as these are specific to a particular antigenic determinant of a particular pathogen.
- Monoclonal antibodies do not affect normal human microbiota.
- Monoclonal antibodies possess the capacity that these can act against various antigens.
- Immune responses are produced up to longer times.

(Boniche et al., 2020; Casadevall et al., 2004; Casadevall and Pirofski, 2012)

7.5.3 Advantages of vaccines as immunotherapy

Vaccines against fungi are active immunotherapies as they amplify the host immune system against fungi by focusing mainly on fungal-specific antigens (Carvalho et al., 2012). The main goal of fungal vaccines is to bring about effective specific immune responses that can lead toward immunological memory to avoid fungal infections (Posch et al., 2017). In a survey, it was described that vaccines avert 6,000,000 deaths per year globally (Leibovitch and Jacobson, 2016). All the three categories of vaccines are being developed against fungal pathogens, i.e., conjugate vaccines, recombinant vaccines, and live-attenuated vaccines (Table 7.3) (Nami et al., 2019). All chemicals and the antigens of the main life-threatening fungal species have been taken into consideration for the development of a successful active vaccination. This is only possible due to whole genome sequencing and proteomics. This concept was used for the development of bacterial vaccines, e.g., group B meningococcus based on the perspective of reverse vaccinology. The fungal vaccines may have broad-spectrum as far as β-glucan-conjugate vaccine is considered, β-glucan is a major fungal cell wall polysaccharide and it also protects them from unfavorable conditions (Cassone, 2008). In the recent years, scientists have made significant progresses in the field of vaccine development, and various antifungal candidate vaccines have been described; some are under preclinical examinations and two of these vaccine candidates are under phase I

TABLE 7.3 Advantages of some major fungal vaccine plans (Nami et al., 2019).

S.No	Vaccine	Advantage
1.	Conjugate vaccines	• Incite reactions in the case of both glycan and protein antigens • Provide more safety in comparison to attenuated vaccines in the case of immunosuppressed individuals
2.	Killed-attenuated vaccine	• Incite different immune reactions • Produce multiple types of antigens for more stimulation of immune reaction
3.	Recombinant vaccine	• Provides more safety in comparison to attenuated vaccines in the case of immunosuppressed individuals

clinical trials. Various experiments have been conducted on the use of sonicated hyphal filtrate, showing some protection degrees. An example of such vaccines is PtiumVac, used to treat equine equinepythiosis in Brazil. It is a licensed immunotherapy for equine pythiosis having a cure rate of 70%–80%, but lacks protective activity (Loreto et al., 2017). Several studies suggest that on using heat-killed yeasts in mice provide positive results against various fungal species including *Candida, Aspergillus, Cryptococcus, Coccidioides and Rhizopus.* There experiments suggested that specific Th1 responses and antibodies responses against glucan and mannan were induced. These results suggest that there is possibility of the formation of a pan-fungal yeast-based vaccine (Liu et al., 2011).

7.6 Future perspectives

Despite having advanced comprehensive knowledge in pathogenesis of fungal infections, there are still high mortalities due to IFIs especially in immunocompromised individuals. Classical antifungals are ineffective due to increase in drug-resistant fungal strains, drug toxicities, etc. There is urgent need of new and specific treatment options for such patients so that fungal infections can be controlled and eradicated. One of the promising treatment against such fungal infections is immunotherapeutic approaches. Immunotherapies exhibit immense potential to upgrade and amplify the host defense mechanisms against various fungal pathogens. Currently various immunotherapeutic approaches are utilized for the treatment of cancers, and they are showing promising results. Likewise, immunotherapies against fungal infections could become promising treatments to combat fungal diseases in future. There are various drawbacks of classical antifungals including ineffectiveness, severe side effects, and drug-resistant fungi. Immune-modulation methods try to overcome such issues and provide a robust method of treatment against these infections. Immunotherapies have the ability to modulate both innate as well as adaptive immune reactions in the form of cytokines, T cells, monoclonal antibodies, antifungal vaccines, etc. These multiple immune therapies are consequently turning out to be interesting treatment therapies to prevent such kind of invasive fungal infections. Out of all the therapeutic approaches, monoclonal antibodies and vaccines are becoming prominent methods. mAbs have been used in cancer therapies, autoimmune diseases, and various infections. The specific and selective behavior of such immunotherapies make them very versatile and in near future immunotherapies will benefit the people and advancement in treatments will be seen.

References

Abassi, M., Boulware, D.R., Rhein, J., 2015. Cryptococcal meningitis: diagnosis and management update. Curr. Trop. Med. Rep. 2, 90–99.

Al-Shami, A., Mahanna, W., Naccache, P.H., 1998. Granulocyte-macrophage colony-stimulating factor-activated signaling pathways in human neutrophils: selective activation of Jak2, Stat3, and Stat5b. J. Biol. Chem. 273, 1058–1063.

Alspaugh, J.A., 2019. Immunomodulators: What Is the Evidence for Use in Mycoses? Antifungal Therapy. CRC Press.

Arendrup, M.C., Sulim, S., Holm, A., Nielsen, L., Nielsen, S.D., Knudsen, J.D., Drenck, N.E., Christensen, J.J., Johansen, H.K., 2011. Diagnostic issues, clinical characteristics, and outcomes for patients with fungemia. J. Clin. Microbiol. 49, 3300–3308.

Aruanno, M., Glampedakis, E., Lamoth, F., 2019. Echinocandins for the treatment of invasive aspergillosis: from laboratory to bedside. Antimicrob. Agents Chemother. 63 e00399–19.

Bandera, A., Trabattoni, D., Ferrario, G., Cesari, M., Franzetti, F., Clerici, M., Gori, A., 2008. Interferon-γ and granulocyte-macrophage colony stimulating factor therapy in three patients with pulmonary aspergillosis. Infection 36, 368–373.

Berraondo, P., Sanmamed, M.F., Ochoa, M.C., Etxeberria, I., Aznar, M.A., Pérez-Gracia, J.L., Rodríguez-Ruiz, M.E., Ponz-Sarvise, M., Castañón, E., Melero, I., 2019. Cytokines in clinical cancer immunotherapy. Br. J. Cancer 120, 6–15.

Bezman, N.A., Kim, C.C., Sun, J.C., MIN-Oo, G., Hendricks, D.W., Kamimura, Y., Best, J.A., Goldrath, A.W., Lanier, L.L., 2012. Molecular definition of the identity and activation of natural killer cells. Nat. Immunol. 13, 1000–1009.

Bhattacharya, S., SAE-Tia, S., Fries, B.C., 2020. Candidiasis and mechanisms of antifungal resistance. Antibiotics 9, 312.

Bicanic, T., Muzoora, C., Brouwer, A.E., Meintjes, G., Longley, N., Taseera, K., Rebe, K., Loyse, A., Jarvis, J., Bekker, L.-G., 2009. Independent association between rate of clearance of infection and clinical outcome of HIV-associated cryptococcal meningitis: analysis of a combined cohort of 262 patients. Clin. Infect. Dis. 49, 702–709.

Boniche, C., Rossi, S.A., Kischkel, B., Vieira Barbalho, F., Nogueira D'aurea Moura, Á., Nosanchuk, J.D., Travassos, L.R., Pelleschi Taborda, C., 2020. Immunotherapy against systemic fungal infections based on monoclonal antibodies. J. Fungi 6, 31.

Bowdish, D.M.E., Davidson, D.J., Scott, M.G., Hancock, R.E.W., 2005. Immunomodulatory activities of small host defense peptides. Antimicrob. Agents Chemother. 49, 1727–1732.

Brandt, M.E., 2002. Candida and candidiasis. Emerg. Infect. Dis. 8, 876.

Brown, G.D., Denning, D.W., Gow, N.A., Levitz, S.M., Netea, M.G., White, T.C., 2012. Hidden killers: human fungal infections. Sci. Transl. Med. 4, 165rv13.

Bruch, A., Kelani, A.A., Blango, M.G., 2021. RNA-based therapeutics to treat human fungal infections. Trends Microbiol.

Carvalho, A., Cunha, C., Iannitti, R.G., Casagrande, A., Bistoni, F., Aversa, F., Romani, L., 2012. Host defense pathways against fungi: the basis for vaccines and immunotherapy. Front. Microbiol. 3, 176.

Casadevall, A., Dadachova, E., Pirofski, L.-A., 2004. Passive antibody therapy for infectious diseases. Nat. Rev. Microbiol. 2, 695–703.

Casadevall, A., Pirofski, L.-A., 2012. Immunoglobulins in defense, pathogenesis, and therapy of fungal diseases. Cell Host Microbe 11, 447–456.

Cassone, A., 2008. Fungal vaccines: real progress from real challenges. Lancet Infect. Dis. 8, 114–124.

Cenci, E., Mencacci, A., Bacci, A., Bistoni, F., Kurup, V.P., Romani, L., 2000. T cell vaccination in mice with invasive pulmonary aspergillosis. J. Immunol. 165, 381–388.

Cerwenka, A., Lanier, L.L., 2016. Natural killer cell memory in infection, inflammation and cancer. Nat. Rev. Immunol. 16, 112–123.

Chabi, M.L., Goracci, A., Roche, N., Paugam, A., Lupo, A., Revel, M.P., 2015. Pulmonary aspergillosis. Diagn. Interven. Imag. 96, 435–442.

Chauhan, P., Nair, A., Patidar, A., Dandapat, J., Sarkar, A., Saha, B., 2021. A primer on cytokines. Cytokine 145, 155458.

Clark, C., Drummond, R.A., 2019. The hidden cost of modern medical interventions: how medical advances have shaped the prevalence of human fungal disease. Pathogens 8, 45.

Dagenais, T.R.T., Keller, N.P., 2009. Pathogenesis of *Aspergillus fumigatus* in invasive aspergillosis. Clin. Microbiol. Rev. 22, 447–465.

Datta, K., Hamad, M., 2015. Immunotherapy of fungal infections. Immunol. Invest. 44, 738–776.

De Lucca, A.J., Walsh, T.J., 2000. Antifungal peptides: origin, activity, and therapeutic potential. Rev. Iberoam. De. Micol. 17, 116–120.

De Pauw, B.E., 2011. What are fungal infections? Mediter. J. Hematol. Infect. Dis. 3.

Deverman, B.E., Patterson, P.H., 2009. Cytokines and CNS development. Neuron 64, 61–78.

Didomenico, B.J.C.O.I.M., 1999. Novel Antifungal Drugs, vol. 2, pp. 509–515.

Ellett, F., Jorgensen, J., Frydman, G.H., Jones, C.N., Irimia, D., 2017. Neutrophil interactions stimulate evasive hyphal branching by *Aspergillus fumigatus*. PLoS Pathogens 13, e1006154.

Farkas, F., Mistrik, M., Batorova, A., 2019. The use of granulocyte colony stimulating factor after autologous hematopoietic stem cell transplantation. Bratisl. Lek. Listy 120, 668–672.

Fisher, J.F., Valencia-Rey, P.A., Davis, W.B., 2016. Pulmonary Cryptococcosis in the Immunocompetent Patient—Many Questions, Some Answers. Oxford University Press, p. ofw167.

Garey, K.W., Rege, M., Pai, M.P., Mingo, D.E., Suda, K.J., Turpin, R.S., Bearden, D.T., 2006. Time to initiation of fluconazole therapy impacts mortality in patients with candidemia: a multi-institutional study. Clin. Infect. Dis. 43, 25–31.

Gellin, B., Modlin, J.F., Casadevall, A., Pirofski, L.-A., 2001. Adjunctive immune therapy for fungal infections. Clin. Infect. Dis. 33, 1048–1056.

Gintjee, T.J., Donnelley, M.A., Thompson, G.R., 2020. Aspiring antifungals: review of current antifungal pipeline developments. J. Fungi 6, 28.

Gonzalez-Lara, M.F., Sifuentes-Osornio, J., Ostrosky-Zeichner, L., 2017. Drugs in clinical development for fungal infections. Drugs 77, 1505–1518.

Goodman, A.R., Cardozo, T., Abagyan, R., Altmeyer, A., Wisniewski, H.-G., Vilček, J., 1996. Long pentraxins: an emerging group of proteins with diverse functions. Cytokine Growth Factor Rev. 7, 191–202.

Goyal, S., Castrillón-Betancur, J.C., Klaile, E., Slevogt, H., 2018. The interaction of human pathogenic fungi with C-type lectin receptors. Front. Immunol. 9, 1261.

Gupta, A., Gupta, R., Singh, R.L., 2017. Microbes and Environment. Principles and Applications of Environmental Biotechnology for a Sustainable Future. Springer.

Hamad, M., 2008. Antifungal immunotherapy and immunomodulation: a double-hitter approach to deal with invasive fungal infections. Scand. J. Immunol. 67, 533–543.

Harmsen, A.G., Stankiewicz, M., 1990. Requirement for $CD4^+$ cells in resistance to *Pneumocystis carinii* pneumonia in mice. J. Exp. Med. 172, 937–945.

Hazel, D.L., Newland, A.C., Kelsey, S.M., 1999. Granulocyte colony stimulating factor increases the efficacy of conventional amphotericin in the treatment of presumed deep-seated fungal infection in neutropenic patients following intensive chemotherapy or bone marrow transplantation for haematological malignancies. Hematology 4, 305–311.

Hebart, H., Bollinger, C., Fisch, P., Sarfati, J., Meisner, C., Baur, M., Loeffler, J.R., Monod, M., Latgé, J.-P., Einsele, H., 2002. Analysis of T-cell responses to *Aspergillus fumigatus* antigens in healthy individuals and patients with hematologic malignancies. Blood 100, 4521–4528.

Hoyt, T.R., Dobrinen, E., Kochetkova, I., Meissner, N., 2015. B cells modulate systemic responses to *Pneumocystis murina* lung infection and protect on-demand hematopoiesis via T cell-independent innate mechanisms when type I interferon signaling is absent. Infect. Immun. 83, 743–758.

Ibrahim, A.S., Luo, G., Gebremariam, T., Lee, H., Schmidt, C.S., Hennessey Jr., J.P., French, S.W., Yeaman, M.R., Filler, S.G., Edwards Jr., J.E., 2013. NDV-3 protects mice from vulvovaginal candidiasis through T-and B-cell immune response. Vaccine 31, 5549–5556.

Irvine, D.J., Hanson, M.C., Rakhra, K., Tokatlian, T., 2015. Synthetic nanoparticles for vaccines and immunotherapy. Chem. Rev. 115, 11109–11146.

Ji, C., Liu, N., Tu, J., Li, Z., Han, G., Li, J., Sheng, C., 2019. Drug repurposing of haloperidol: discovery of new benzocyclane derivatives as potent antifungal agents against cryptococcosis and candidiasis. ACS Infect. Dis. 6, 768–786.

Karavalakis, G., Yannaki, E., Papadopoulou, A., 2021. Reinforcing the immunocompromised host defense against fungi: progress beyond the current state of the art. J. Fungi 7, 451.

Khan, H.A., Baig, F.K., Mehboob, R., 2017. Nosocomial infections: epidemiology, prevention, control and surveillance. Asian Pacific J. Trop. Biomed. 7, 478–482.

Kuderer, N.M., Dale, D.C., Crawford, J., Lyman, G.H., 2007. Impact of Primary Prophylaxis with Granulocyte Colony-Stimulating Factor on Febrile Neutropenia and Mortality in Adult Cancer Patients Receiving Chemotherapy: A Systematic Review. Database of Abstracts of Reviews of Effects (DARE): Quality-Assessed Reviews [Internet].

Kullberg, B.J., Arendrup, M.C., 2015. Invasive candidiasis. N. Engl. J. Med. 373, 1445–1456.

Kumaresan, P.R., Manuri, P.R., Albert, N.D., Maiti, S., Singh, H., Mi, T., Roszik, J., Rabinovich, B., Olivares, S., Krishnamurthy, J., 2014. Bioengineering T cells to target carbohydrate to treat opportunistic fungal infection. Proc. Natl. Acad. Sci. USA 111, 10660–10665.

Kwon-Chung, K.J., Bennett, J.E., Wickes, B.L., Meyer, W., Cuomo, C.A., Wollenburg, K.R., Bicanic, T.A., Castañeda, E., Chang, Y.C., Chen, J., 2017. The case for adopting the "species complex" nomenclature for the etiologic agents of cryptococcosis. mSphere 2 e00357–16.

Laniado-Laborín, R., Cabrales-Vargas, M.N., 2009. Amphotericin B: side effects and toxicity. Rev. Iberoam. De. Micol. 26, 223–227.

Latgé, J.-P., Chamilos, G., 2019. *Aspergillus fumigatus* and aspergillosis in 2019. Clin. Microbiol. Rev. 33 e00140–18.

Ledoux, M.-P., Guffroy, B., Nivoix, Y., Simand, C., Herbrecht, R., 2020. Invasive Pulmonary Aspergillosis. Thieme Medical Publishers, pp. 080–098.

Leibovitch, E.C., Jacobson, S., 2016. Vaccinations for neuroinfectious disease: a global health priority. Neurotherapeutics 13, 562–570.

Li, Y., Zhang, Y., Lu, L., 2019. Calcium signaling pathway is involved in non-CYP51 azole resistance in *Aspergillus fumigatus*. Med. Mycol. 57, S233–S238.

Lin, S.-J., Schranz, J., Teutsch, S.M., 2001. Aspergillosis case-fatality rate: systematic review of the literature. Clin. Infect. Dis. 32, 358–366.

Liu, M., Clemons, K.V., Bigos, M., Medovarska, I., Brummer, E., Stevens, D.A., 2011. Immune responses induced by heat killed *Saccharomyces cerevisiae*: a vaccine against fungal infection. Vaccine 29, 1745–1753.

Loeffler, J., Stevens, D.A., 2003. Antifungal drug resistance. Clin. Infect. Dis. 36, S31–S41.

Lone, S.A., Ahmad, A., 2019. *Candida auris*—the growing menace to global health. Mycoses 62, 620–637.

Loreto, É., Tondolo, J.S.M., Alves, S.H., Santurio, J.M., 2017. Immunotherapy—Myths, Reality, Ideas, Future. InTech.

Louie, A., Stein, D.S., Zack, J.Z., Liu, W., Conde, H., Fregeau, C., Vanscoy, B.D., Drusano, G.L., 2011. Dose range evaluation of Mycograb C28Y variant, a human recombinant antibody fragment to heat shock protein 90, in combination with amphotericin B-desoxycholate for treatment of murine systemic candidiasis. Antimicrob. Agents Chemother. 55, 3295–3304.

Ma, L.L., Spurrell, J.C.L., Wang, J.F., Neely, G.G., Epelman, S., Krensky, A.M., Mody, C.H., 2002. CD8 T cell-mediated killing of *Cryptococcus neoformans* requires granulysin and is dependent on CD4 T cells and IL-15. J. Immunol. 169, 5787–5795.

Macdougall, L., Fyfe, M., Romney, M., Starr, M., Galanis, E., 2011. Risk factors for *Cryptococcus gattii* infection, British Columbia, Canada. Emerg. Infect. Dis. 17, 193.

Marquez, L., Quave, C.L., 2020. Prevalence and therapeutic challenges of fungal drug resistance: role for plants in drug discovery. Antibiotics 9, 150.

Matthews, R.C., Burnie, J.P., 2004. Recombinant antibodies: a natural partner in combinatorial antifungal therapy. Vaccine 22, 865–871.

May, R.C., Stone, N.R.H., Wiesner, D.L., Bicanic, T., Nielsen, K., 2016. Cryptococcus: from environmental saprophyte to global pathogen. Nat. Rev. Microbiol. 14, 106–117.

Maziarz, E.K., Perfect, J.R., 2016. Cryptococcosis. Infect. Dis. Clin. 30, 179–206.

Mccarty, T.P., Pappas, P.G., 2016. Invasive candidiasis. Infect. Dis. Clin. 30, 103–124.

Medici, N.P., DEL Poeta, M., 2015. New insights on the development of fungal vaccines: from immunity to recent challenges. Mem. Inst. Oswaldo Cruz 110, 966–973.

Mehraj, U., Nisar, S., Bhat, B.A., Sheikh, B.A., Hamdani, S.S., Qayoom, H., Mir, M.A. Organs of the immune system. Basics and Fundamentals of Immunology, 53.

Mehraj, U., Nisar, S., Qayoom, H., Mir, M.A. Monoclonal antibodies IN therapeutics. Immunoglob. Magic Bull. Therap. Antibodies, 155.

Mir, M.A., AL-Baradie, R., 2013. Tuberculosis time bomb-A global emergency: need for alternative vaccines. Majmaah J. Health Sci. 1, 77–82.

Mir, M.A., Bhat, B.A., Sheikh, B.A., Rather, G.A., Mehraj, S., Mir, W.R., 2021. Nanomedicine in Human Health Therapeutics and Drug Delivery: Nanobiotechnology and Nanobiomedicine. Applications of Nanomaterials in Agriculture, Food Science, and Medicine. IGI Global.

Mir, M.A., Mehraj, U., Sheikh, B.A., Hamdani, S.S., 2020. Nanobodies: the "magic bullets" in therapeutics, drug delivery and diagnostics. Hum. Antibodies 28, 29–51.

Morrell, M., Fraser, V.J., Kollef, M.H., 2005. Delaying the empiric treatment of Candida bloodstream infection until positive blood culture results are obtained: a potential risk factor for hospital mortality. Antimicrob. Agents Chemother. 49, 3640–3645.

Nami, S., Aghebati-Maleki, A., Morovati, H., Aghebati-Maleki, L., 2019. Current antifungal drugs and immunotherapeutic approaches as promising strategies to treatment of fungal diseases. Biomed. Pharmacother. 110, 857–868.

Olli, S., Kirti, P.B., 2006. Cloning, characterization and antifungal activity of defensin Tfgd1 from *Trigonella foenum-graecum* L. BMB Rep. 39, 278–283.

Ostrosky-Zeichner, L., Shoham, S., Vazquez, J., Reboli, A., Betts, R., Barron, M.A., Schuster, M., Judson, M.A., Revankar, S.G., Caeiro, J.P., 2014. MSG-01: a randomized, double-blind, placebo-controlled trial of caspofungin prophylaxis followed by preemptive therapy for invasive candidiasis in high-risk adults in the critical care setting. Clin. Infect. Dis. 58, 1219–1226.

Pachl, J., Svoboda, P., Jacobs, F., Vandewoude, K., Van Der Hoven, B., Spronk, P., Masterson, G., Malbrain, M., Aoun, M., Garbino, J., 2006. A randomized, blinded, multicenter trial of lipid-associated amphotericin B alone versus in combination with an antibody-based inhibitor of heat shock protein 90 in patients with invasive candidiasis. Clin. Infect. Dis. 42, 1404–1413.

Pappas, P.G., Lionakis, M.S., Arendrup, M.C., Ostrosky-Zeichner, L., Kullberg, B.J., 2018. Invasive candidiasis. Nat. Rev. Dis. Prim. 4, 1–20.

Park, S.J., Hughes, M.A., Burdick, M., Strieter, R.M., Mehrad, B., 2009. Early NK cell-derived IFN-γ is essential to host defense in neutropenic invasive aspergillosis. J. Immunol. 182, 4306–4312.

Pathakumari, B., Liang, G., Liu, W., 2020. Immune defence to invasive fungal infections: a comprehensive review. Biomed. Pharmacother. 130, 110550.

Patterson, T.F., Thompson Iii, G.R., Denning, D.W., Fishman, J.A., Hadley, S., Herbrecht, R., Kontoyiannis, D.P., Marr, K.A., Morrison, V.A., Nguyen, M.H., 2016. Practice guidelines for the diagnosis and management of aspergillosis: 2016 update by the Infectious Diseases Society of America. Clin. Infect. Dis. 63, e1–e60.

Perlin, D.S., Rautemaa-Richardson, R., Alastruey-Izquierdo, A., 2017. The global problem of antifungal resistance: prevalence, mechanisms, and management. Lancet Infect. Dis. 17, e383–e392.

Posch, W., Steger, M., Wilflingseder, D., Lass-Flörl, C., 2017. Promising immunotherapy against fungal diseases. Expet Opin. Biol. Ther. 17, 861–870.

Prasad, R., Shah, A.H., Rawal, M.K., 2016. Antifungals: mechanism of action and drug resistance. Yeast Membrane Transp. 327–349.

Puccetti, P., Romani, L., Bistoni, F., 1995. A TH1-TH2-like switch in candidiasis: new perspectives for therapy. Trends Microbiol. 3, 237–240.

Qadri, H., Haseeb, A., Mir, M., 2021a. Novel strategies to combat the emerging drug resistance in human pathogenic microbes. Curr. Drug Targets 22, 1–13.

Qadri, H., Qureshi, M.F., Mir, M.A., Shah, A.H., 2021b. Glucose-The X Factor for the Survival of Human Fungal Pathogens and Disease Progression in the Host. Microbiological Research, p. 126725.

Rachini, A., Pietrella, D., Lupo, P., Torosantucci, A., Chiani, P., Bromuro, C., Proietti, C., Bistoni, F., Cassone, A., Vecchiarelli, A., 2007. An anti-β-glucan monoclonal antibody inhibits growth and capsule formation of *Cryptococcus neoformans* in vitro and exerts therapeutic, anticryptococcal activity in vivo. Infect. Immun. 75, 5085–5094.

Rhodes, J., 2019. Rapid worldwide emergence of pathogenic fungi. Cell Host Microbe 26, 12–14.

Rigby, A.L., Glanville, A.R., 2012. Miliary pulmonary cryptococcosis in an HIV-positive patient. Am. J. Respir. Crit. Care Med. 186, 200–201.

Roilides, E., Lamaignere, C.G., Farmaki, E., 2002. Cytokines in immunodeficient patients with invasive fungal infections: an emerging therapy. Int. J. Infect. Dis. 6, 154–163.

Rosen, L.B., Freeman, A.F., Yang, L.M., Jutivorakool, K., Olivier, K.N., Angkasekwinai, N., Suputtamongkol, Y., Bennett, J.E., Pyrgos, V., Williamson, P.R., 2013. Anti−GM-CSF autoantibodies in patients with cryptococcal meningitis. J. Immunol. 190, 3959−3966.

Sanglard, D., Ischer, F., Parkinson, T., Falconer, D., Bille, J., 2003. *Candida albicans* mutations in the ergosterol biosynthetic pathway and resistance to several antifungal agents. Antimicrob. Agents Chemother. 47, 2404−2412.

Segal, B.H., Kwon-Chung, J., Walsh, T.J., Klein, B.S., Battiwalla, M., Almyroudis, N.G., Holland, S.M., Romani, L., 2006. Immunotherapy for fungal infections. Clin. Infect. Dis. 42, 507−515.

Setianingrum, F., Rautemaa-Richardson, R., Denning, D.W., 2019. Pulmonary cryptococcosis: a review of pathobiology and clinical aspects. Med. Mycol. 57, 133−150.

Shaw, K.J., Ibrahim, A.S., 2020. Fosmanogepix: a review of the first-in-class broad spectrum agent for the treatment of invasive fungal infections. J. Fungi 6, 239.

Shorr, A.F., Chung, K., Jackson, W.L., Waterman, P.E., Kollef, M.H., 2005. Fluconazole prophylaxis in critically ill surgical patients: a meta-analysis. Crit. Care Med. 33, 1928−1935.

Stuehler, C., Khanna, N., Bozza, S., Zelante, T., Moretti, S., Kruhm, M., Lurati, S., Conrad, B., Worschech, E., Stevanović, S., 2011. Cross-protective TH1 immunity against *Aspergillus fumigatus* and *Candida albicans*. Blood 117, 5881−5891.

Tan, S., Li, D., Zhu, X., 2020. Cancer immunotherapy: pros, cons and beyond. Biomed. Pharmacother. 124, 109821.

Tarang, S., Kesherwani, V., Latendresse, B., Lindgren, L., Rocha-Sanchez, S.M., Weston, M.D., 2020. In silico design of a multivalent vaccine against *Candida albicans*. Sci. Rep. 10, 1−7.

Thullen, T.D., Ashbaugh, A.D., Daly, K.R., Linke, M.J., Steele, P.E., Walzer, P.D., 2003. New rat model of Pneumocystis pneumonia induced by anti-$CD4^+$ T-lymphocyte antibodies. Infect. Immun. 71, 6292−6297.

Torosantucci, A., Bromuro, C., Chiani, P., DE Bernardis, F., Berti, F., Galli, C., Norelli, F., Bellucci, C., Polonelli, L., Costantino, P., 2005. A novel glyco-conjugate vaccine against fungal pathogens. J. Exp. Med. 202, 597−606.

Tugume, L., Rhein, J., Hullsiek, K.H., Mpoza, E., Kiggundu, R., Ssebambulidde, K., Schutz, C., Taseera, K., Williams, D.A., Abassi, M., 2019. HIV-associated cryptococcal meningitis occurring at relatively higher CD4 counts. J. Infect. Dis. 219, 877−883.

Ueno, K., Kinjo, Y., Okubo, Y., Aki, K., Urai, M., Kaneko, Y., Shimizu, K., Wang, D.-N., Okawara, A., Nara, T., 2015. Dendritic cell-based immunization ameliorates pulmonary infection with highly virulent *Cryptococcus gattii*. Infect. Immun. 83, 1577−1586.

Valedkarimi, Z., Nasiri, H., Aghebati-Maleki, L., Majidi, J., 2017. Antibody-cytokine fusion proteins for improving efficacy and safety of cancer therapy. Biomed. Pharmacother. 95, 731−742.

Van Daele, R., Spriet, I., Wauters, J., Maertens, J., Mercier, T., Van Hecke, S., Brüggemann, R., 2019. Antifungal drugs: what brings the future? Med. Mycol. 57, S328−S343.

Vandeputte, P., Ferrari, S., Coste, A.T., 2012. Antifungal resistance and new strategies to control fungal infections. Int. J. Microbiol.

Viejo-Díaz, M., Andrés, M.T., Fierro, J.F., 2005. Different anti-Candida activities of two human lactoferrin-derived peptides, Lfpep and kaliocin-1. Antimicrob. Agents Chemother. 49, 2583−2588.

Vila, T., Sultan, A.S., Montelongo-Jauregui, D., Jabra-Rizk, M.A., 2020. Oral candidiasis: a disease of opportunity. J. Fungi 6, 15.

Williams, T.J., Harvey, S., Armstrong-James, D., 2020. Immunotherapeutic approaches for fungal infections. Curr. Opin. Microbiol. 58, 130−137.

Xu, H., Fang, T., Omran, R.P., Whiteway, M., Jiang, L., 2020. RNA sequencing reveals an additional Crz1-binding motif in promoters of its target genes in the human fungal pathogen *Candida albicans*. Cell Commun. Signal. 18, 1−14.

Zhou, Q., Gault, R.A., Kozel, T.R., Murphy, W.J., 2006. Immunomodulation with CD40 stimulation and interleukin-2 protects mice from disseminated cryptococcosis. Infect. Immun. 74, 2161−2168.

CHAPTER

8

Combinatorial approach to combat drug resistance in human pathogenic bacteria

Manzoor Ahmad Mir[1], *Manoj Kumawat*[2], *Bilkees Nabi*[3] and *Manoj Kumar*[2]

[1]Department of Bioresources, School of Biological Sciences, University of Kashmir, Srinagar, Jammu and Kashmir, India; [2]Department of Microbiology, ICMR-NIREH, Bhopal, Madhya Pradesh, India; [3]Department of Biochemistry and Biochemical Engineering, Sam Higginbottom University of Agriculture, Technology and Sciences, Prayagraj, Uttar Pradesh, India

8.1 Introduction

Pathogenic bacteria found in food pose a significant threat to human health. The contaminated food and water that millions of people consume every year cause diarrhea. Antibiotics are of vital medicinal value which can be customary organic compounds; relatively new synthetic organic agents, natural products, peptides, and proteins, are widely used as antimicrobial agents to treat, control, or prevent diseases caused by microorganisms (*Campylobacter* spp., *Escherichia coli, Listeria* spp., and *Salmonella*), but their use to large extent can result in antimicrobial-resistant bacteria (Balta et al., 2020). Antibiotic use in animals promotes the formation of a wide spectrum of resistant zoonotic infections, such as *Salmonella*, which limits the performance of antibiotic therapies used in humans when contracting infection, according to mounting evidence. At any point during processing or storage, cross-contamination can introduce bacteria that are resistant to antimicrobials. Antibiotics can be used in a variety of settings, including treating diseases and promoting growth in livestock, leading to the spread of antibiotic-resistant bacteria. Antibiotic resistance has been identified as one of the most critical challenges confronting our national and global health systems, according to the Centers for Disease Control (CDC) and Prevention and the World Health Organization (WHO).

Human Pathogenic Microbes
https://doi.org/10.1016/B978-0-323-96127-1.00014-0

Between 2000 and 2015, global antibiotic usage climbed by 65%, with the rate of consumption increasing by 39%, from 113 defined daily doses (DDDs) per 1000 people to 157 DDDs per 1000 people (Laxminarayan et al., 2016). Population growth could result in the total consumption of goods and services increasing even if per capita consumption remains constant. As the human population grows slowly compared to antibiotic usage, both per capita and total consumption should be considered for resistance monitoring. Global usage of broadspectrum antibiotics (such as glycyl cyclines, oxazolidinones, carbapenems, and polymyxins) increased from 2000 to 2015, thus raising concerns about antibiotic resistance in last-resort antibiotics. The worldwide increase of foodborne infections associated with antibiotic-resistant pathogens and the spread of antibiotic resistance is one of major concerns in developed as well as developing countries. Due to easy availability of drugs and insufficient health services in developing countries like India, there is an increase in the number of self-medications as opposed to prescriptions, leading to impending health problems and antimicrobial resistance.

Salmonella and *Campylobacter* resistant to fluoroquinolones were found to survive in food animals during the mid-1990s (Aarestrup and Wegener, 1999). Every year, *Salmonella* causes a third of all foodborne illness outbreaks in humans (Ramsay et al., 2002). *Salmonella* bacteria are isolated more frequently from poultry and their products than any other type of food animal. Contaminated poultry products are among the most important causes of outbreaks of foodborne illness in humans (Braden, 2006; Yohannes et al., 2014). To make the search for new antibiotic agents more efficient, there are still knowledge gaps regarding the chemical and microbiological mechanisms involved. In addition to improving the application of antimicrobial active films and coatings to extend the shelf life of products, it is crucial to improve the technical implementation of antimicrobial.

8.2 Food-borne bacterial infections

More than 250 foodborne diseases are known till date, which occur by a number of bacteria, viruses, and parasites. The foodborne illnesses are caused by the microorganisms in contaminated foods, by the harmful toxins and chemicals secreted by the microorganisms. The most significant zoonotic pathogens responsible for foodborne diseases in humans include *Campylobacter, Salmonella, Staphylococcus, Escherichia coli, Clostridium,* and *Listeria* (Jansen et al., 2019). Foodborne infections cause 48 million people to become unwell each year, 128,000 patients to be hospitalized, and 3000 people to die, according to the CDC (Bondi et al., 2017). Raw foods are more likely than others to cause foodborne diseases including food poisoning. They have high chance to carry harmful germs that can make the food contaminated. Among raw foods of animal origin, raw meat and dairy products are most susceptible to contamination. Undercooked poultry, raw eggs, unpasteurized milk, and raw shellfish are all foods that should be avoided. In addition, fruits and vegetables can be contaminated by water contaminated with pesticides, during processing and during other stages in the food production chain (González-Rodríguez et al., 2011). Additionally, raw meat and poultry products in kitchens may be contaminated by cross-contamination with other food items.

8.2.1 Campylobacter

Theodor Escherich discovered and described nonculturable spiral-shaped bacteria in 1886, and campylobacteriosis was the first illness caused by eating food or drinking water contaminated with the bacteria *Campylobacter jejuni.* Each year, 400,000,000 cases are reported, 13,000 hospitalizations result, and 120 deaths occur because of them. Another 310,000 cases of infection are due to a drug-resistant organisms (Centers for Disease and Prevention, 2013b), resulting in 15% of hospitalization and 6% fatalities due to foodborne disease, estimated due to *Campylobacter* infection. Animal reservoirs, including birds, are mainly known to be carriers of *Campylobacter* organisms, which are zoonotic pathogens that can spread to humans. As a zoonotic disease, campylobacteriosis is important both clinically and economically. Humans are susceptible to campylobacteriosis due to a wide range of risk factors. As a risk factor, traveling is the most important factor. Undercooked chicken, exposure to the environment, and contact with farm animals are also significant factors. There are no known mortality rates for *campylobacter* outbreaks, but secondary complications can occur: costs related to treating *Campylobacter* infections, loss of productivity due to the infection, and costs related to controlling the pathogen account for the majority of the overall economic losses (Neimann et al., 2003). Raw and insufficiently cooked animal items, as well as nonchlorinated water, are among the most common sources of *Campylobacter* infection. Humans can also become infected by eating contaminated seafood and meat, as well as drinking unclean water (Lanzas et al., 2020). Pasteurizing milk; preventing postpasteurization contamination; cooking raw meat, poultry, and fish; and preventing cross-contamination between raw and cooked food can all aid in the prevention of *Campylobacter* infections.

8.2.2 Salmonella

Salmonellosis is among the most prevalent pathogenic illnesses transmitted through the food chain. Typhoidal salmonellosis is caused by the host-restricted serotype *S. typhi* in humans, whereas nontyphoidal salmonellosis is caused by other *Salmonella* serotypes. In many parts of the world, including in India, *S. typhimurium* is the most common nontyphoidal *Salmonella* isolated from food poisoning cases (Bakshi et al., 2003; Helke et al., 2017). *S. typhimurium* and *S. enteritidis* are the most frequently isolated serovar from cases of human food poisoning associated with chicken products (Dias de Oliveira et al., 2005). There are over 2300 serotypes in the *Salmonella* family, but two of them, *Salmonella enteritidis* and *Salmonella typhimurium*, are the most widespread in the world and are devastating microbes in humans as well as in animals. The *Salmonella* family includes Gram-negative bacteria that are relatively anaerobic, do not produce spores, and grow as straight rods. More than 150 varieties of *Salmonella* bacteria cause foodborne salmonellosis, which is spread by contact with the bowel contents or feces of animals, including humans (Velge et al., 2005). The bacteria may continue to live and proliferate in the intestine after being ingested, causing an infection and disease. *Salmonella* species that are not typhoidal are zoonotic, and dietary products derived from animals are the primary means of transmission in humans. The most common food items linked to human salmonellosis outbreaks are chickens, pigs, and cattle, as well as their byproducts such as eggs, milk, and meat (Kumawat, 2016). *Salmonella* is widely recognized to reside in poultry products, especially eggs, which are a major source of human cases of salmonellosis.

Foodborne diseases like salmonellosis are most common in both poor and industrialized nations, though the incidence varies by location (Silva et al., 2014). Eggs and egg products, poultry, meats, and meat products are the primary vehicles of transmission, according to WHO. *S. typhimurium* and *S. enteritidis* can cause gastroenteritis as well as other systemic disorders. *S. typhimurium* is the most frequent pathogen usually found in all major food animal species, as well as the leading cause of salmonellosis in people around world (Centers for Disease and Prevention, 2013a). There were 93.8 million reported foodborne *Salmonella* cases worldwide in 2010 and 0.15 million deaths were attributed to the disease (Majowicz et al., 2010). Food derived from animals, especially meat, eggs, and poultry products, is the most common way for *Salmonella* to spread to humans (Kumar et al., 2013). Diarrhea, fever, and stomach pains are symptoms of salmonellosis, which emerge 12–72 h after infection. Substantial outbreaks of salmonellosis connected to diet have been attributed to poultry meat and eggs. Despite the presence of infection in these animals, poultry and pigs do not show any clinical symptoms of illness (Leach et al., 1999). Poultry meat is more popular in India due to benefits like ease of digestion and widespread acceptance (Yashoda et al., 2001). Poultry is a prime source for *Salmonella* and a leading source of human sickness. *Salmonella* control is a source of concern for health officials due to their link to a number of clinical disorders, including typhoidal and nontyphoidal ailments. It has been found that the majority of outbreaks are caused by raw poultry or poultry that has been undercooked. Infections with *Salmonella* cause a number of acute and chronic diseases in chicken. These diseases have been linked to major financial losses for poultry producers, prompting the implementation of comprehensive testing and control programs. Antibiotic-resistant human and animal *Salmonella* infections are a worldwide issue, especially in impoverished nations. It is due to the rise and spread of antimicrobial-resistant (AMR) strains, the *Salmonella* enteric which is the causative agent of salmonellosis is becoming a rising public health concern. Antibiotic-resistant human and animal *Salmonella* infections are a worldwide issue, especially in impoverished nations (Nair et al., 2018). *S. typhimurium* DT104, a multidrug-resistant definitive variety that has recently made its presence, is primarily spread through the eating of contaminated beef (Ngwai et al., 2006).

8.2.3 Staphylococcus

Staphylococcus bacteria often are found on the skin, in the nose, and in the throat of people who have colds or sinus infections. As one of the most frequent foodborne pathogens, *Staphylococci* are commonly detected in infected wounds, boils, pimples, and acne (Su et al., 2005). Besides being prevalent in raw milk, untreated water, and sewage, *Staphylococci* can be the reason for a wide range of diseases in people and animals. Healthy people are thought to carry the microbes in their noses, throats, and on body surfaces, and naturally infected cows are a major source of *S. aureus* (Leonard and Markey, 2008).

Staphylococcus is a common source of food poisoning, which is caused by eating contaminated food containing staphylococcal enterotoxins. Food deterioration, reduced shelf life, and food poisoning are all caused by the toxins (Thomsen, 2018). It has received widespread attention as a result of rising mortality linked to *Staphylococcus* multidrug resistance. Food can become contaminated with *Staphylococcus* when it is produced by animals that have been infected or when it is manufactured or stored in an unsanitary manner. Unless the food is stored at room temperature for prolonged time, the organism present in food can develop the toxin (Schwan, 2019). While proliferating in food, the bacteria generate enterotoxin. This enterotoxin stimulates the vomiting center of the nervous system and inhibits the absorption of sodium and water by the small intestine. Abdominal pains, nausea, and vomiting may occur in people who are more vulnerable, such as children and the elderly. MRSA (Methicillin-resistant *S. aureus*) was produced as a result of extensive and excessive antibiotic use, as well as bacteria's ability to establish and acquire antimicrobial resistance quickly. MRSA variants are now recognized as a prominent inpatient infection worldwide, according to records dating back to the early 1960s. MRSA is a well familiar pathogen in both human as well veterinary medicines (Smith et al., 2013). Multidrug-resistant *Staphylococcus aureus* (MRSA) strains are frequently associated with nosocomial infections that can lead to life-threatening diseases like necrotizing fasciitis, endocarditis, osteomyelitis, and toxic shock syndrome (Rantala, 2014; Qadri et al., 2021).

8.2.4 Escherichia coli

Escherichia coli (*E. coli*) are bacteria that are often detected in the environment, foods, and the gut of humans and animals. *E. coli* is a vast and diversified genus of bacteria that can be used to assess contamination from manure, soil, and contaminated water. According to research, while most strains of *E. coli* are safe, others can cause human illness (Khan and Steiner, 2002). Food products derived from cows, such as milk, milk products, tainted beef, and ground beef, are clearly linked with outbreaks of bovine diseases. Humans and animals can get diarrhea only in certain strains of *E. coli*, while others can get respiratory problems, pneumonia, and urinary tract infections (Mazumdar et al., 2006).

A major outbreak of bloody diarrhea in the United States linked with undercooked hamburgers in 1982 first identified *E. coli* as a serious foodborne pathogen. *E. coli* O157:H7 is the most distinguished serotype for containing pathogens which infect humans through food (De la Pomelie et al., 2021). Throughout the world, outbreaks of this strain have been reported. Some strains of *E. coli* produce shiga toxin, which is an important foodborne and zoonotic pathogen. Colonic epithelial cells are killed by toxins and secreted. In animals, the toxin induces regional fluid accumulation and a colonic abscess characterized by sloughing of the surface and crypt epithelial cells, which can lead to serious life-threatening illnesses such as bacteremia, meningitis, meningoencephalitis, pregnancy loss, stillbirth, or fetal infection in high-risk groups (Kourtis et al., 2020).

8.2.5 Clostridium

The *Clostridium* infection is also among the most serious healthcare-associated diseases, and community-associated infections that have emerged in the recent years. *Clostridium perfringens* is a genus of bacteria related to the botulinum organism. The bacterium *Clostridium difficile* produces spores that cause diarrhea in both humans and animals (Wilcox and Fawley, 2000). It is an anaerobic bacterium that is transmitted through the release of feces and oral feces. People who have recently obtained antibiotic treatment or medical attention are most likely to contract an infection (Halabi et al., 2013). It has not been shown that *C. difficile* infection in humans spread directly from animals, food, or environment.

C. difficile has been detected in a number of foods including raw meat, unpasteurized milk, uncooked vegetables, and seafood. Several investigations support the idea that the spore-contaminated foods might be the reason of *C. difficile* exposure and transmission (Romano et al., 2018). The *Clostridium* bacterium is the most common cause of healthcare-associated diseases in the United States, with an estimation of approximately 500,000 illness and 29,000 fatalities per year (Turner et al., 2019). The asymptomatic animals which shed the bacterium into the environment and also transport, infecting people or communities either directly or through the food chain are increasing and their frequency is a serious public health issue. Besides this the bacterium's invasion of healthy sheep and goats increases both animal-to-animal and zoonotic transmission.

8.2.6 Listeria

Listeria monocytogenes is naturally occurring foodborne microbe that is a prominent source of sickness in both animals and people. It's a significant foodborne zoonotic pathogen that can cause substantial public health problems if you eat infected animal products (Marrakchi et al., 1993). Since 1980, when several outbreaks and rare instances of listeriosis have been linked to the eating of contaminated food, *Listeria* has emerged as a foodborne disease. The most common pathogen of animals and humans, *Listeria*, is a bacterium able to thrive in diverse environments including those with varying pH levels, temperature changes, and salt concentrations. As a result, *Listeria* is a major foodborne pathogen (Tkacikova et al., 2000).

Despite the fact that raw milk is pasteurized to kill *Listeria*, the practice does not totally remove the risk of infection in dairy products. Humans can contract *Listeria* from animal sources as an occupational hazard, which is particularly dangerous for farmers, butchers, poultry workers, and veterinarians (Chahad et al., 2012). The most vulnerable to infection include pregnant ladies, newborns, the aged, and those with weakened immune systems (Mir et al., n.d.).

8.3 Antibiotic treatment and their efficiency analysis

Each year, contaminated food products cause approximately 600 million cases of sickness and 420,000 fatalities as a result of microbiological contamination (Chahad et al., 2012). It is estimated that foodborne transmission is involved in approximately 15% of newly emerging infectious diseases (EIDs) (Nesbitt et al., 2012). Due to the interconnection of global food systems, the spread of antibiotic-resistant foodborne diseases is increasing. Antibiotic-resistant bacteria are becoming a serious public health issue, and there is a need for effective detection methods for identifying them and preventing them from spreading throughout our food systems and the environment (Witte et al., 2008). Regardless of the fact that many different compounds with antimicrobial capabilities have been found, there has recently been a surge in interest in the usage of potential antimicrobial food preservatives. Natural antimicrobials are typically thought to be better than their synthetic counterparts, which are more commonly accepted by consumers and may be effective against infections resistant to a variety of synthetic antibiotics (Pisoschi et al., 2018).

Human medicine is plagued with the problem of inappropriate antibiotic use. Inappropriate antibiotic dosages, over-the-counter accessibility, poor patient compliance with recommended treatments, usage of substandard medications, and self-medication with previously unused antibiotics—all contribute to the spread of antibiotic resistance (Lushniak, 2014). Antibiotics' safety for human and veterinary health has been questioned. Antibiotics disrupt human and animal health in addition to killing germs. Antibiotics were licensed for human use after extensive clinical study demonstrated their efficacy and safety. However, from the standpoint of bacteria, they pose a health danger to people (Muller et al., 2015).

Fluoroquinolones are antibiotics that are prescribed by doctors all over the world, since they are generally safe for most people. Antibiotics' well-known side effects have prompted health professionals and regulatory agencies to reconsider their usage in biomedical and veterinary applications. Fluoroquinolone side effects, such as muscle and tendon injuries, neuropsychiatric problems, and mitochondrial toxicity, are convincing instances of what antibiotics do to humans other than cure infections.

Because microorganisms might develop resistance to novel antimicrobial agents, ongoing research to test more novel antimicrobial agents that can aid the agriculture and food industries in combating antimicrobial resistance will be required. As previously said, excessive use of medications, particularly antibiotics, is the leading cause of drug resistance today. As a result, several research studies have explored the use of natural and effective antimicrobial agents as complementary therapies strategy (Kumar et al., 2021). The efficacy of these chemicals is governed by their genotypes, agronomic properties, chemical composition, and environmental conditions. Combining essential oils with regular antibiotics, on the other hand, can result in distinct inhibitory processes in resistant bacteria, and this could be an alternative option for combating microbial resistance (Owen and Laird, 2018). Nanocarriers, on the other hand, play a functional role to improve medicine intake by boosting solubility, limiting drug leakage, controlling intracellular penetration, and so on. Identifying the evolution of resistance mechanisms and understanding pathways by which these strategies propagate across bacteria could contribute to a better understanding of molecular level of antimicrobial processes (Allahverdiyev et al., 2011; Mir et al., 2021). Figs. 8.1 and 8.2 show how antibiotic resistance happens with the cells (Table 8.1).

8.4 Antibiotic resistance and antimicrobial resistance shown by *Salmonella*

Pathogenic bacteria's resistance to a variety of antimicrobial treatments has arisen as a significant global issue. *Salmonella,* a Gram-negative bacillus is found in pigs, cattle, and poultry, among other animals (Wigley, 2004). Salmonella infections generate significant patient mortality and morbidity, despite the major medical improvements made possible by antibiotics over the previous 50 years, and rising antibiotic resistance poses a serious threat. The methods of food productions including transport, storage, handling, preparation are all taken into consideration in preventive and control measures for foodborne illness, besides ensuring good health and safe food for human use (Kamboj et al., 2020). Resistant foodborne infections are by far the most serious public health hazards associated to the possibility of antimicrobial resistance growing in the food supply chain. The CDC have identified four priority action areas for the *Salmonella* outbreak: governance, infection prevention,

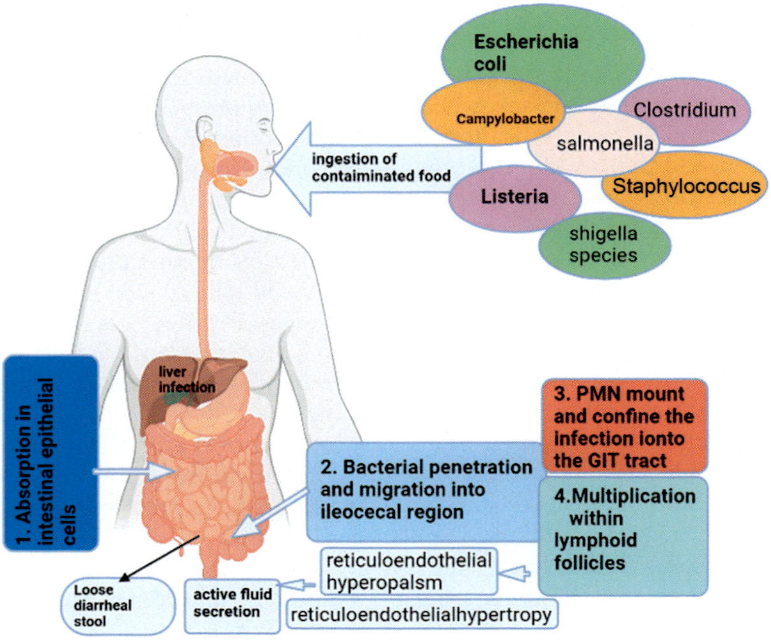

FIGURE 8.1 Diagrammatic representation of different types of bacteria-contaminated food, which is ingested by humans and shows the consequence of the bacterial contaminated food.

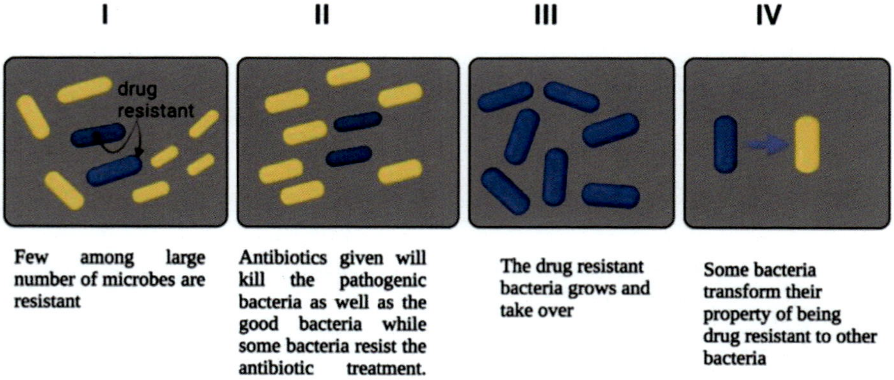

FIGURE 8.2 Represents the development of antibiotic resistance among microbes. In I-cell few microbes are resistant, II-cell shows that on introducing the antibiotics both pathogenic and good bacteria are killed except resistant bacteria which withstand the antibiotics, in III-cell the resistant bacteria are free to show growth even in presence of antibiotics, and IV-cell shows resistant bacteria transform the normal bacteria into resistant bacteria.

resistance mapping, and the development of novel treatments and diagnostic testing (Newell et al., 2010). *Salmonella* is one of the foodborne infections that poses a significant risk to human health and has a global distribution. *Salmonella* can pass through the stomach's defense and reach into the intestines, where they adhere to nonphagocytic cells such the epithelial cells of the small intestine. Local damage without septicemia is a hallmark of

TABLE 8.1 A few foodborne pathogens, disease, antimicrobials.

S.No.	Bacterial pathogen	Disease	Antimicrobial	Resistance mechanism
1.	*Mycobacterium tuberculosis*	Tuberculosis	Fluoroquinolones	It works by modifying enzymes and imitating the target.
2.	*Streptococcus*	Pneumococcal meningitis	Penicillin	Shows the genetic alterations in the penicillin-binding protein
3.	*Vibrio cholerae*	Severe watery diarrhea	Sulfonamides	Chromosomal mutations in encoding dihydropteroate synthase
4.	*Shigella dysenteriae*	Severe diarrhea	Chloramphenicol, tetracycline	Results in decreased permeability, efflux
5.	*Salmonella typhi*	Typhoid	Chloramphenicol	The enzyme chloramphenicol acetyltransferase is produced, and active efflux is observed
6.	*Campylobacter jejuni*	Gastrointestinal illnesses	Tetracycline	Shows target protection, alters the ribosomal conformation
7.	*Streptococcus* spp.	Sore throat, scarlet fever	Tetracycline	Shows target protection, and also change in ribosomal conformation
8.	*Enterobacter*	Sore throat, scarlet fever	B-lactam antibiotics	Reduces cell membrane permeability
9.	*Enterococci*	Soft tissue infections	Vancomycin	Antibiotic target are circumvented
10.	*Staphylococcus aureus*	Vomiting, diarrhea, dehydration	Methicillin	It has a link to the mecA gene and decreases the affinity of all b-lactams

Salmonella infection in Peyer's patches' microfold cell, which is aided by fimbrial adhesions. Ruffling the target cell membrane causes bacteria to internalize into the membrane-bound vacuoles. Ruffles also assist bacterial reception in membrane-bound vacuoles or vesicles, which tend to congeal. The most common gastrointestinal symptoms for this infection are headache; vomiting; abdominal discomfort; and watery, greenish, foul-smelling diarrhea or bloody diarrhea with mucus (Mushin, 1948).

A variety of *Salmonella* spp. have been reported to exhibit multidrug resistance (Olsen et al., 2004). In 2012, 3% of nontyphoid *Salmonella* isolates tested and evaluated by the National Antimicrobial Resistance Monitoring System (NARMS) were showing resistance to nalidixic acid, a quinolone used to screen for ciprofloxacin resistance. In addition, approximately 50% of *Salmonella enteritidis* strains, the most common *Salmonella* serotype, were resistant to nalidixic acid (Karp et al., 2017). *Salmonella* isolates resistant to at least ampicillin, chloramphenicol, streptomycin, sulfonamides, and tetracycline were found in 4% of nontyphoid *Salmonella* isolates in the NARMS 2012 human isolates final report (Voss-Rech et al., 2017). Multiple drug-resistant *S. typhimurium* cases are associated with invasive infections, longer and more frequent hospitalizations, and a higher mortality rate.

Salmonella that are nontyphoidal are mainly transmitted through the utilization of animal-derived foods. These include poultry, eggs and their products, pork, and cattle, and are frequently identified as food sources. Nontyphoidal *Salmonella* has over 150 serotypes and is most common zoonotic bacterial foodborne disease causing agent in humans. The *Salmonellae* are found in both people and animals and are widely spread in nature. Every year, 100,000 cases of drug-resistant diseases are reported. Further nontyphoidal *Salmonella* are responsible for roughly 155,000 deaths worldwide, with an estimated 94 million infections (Majowicz et al., 2010). Foodborne sickness and the spread of antimicrobial-resistant strains of *Salmonella* caused by this pathogen are serious public health problems for humans and animals alike, although incidence rates differ by country. Poultry products, notably eggs, are commonly regarded as a substantial *Salmonella* reservoir and have been linked to occasional instances and outbreaks of human salmonellosis. Salmonella infection in the country could be caused by the presence of carrier farm animals, unlawful animal slaughter in open areas, unsanitary techniques used for slaughtering in abattoirs, and eating uncooked meat by using prevalent traditional methods. *S. typhimurium* DT104, a multidrug-resistant definitive variety that is recently seen is primarily spread through the eating of contaminated beef. *Salmonella* infections with antibiotic resistance occurring in humans and animals are a worldwide concern, especially in the developing countries which face high risk of infection due to unsanitary living styles, close contact and sharing of homes of humans with animals, and consumption of raw or undercooked animal-based foods (Fig. 8.3).

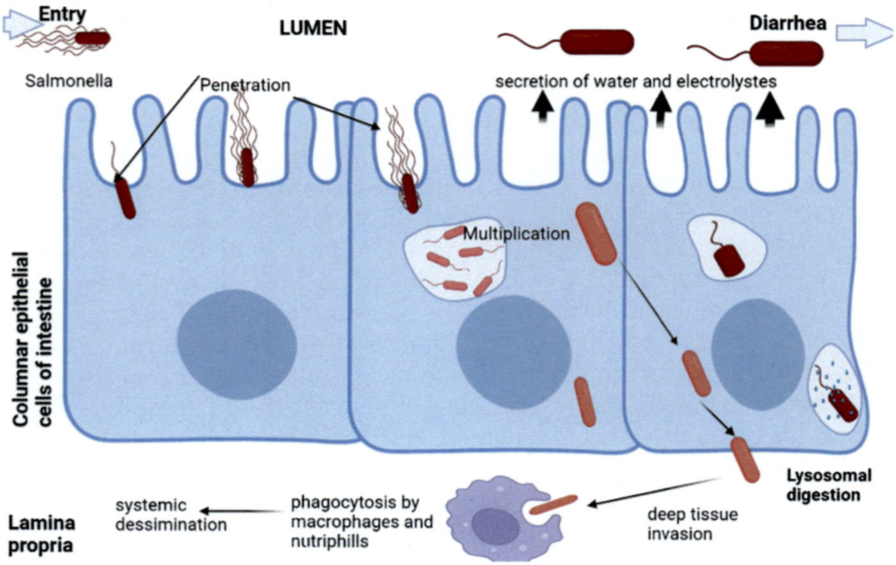

FIGURE 8.3 Showing the internalization of *Salmonella* and multiplication within the cells and invasion from lumen to tissue through the columnar epithelial cells of intestine which results in secretion of water and electrolytes and ultimately causes diarrhea as one of the symptoms.

8.5 Future approach to fight salmonellosis

The use of antimicrobial drugs for growth promotion and disease prevention in food animals is linked to the global health crisis of antibiotic resistance. With new scientific and commercial tools, the investigation of antibiotic-resistant foodborne bacteria should continue. *S. Typhimurium* is well-known as one among most common causes of human and animal food toxicity and NTS illness. Immunizations have been considered as a potential solution to the emerging antimicrobial resistance crisis by health authorities, clinicians, and drug developers. The vaccines much promise for the eventual control of *Salmonella* disease; vaccines can also provide cross-protective immunity against other serovars of *Salmonella* (Baliban et al., 2020). New recombinant vaccine technology has played a key role in lowering direct and indirect *Salmonella* infections, which would have otherwise demanded the use of antibiotics. Antibiotic-active natural compounds (penicillin, cephalosporins, vancomycin, tetracycline, and aminoglycosides) as well as synthetic antimicrobials can be used to formulate new antibiotics (Demain, 2009).

Another option is to utilize immunotherapeutics, which enhance the immune system of the patient and provide tolerance against infectious *Salmonella* (Pangilinan and Lee, 2019). More research into the molecular biology of bacterial cell death, including genetic regulatory programs and biochemical mechanisms, is desperately needed. Antibiotic-induced cell death in bacteria is a complicated process involving numerous genetic and metabolic pathways (Kohanski et al., 2010). Therefore, high-throughput approaches for cell-based screening must be developed, expanding beyond growth inhibition to whole-genome sequencing of resistant organisms. New natural compounds and biochemical pathways have been discovered, thanks to the field of metagenomics. Many foodborne infections and antibiotic resistance challenges require additional information, data, or awareness, and a One Health strategy will be advantageous in these cases (World Health Organization, 2017). Improved surveillance, standardized methodologies, and data sharing are all critical. Monitoring shared human and animal environments necessitates effort.

8.6 Combinatorial approach in tuberculosis

Tuberculosis (TB) is caused by bacteria *Mycobacterium tuberculosis* (*M. tuberculosis*), which is the largest cause of death from infectious illness globally, resulting in the deaths of 1.5 million people each year. The emergence and dissemination of multidrug-resistant *Mycobacterium* TB strains is one of the most serious public health challenges (Merker et al., 2015; Sheikh et al., 2021). As a result, developing more efficient and effective antimycobacterial medicines for the treatment of *Mycobacterium* has been a challenge. *M. tuberculosis* is a facultative intracellular bacterium that may survive and reproduce inside phagocytes like macrophages by hijacking their effector activities. The lungs are the central area of *M. tuberculosis* infection, which results in symptoms including fever, night sweats, weight loss, and sputum and/or blood coughing. *M. tuberculosis* can live for a long time in the host, increasing the chance of reactivation, even decades after infection (Koegelenberg et al., 2021).

The WHO designated tuberculosis a Global Emergency in 1994, and it is now the biggest cause of mortality, with an estimated 1.5 million deaths globally in 2015. A combination of four antimicrobial drugs must be used over the period of 6 months to treat tuberculosis. Efforts to discover innovative therapies for Mtb strains that are multidrug resistant (MDR) and also those showing extensive drug resistance (XDR) have been considerable over the last decade (Kinnings et al., 2009). The first combination medicine therapy for TB treatment was born after years of exploration in two separate but concurrent areas of inquiry. The first antibiotic to successfully treat TB was penicillin (streptomycin). Importantly, the discoveries that led to the first combination therapy and subsequent tuberculosis treatment success were made across a broad spectrum of science, from basic soil microbiology to rigorous clinical trials, demonstrating the importance of investing in all aspects of science to advance human disease treatment. Despite considerable advances in the TB drug discovery pipeline over the last decade, there is still a continuing demand for better therapeutic interventions (Kinnings et al., 2009; Mir and Al-Baradie, 2013). It's worth reflecting on the lessons learned during decades of fundamental drug design, preclinical research, and clinical testing that led to the first TB prevention triumphs. As we go forward in the future years to reduce the global incidence of TB and certain other ailments that need combination therapy, continued support is needed (Raviglione and Sulis, 2016).

8.7 Therapeutics and their efficiency analysis

Tuberculosis is an infectious disease spread through the air that is treated with a combination of drugs. Persistence to long-term antituberculosis medication is required to maintain an adequate blood drug level. It is mainly due to poor medical management of patients that drug-resistant *M. tuberculosis* strains emerge and spread (Migliori et al., 2009). *M. tuberculosis* strains which are resistant to regular anti-TB therapy are emerging in almost every country that reports to the WHO. Treatment with antibiotics, which reduces the bacterial load in the lungs, as well as taking other public health measures like isolation and cough etiquette, can assist to lower the likelihood of transmission. Besides, due to the poor, expensive, inefficient, and toxic alternatives to first-line medications, treating drug-resistant tuberculosis is challenging (Tiberi et al., 2019).

The earliest experimental indication of the potential efficacy of novel antituberculosis medications came in 1940, when a sulfonamide dapsone-derivative chemical known as promin was given to guinea pigs for the first time, but it was never given to humans. Because of its bactericidal properties, Schatz and Waksman declared in 1944 that the medication might be given for tuberculosis treatment. Ethambutol and rifampicin were first used in 1961 and 1963, respectively, with treatments lasting 1 to 2 years. In 1970, experiments on rifampicin-containing regimens yielded positive results after 9 months of treatment, while in 1974, the addition of rifampicin and pyrazinamide at low doses established the usefulness of the combination of a 6-month treatment. The complete historical events for antituberculosis agents are shown in Fig. 8.2. Even while the newly available drugs' adherence, efficacy, safety, and tolerability profiles (especially delamanid and bedaquiline) are encouraging, still we cannot predict their long-term efficacy or affordability in resource-limited settings at this time (Schaberg, 1995) (Fig. 8.4).

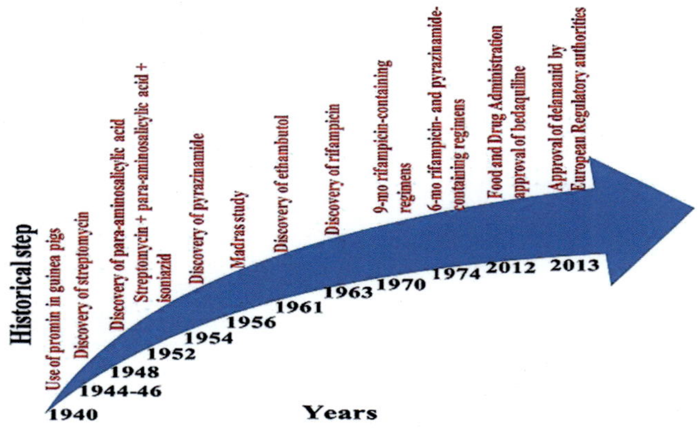

FIGURE 8.4 Represents the year-wise introduction of antituberculosis drugs.

8.8 Advanced therapeutics for tuberculosis

Fixed-dose combination (FDC) formulations are historical antituberculosis therapeutic phases that are now prescribed for the treatment of active TB. FDC tablets, each containing two or more anti-TB medications, have been designed since the 1980s to simplify TB therapy and enable physician and patient compliance with treatment protocols due to the massive number of pills used in TB treatment regimens. The daily recommended dosage of different antituberculosis drugs (mg/kg) is shown in Fig. 8.5. It is necessary to enhance the existing therapeutic treatment of drug-susceptible and drug-resistant strains. The existing regimens have a high pill load, a long duration, and effectiveness, safety, and durability that vary (Tang et al., 2015). The overall treatment success rate falls short of the WHO's recommended 85%, resulting in an increase in medication resistance. Research efforts need to be pursued in order to determine the potential of the new drugs and to learn how to combine them more effectively (Stoll et al., 2002).

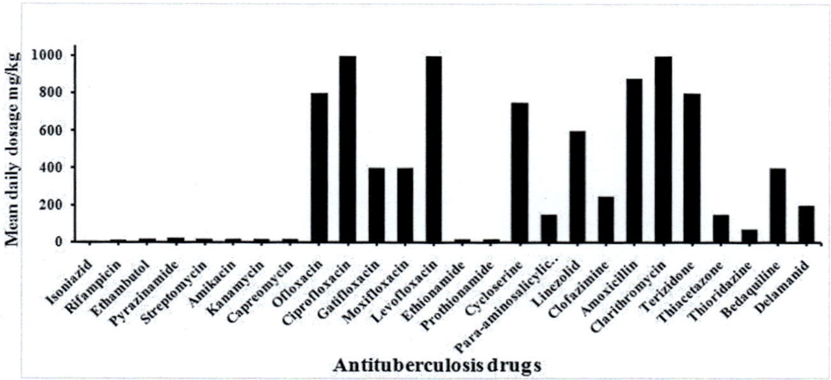

FIGURE 8.5 The graph shows the recommended daily dose of different antituberculosis drugs as per mg/body weight.

The introduction and usage of novel antibiotics to treat tuberculosis that are effective against both stimulant and drug-resistant strains while not interacting with antiretrobacterial medications enable a more constructive strategy to tuberculosis patients. New research and development activities, as well as the preservation of current therapeutic choices, are requested. To avoid a severe increase in drug-resistant forms of tuberculosis, training and educational initiatives centered on the rational use of antituberculosis medications are required.

8.9 Current therapeutics for treatment of cholera

Vibrio cholerae causes cholera in human as an acute intestinal infectious disease. Diarrhea and vomiting are usually accompanied by severe fluid and electrolyte losses, which lead to hypovolemia and hypovolemic shock. In this potentially fatal condition, aqueous feces are excreted in large quantities, resulting in rapid dehydration that can lead to hypovolaemia and metabolic acidosis (Cottingham et al., 2003). During outbreaks, some of the deaths may be fatal in vulnerable groups, but this number could be 1% if properly treated. Infection with *cholerae* is primarily because of the utilization of contaminated water or food. Presence of *cholerae* in the water can act as a prevalent source or reservoir for the organism in endemic areas. As a result, cholera usually occurs in places with insufficient access to clean water and sanitation. Around 50 nations report cholera cases every 3 to 5 years, commonly in African and Asian regions (Ali et al., 2012). It is reported that the cholera spread over the world from its initial reservoir in India's Ganges delta during the 19th and 20th centuries. Individuals with severe cholera can expel between 1010 and 1012 organisms per liter of stool, regardless of whether they are exposed to environmental *cholerae* or not (Ali et al., 2015).

Severe cholera diarrhea is coupled with fast fluid and electrolyte losses, leading to hypovolemia and electrolyte imbalances. Severe hypovolemia may occur within hours of symptoms (Guerrant et al., 2003). Cholera has become a treatable and restricted disease in developed countries as sanitary systems have improved. However, the number of antibiotic-resistant strains is increasing along with severe weather events. During the infection, the cholera toxin (CT) is secreted in the small intestine, which causes the pathological effects associated with cholera such as dehydration (Atia and Buchman, 2009).

Cholera is treated with vigorous volume replacement. Unless there is significant volume depletion or shock, when rapid fluid replacement is essential and intravenous fluids should be administered, oral fluid replacement is typically sufficient (Lazzerini, 2012). A proper diet is important for all patients, and certain micronutrients may also be beneficial for children. Antibiotics are used as a supplement to other treatments for individuals with mild to severe volume depletion, especially in epidemic situations. The use of antibiotics is an adjunctive therapy for patients suffering from cholera and moderate to severe volume depletion. Several investigations have demonstrated that effective cholera antibiotics are also beneficial in these patients (Holmgren, 1981). A number of studies have also demonstrated the efficacy of tetracyclines, the antibiotic class with the most clinical experience. A global study found that two oral cholera vaccines, also recommended by WHO as part of cholera control programs, are effective in halting outbreaks in areas of high risk by 60%–80% in endemic areas, along with other prevention and control measures. To prevent malnutrition and facilitate recovery from cholera, patients with the disease should receive adequate nutrition, as they would be suffering with any other cause of acute diarrhea. Proper sanitation and access to clean water are important to cholera prevention.

8.10 Advanced therapies against human pathogenic bacteria

Antibiotic misuse and the emergence of multidrug-resistant pathogenic bacteria have altered the way we think about antibiotics and bacteria-human interactions. AMR is on the rise, posing a severe threat to worldwide health sector. In the combat against drug-resistant microorganisms, conventional antibiotics have been defeated. The clinical failures of current antibiotic therapy are connected to long detection procedures, poor penetration at infection sites, alteration of the indigenous microbiota, and a high risk of mutational resistance (Lee et al., 2017). Alternative treatment solutions are now needed to address the AMR epidemic. Antibiotic-resistant microbes kill 700,000 people worldwide each year, with the death rates because of AMR expected to climb to 10 million by 2050, costing the global economy $100 trillion. Although a broadspectrum antibiotic strategy to bacterial therapy has had a lot of success, new and recurring bacterial infections continue to be a major problem for global health care (Devi et al., 2017). Antibiotic resistance has been the focus of research in recent decades, with the goal of creating quick detection and effective medication delivery methods. Inorganic materials, polymers, peptides, dendrimers, and liposomes are among the delivery agents that can be used.

POT (pathogen-oriented therapy) is emerging as a promising strategy for combating antibiotic resistance. Antibacterial chemicals or materials are applied directly to infected areas using POT to treat specific bacteria strains or species, improving treatment effectiveness, reducing nontargeting, and preventing drug resistance (Shang et al., 2020). Combining antibiotics with micro or nanodelivery materials is another possible technique for increasing antibiotic efficacy. Antibiotics might be protected against enzyme degradation using certain materials/vehicles, improving their therapeutic efficacy. In addition to detection, nano-vehicles are used to heal bacterial diseases. Chemical conjugation, adsorption, and encapsulation are some of the strategies that may be used to load medications onto the outside or inside of a vehicle for delivery to an infection site. Antibiotics administered in this manner safeguard the drug from the pathogen's resistance mechanisms (Van Giau et al., 2019; Mir et al., 2021). The discovery of novel antibiotics or alternative therapeutic techniques is required to prevent major drug-resistant illnesses.

8.11 Future perspectives

The majority of fast spreading lethal infectious diseases affecting humans are caused by microbial pathogens. Pathogens have turned the planet into a playground for developing drug resistance. As a result, effective and rapid detection methods for pathogenic micro-organisms are critical. The development of novel antimicrobials should include not only the identification of new cellular targets but also the definition of the organism's inhibitory activity profile—species-specific antibiotics or combinations. If the germs are responsive to antibiotics, the host may be able to recover. Antibiotic overuse, on the other hand, has led to a rapid surge in novel mutant pathogenic strains with widespread resistance to currently available antibiotics. In present times the leading cause of human illness is mainly the food-borne zoonotic infection worldwide, with developing countries bearing disproportionately higher cases, resulting in a major economic loss and public health problems. The most common bacteria detected in animal products are *Staph aureus*, *Salmonella* species, *Campylobacter* species, *Listeria monocytogenes*, and *E. coli*. Bacteria can infiltrate the food chain at any point along the route, from food animal killing through final absorption of animal products. Due to the rise of multidrug-resistant forms, these bacterial infections have become a serious public health issue. As the incidence of drug-resistant bacteria, particularly Gram-negative bacteria, grows, new techniques to address illnesses caused by these germs are urgently needed. Traditional antibiotics' enormous success and, of course, widespread use have considerably enhanced people's quality of life, but drug resistance today poses a severe danger to public health. But traditional antibacterial treatment has a number of drawbacks, including a scarcity of treatment regimens for MDR infections and a lack of antibiotic-specific action. Finally, antibiotic resistance can strike humans and animals at any time during their lifetimes. To lessen reliance on pharmaceutical therapies, alternative medicines should be developed. Antibiotic effectiveness has been dropping since they were first used in medical advancements more than seven decades ago. Experts from a number of fields, such as clinical research, microbiology, genetics, computational engineering, imaging, and modeling, should work together to find new treatments and strategies to address the problem.

References

Aarestrup, F.M., Wegener, H.C., 1999. The effects of antibiotic usage in food animals on the development of antimicrobial resistance of importance for humans in *Campylobacter* and *Escherichia coli*. Microb. Infect. 1, 639–644.

Ali, M., Lopez, A.L., You, Y.A., Kim, Y.E., Sah, B., Maskery, B., Clemens, J., 2012. The global burden of cholera. Bull. World Health Organ. 90, 209–218A.

Ali, M., Nelson, A.R., Lopez, A.L., Sack, D.A., 2015. Updated global burden of cholera in endemic countries. PLoS Neglected Trop. Dis. 9, e0003832.

Allahverdiyev, A.M., Kon, K.V., Abamor, E.S., Bagirova, M., Rafailovich, M., 2011. Coping with antibiotic resistance: combining nanoparticles with antibiotics and other antimicrobial agents. Expert Rev. Anti Infect. Ther. 9, 1035–1052.

Atia, A.N., Buchman, A.L., 2009. Oral rehydration solutions in non-cholera diarrhea: a review. Am. J. Gastroenterol. 104, 2596–2604. Quiz 2605.

Bakshi, C.S., Singh, V.P., Malik, M., Singh, R.K., Sharma, B., 2003. 55 kb plasmid and virulence-associated genes are positively correlated with *Salmonella enteritidis* pathogenicity in mice and chickens. Vet. Res. Commun. 27, 425–432.

Baliban, S.M., Lu, Y.J., Malley, R., 2020. Overview of the nontyphoidal and paratyphoidal *Salmonella* vaccine pipeline: current status and future prospects. Clin. Infect. Dis. 71, S151–S154.

Balta, I., Linton, M., Pinkerton, L., Kelly, C., Stef, L., Pet, I., Stef, D., Criste, A., Gundogdu, O., Corcionivoschi, N., 2020. The s. Food Contr. 107745.

Bondi, M., Lauková, A., DE Niederhausern, S., Messi, P., Papadopoulou, C., 2017. Natural Preservatives to Improve Food Quality and Safety. Hindawi.

Braden, C.R., 2006. *Salmonella enterica* serotype Enteritidis and eggs: a national epidemic in the United States. Clin. Infect. Dis. 43, 512–517.

Centers for Disease Control and Prevention, 2013a. Notes from the field: multistate outbreak of human *Salmonella typhimurium* infections linked to contact with pet hedgehogs - United States, 2011–2013. MMWR Morb. Mortal. Wkly. Rep. 62, 73.

Centers for Disease Control and Prevention, 2013b. Recurrent outbreak of *Campylobacter jejuni* infections associated with a raw milk dairy–Pennsylvania, April-May 2013. MMWR Morb. Mortal. Wkly. Rep. 62, 702.

Chahad, O.B., EL Bour, M., Calo-Mata, P., Boudabous, A., BARROS-Velazquez, J., 2012. Discovery of novel biopreservation agents with inhibitory effects on growth of food-borne pathogens and their application to seafood products. Res. Microbiol. 163, 44–54.

Cottingham, K.L., Chiavelli, D.A., Taylor, R.K., 2003. Environmental microbe and human pathogen: the ecology and microbiology of *Vibrio cholerae*. Front. Ecol. Environ. 1, 80–86.

De la Pomelie, D., Leroy, S., Talon, R., Ruiz, P., Gatellier, P., Sante-Lhoutellier, V., 2021. Investigation of *Escherichia coli* O157:H7 survival and interaction with meal components during gastrointestinal digestion. Foods 10.

Demain, A.L., 2009. Antibiotics: natural products essential to human health. Med. Res. Rev. 29, 821–842.

Devi, V.D., Kalpana, G., Saranraj, P., 2017. Antibacterial activity of Essential oils against human pathogenic bacteria. Adv. Biol. Res. 11, 357–364.

Dias de Oliveira, S., Siqueira Flores, F., Dos Santos, L.R., Brandelli, A., 2005. Antimicrobial resistance in *Salmonella enteritidis* strains isolated from broiler carcasses, food, human and poultry-related samples. Int. J. Food Microbiol. 97, 297–305.

González-Rodríguez, R., Rial-Otero, R., Cancho-Grande, B., Gonzalez-Barreiro, C., Simal-Gándara, J., 2011. A review on the fate of pesticides during the processes within the food-production chain. Crit. Rev. Food Sci. Nutr. 51, 99–114.

Guerrant, R.L., Carneiro-Filho, B.A., Dillingham, R.A., 2003. Cholera, diarrhea, and oral rehydration therapy: triumph and indictment. Clin. Infect. Dis. 37, 398–405.

Halabi, W.J., Nguyen, V.Q., Carmichael, J.C., Pigazzi, A., Stamos, M.J., Mills, S., 2013. *Clostridium difficile* colitis in the United States: a decade of trends, outcomes, risk factors for colectomy, and mortality after colectomy. J. Am. Coll. Surg. 217, 802–812.

Helke, K.L., Mccrackin, M.A., Galloway, A.M., Poole, A.Z., Salgado, C.D., Marriott, B.P., 2017. Effects of antimicrobial use in agricultural animals on drug-resistant foodborne salmonellosis in humans: a systematic literature review. Crit. Rev. Food Sci. Nutr. 57, 472–488.

Holmgren, J., 1981. Actions of cholera toxin and the prevention and treatment of cholera. Nature 292, 413–417.

Jansen, W., Muller, A., Grabowski, N.T., Kehrenberg, C., Muylkens, B., AL Dahouk, S., 2019. Foodborne diseases do not respect borders: zoonotic pathogens and antimicrobial resistant bacteria in food products of animal origin illegally imported into the European Union. Vet. J. 244, 75–82.

Kamboj, S., Gupta, N., Bandral, J.D., Gandotra, G., Anjum, N., 2020. Food safety and hygiene: a review. Int. J. Chem. Stud. 8, 358–368.

Karp, B.E., Tate, H., Plumblee, J.R., Dessai, U., Whichard, J.M., Thacker, E.L., Hale, K.R., Wilson, W., Friedman, C.R., Griffin, P.M., 2017. National antimicrobial resistance monitoring system: two decades of advancing public health through integrated surveillance of antimicrobial resistance. Foodborne Pathog. & Dis. 14, 545–557.

Khan, M.A., Steiner, T.S., 2002. Mechanisms of emerging diarrheagenic *Escherichia coli* infection. Curr. Infect. Dis. Rep. 4, 112–117.

Kinnings, S.L., Liu, N., Buchmeier, N., Tonge, P.J., Xie, L., Bourne, P.E., 2009. Drug discovery using chemical systems biology: repositioning the safe medicine Comtan to treat multi-drug and extensively drug resistant tuberculosis. PLoS Comput. Biol. 5, e1000423.

Koegelenberg, C.F.N., Schoch, O.D., Lange, C., 2021. Tuberculosis: the past, the present and the future. Respiration 100, 553–556.

Kohanski, M.A., Dwyer, D.J., Collins, J.J., 2010. How antibiotics kill bacteria: from targets to networks. Nat. Rev. Microbiol. 8, 423–435.

Kourtis, A.P., Sheriff, E.A., Weiner-Lastinger, L.M., Elmore, K., Preston, L.E., Dudeck, M., Mcdonald, L.C., 2020. Antibiotic multi-drug-resistance of *Escherichia coli* causing device- and procedure-related infections in the United States reported to the National Healthcare Safety Network (NHSN), 2013–2017. Clin. Infect. Dis. e4552–e4559.

Kumar, M., Sarma, D.K., Shubham, S., Kumawat, M., Verma, V., Nina, P.B., Jp, D., Kumar, S., Singh, B., Tiwari, R.R., 2021. Futuristic non-antibiotic therapies to combat antibiotic resistance: a review. Front. Microbiol. 12, 609459.

Kumar, P.P., Milton, A., Reddy, D.A., Agarwal, R., 2013. Salmonellosis: Are We Living with It?.

Kumawat, M., 2016. Identification and Characterization of *Salmonella* Typhimurium Peptidyl-Prolyl Cis-Trans Isomerase Encoded by Ppi Gene. Sam Higginbottom Institute of Agriculture, Technology & Sciences (SHIATS).

Lanzas, C., Davies, K., Erwin, S., Dawson, D., 2020. On modelling environmentally transmitted pathogens. Interf. Focus 10, 20190056.

Laxminarayan, R., Matsoso, P., Pant, S., Brower, C., Rottingen, J.A., Klugman, K., Davies, S., 2016. Access to effective antimicrobials: a worldwide challenge. Lancet 387, 168–175.

Lazzerini, M., 2012. Evidence Based Treatment of Cholera: A Review of Existing Literature. Cholera, p. 203.

Leach, S.A., Williams, A., Davies, A.C., Wilson, J., Marsh, P.D., Humphrey, T.J., 1999. Aerosol route enhances the contamination of intact eggs and muscle of experimentally infected laying hens by *Salmonella typhimurium* DT104. FEMS Microbiol. Lett. 171, 203–207.

Lee, W.X., Basri, D.F., Ghazali, A.R., 2017. Bactericidal effect of pterostilbene alone and in combination with gentamicin against human pathogenic bacteria. Molecules 22.

Leonard, F.C., Markey, B.K., 2008. Meticillin-resistant *Staphylococcus aureus* in animals: a review. Vet. J. 175, 27–36.

Lushniak, B.D., 2014. Antibiotic resistance: a public health crisis. Publ. Health Rep. 129, 314–316.

Majowicz, S.E., Musto, J., Scallan, E., Angulo, F.J., Kirk, M., O'brien, S.J., Jones, T.F., Fazil, A., Hoekstra, R.M., International Collaboration on Enteric Disease "Burden of Illness" Studies, 2010. The global burden of nontyphoidal *Salmonella* gastroenteritis. Clin. Infect. Dis. 50, 882–889.

Marrakchi, A.E., Hamama, A., Othmani, F.E., 1993. Occurrence of *Listeria monocytogenes* in milk and dairy products produced or imported into Morocco. J. Food Protect. 56, 256–259.

Mazumdar, K., Dutta, N.K., Dastidar, S.G., Motohashi, N., Shirataki, Y., 2006. Diclofenac in the management of *E. coli* urinary tract infections. In Vivo 20, 613–619.

Merker, M., Blin, C., Mona, S., Duforet-Frebourg, N., Lecher, S., Willery, E., Blum, M.G., Rüsch-Gerdes, S., Mokrousov, I., Aleksic, E., 2015. Evolutionary history and global spread of the *Mycobacterium tuberculosis* Beijing lineage. Nat. Genet. 47, 242–249.

Migliori, G.B., D'arcy Richardson, M., Sotgiu, G., Lange, C., 2009. Multidrug-resistant and extensively drug-resistant tuberculosis in the West. Europe and United States: epidemiology, surveillance, and control. Clin. Chest Med. 30, 637–665 (vii).

Mir, M.A., AL-Baradie, R., 2013. Tuberculosis time bomb-A global emergency: need for alternative vaccines. Majmaah J. Health Sci. 1, 77–82.

Mir, M.A., Albaradeh, R., Agrewala, J.N., n.d. Innate–Effector Immune Response Elicitation Against Tuberculosis Through Anti-B7-1 (Cd80) And Anti-B7-2 (Cd86) Signaling in Macrophages.

Mir, M.A., Bhat, B.A., Sheikh, B.A., Rather, G.A., Mehraj, S., Mir, W.R., 2021. Nanomedicine in human health therapeutics and drug delivery: nanobiotechnology and nanobiomedicine. In: Applications of Nanomaterials in Agriculture, Food Science, and Medicine. IGI Global.

Muller, A., Theuretzbacher, U., Mouton, J., 2015. Use of old antibiotics now and in the future from a pharmacokinetic/pharmacodynamic perspective. Clin. Microbiol. Infect. 21, 881–885.

Mushin, R., 1948. An outbreak of gastro-enteritis due to *Salmonella* derby. Epidemiol. Infect. 46, 151–157.

Nair, D.V.T., Venkitanarayanan, K., Kollanoor johny, A., 2018. Antibiotic-resistant *Salmonella* in the food supply and the potential role of antibiotic alternatives for control. Foods 7.

Neimann, J., Engberg, J., Molbak, K., Wegener, H.C., 2003. A case-control study of risk factors for sporadic campylobacter infections in Denmark. Epidemiol. Infect. 130, 353–366.

Nesbitt, A., Ravel, A., Murray, R., Mccormick, R., Savelli, C., Finley, R., Parmley, J., Agunos, A., Majowicz, S.E., Gilmour, M., Canadian Integrated Program for Antimicrobial Resistance Surveillance Public Health Partnership; Canadian Public Health Laboratory Network, 2012. Integrated surveillance and potential sources of *Salmonella enteritidis* in human cases in Canada from 2003 to 2009. Epidemiol. Infect. 140, 1757–1772.

Newell, D.G., Koopmans, M., Verhoef, L., Duizer, E., Aidara-Kane, A., Sprong, H., Opsteegh, M., Langelaar, M., Threfall, J., Scheutz, F., 2010. Food-borne diseases—the challenges of 20 years ago still persist while new ones continue to emerge. Int. J. Food Microbiol. 139, S3–S15.

Ngwai, Y.B., Adachi, Y., Ogawa, Y., Hara, H., 2006. Characterization of biofilm-forming abilities of antibiotic-resistant *Salmonella typhimurium* DT104 on hydrophobic abiotic surfaces. J. Microbiol. Immunol. Infect. 39, 278–291.

Olsen, S.J., Ying, M., Davis, M.F., Deasy, M., Holland, B., Iampietro, L., Baysinger, C.M., Sassano, F., Polk, L.D., Gormley, B., Hung, M.J., Pilot, K., Orsini, M., Van Duyne, S., Rankin, S., Genese, C., Bresnitz, E.A., Smucker, J., Moll, M., Sobel, J., 2004. Multidrug-resistant *Salmonella* Typhimurium infection from milk contaminated after pasteurization. Emerg. Infect. Dis. 10, 932–935.

Owen, L., Laird, K., 2018. Synchronous application of antibiotics and essential oils: dual mechanisms of action as a potential solution to antibiotic resistance. Crit. Rev. Microbiol. 44, 414–435.

Pangilinan, C.R., Lee, C.H., 2019. *Salmonella*-based targeted cancer therapy: updates on a promising and innovative tumor immunotherapeutic strategy. Biomedicines 7.

Pisoschi, A.M., Pop, A., Georgescu, C., Turcus, V., Olah, N.K., Mathe, E., 2018. An overview of natural antimicrobials role in food. Eur. J. Med. Chem. 143, 922–935.

Qadri, H., Haseeb, A., Mir, M., 2021. Novel strategies to combat the emerging drug resistance in human pathogenic microbes. Curr. Drug Targets 22, 1–13.

Ramsay, E.C., Daniel, G.B., Tryon, B.W., Merryman, J.I., Morris, P.J., Bemis, D.A., 2002. Osteomyelitis associated with *Salmonella enterica* SS arizonae in a colony of ridgenose rattlesnakes (*Crotalus willardi*). J. Zoo Wildl. Med. 33, 301–310.

Rantala, S., 2014. Streptococcus dysgalactiae subsp. equisimilis bacteremia: an emerging infection. Eur. J. Clin. Microbiol. Infect. Dis. 33, 1303–1310.

Raviglione, M., Sulis, G., 2016. Tuberculosis 2015: burden, challenges and strategy for control and elimination. Infect. Dis. Rep. 8, 6570.

Romano, V., Pasquale, V., Lemee, L., El Meouche, I., Pestel-Caron, M., Capuano, F., Buono, P., Dumontet, S., 2018. Clostridioides difficile in the environment, food, animals and humans in Southern Italy: occurrence and genetic relatedness. Comp. Immunol. Microbiol. Infect. Dis. 59, 41–46.

Schaberg, T., 1995. The dark side of antituberculosis therapy: adverse events involving liver function. Eur. Respir. J. 8, 1247–1249.

Schwan, W.R., 2019. *Staphylococcus aureus* Toxins: Armaments for a Significant Pathogen. Toxins, Basel), p. 11.

Shang, Z., Chan, S.Y., Song, Q., Li, P., Huang, W., 2020. The Strategies of Pathogen-Oriented Therapy on Circumventing Antimicrobial Resistance. Research (Wash D C), p. 2016201.

Sheikh, B.A., Bhat, B.A., Mehraj, U., Mir, W., Hamadani, S., Mir, M.A., 2021. Development of new therapeutics to meet the current challenge of drug resistant tuberculosis. Curr. Pharmaceut. Biotechnol. 22, 480–500.

Silva, C., Calva, E., Maloy, S., 2014. One health and food-borne disease: *Salmonella* transmission between humans, animals, and plants. Microbiol. Spectr. 2 (2), 1–08.

Smith, I.D., Winstanley, J.P., Milto, K.M., Doherty, C.J., Czarniak, E., Amyes, S.G., Simpson, A.H., Hall, A.C., 2013. Rapid in situ chondrocyte death induced by *Staphylococcus aureus* toxins in a bovine cartilage explant model of septic arthritis. Osteoarthri. Cartil. 21, 1755–1765.

Stoll, B.J., Hansen, N., Fanaroff, A.A., Wright, L.L., Carlo, W.A., Ehrenkranz, R.A., Lemons, J.A., Donovan, E.F., Stark, A.R., Tyson, J.E., Oh, W., Bauer, C.R., Korones, S.B., Shankaran, S., Laptook, A.R., Stevenson, D.K., Papile, L.A., Poole, W.K., 2002. Changes in pathogens causing early-onset sepsis in very-low-birth-weight infants. N. Engl. J. Med. 347, 240–247.

Su, H.P., Chiu, S.I., Tsai, J.L., Lee, C.L., Pan, T.M., 2005. Bacterial food-borne illness outbreaks in Northern Taiwan, 1995–2001. J. Infect. Chemother. 11, 146–151.

Tang, S., Yao, L., Hao, X., Zhang, X., Liu, G., Liu, X., Wu, M., Zen, L., Sun, H., Liu, Y., 2015. Efficacy, safety and tolerability of linezolid for the treatment of XDR-TB: a study in China. Eur. Respir. J. 45, 161–170.

Thomsen, I.P., 2018. Antibody-based intervention against the pore-forming toxins of *Staphylococcus aureus*. Virulence 9, 645–647.

Tiberi, S., Zumla, A., Migliori, G.B., 2019. Multidrug and extensively drug-resistant tuberculosis: epidemiology, clinical features, management and treatment. Infect. Dis. Clin. 33, 1063–1085.

Tkacikova, L., Kantikova, M., Dmitriev, A., Mikula, I., 2000. Use of the molecular typing methods to evaluate the control of *Listeria monocytogenes* contamination in a raw milk and dairy products. Folia Microbiol. 45, 157–160.

Turner, N.A., Smith, B.A., Lewis, S.S., 2019. Novel and emerging sources of *Clostridioides difficile* infection. PLoS Pathog. 15, e1008125.

Van Giau, V., An, S.S.A., Hulme, J., 2019. Recent advances in the treatment of pathogenic infections using antibiotics and nano-drug delivery vehicles. Drug Design Devel. & Ther. 13, 327.

Velge, P., Cloeckaert, A., Barrow, P., 2005. Emergence of *Salmonella* epidemics: the problems related to *Salmonella enterica* serotype Enteritidis and multiple antibiotic resistance in other major serotypes. Vet. Res. 36, 267–288.

Voss-Rech, D., Potter, L., Vaz, C.S., Pereira, D.I., Sangioni, L.A., Vargas, A.C., DE Avila Botton, S., 2017. Antimicrobial resistance in nontyphoidal *Salmonella* isolated from human and poultry-related samples in Brazil: 20-year meta-analysis. Foodb. Pathog. Dis. 14, 116–124.

Wigley, P., 2004. Genetic resistance to *Salmonella* infection in domestic animals. Res. Vet. Sci. 76, 165–169.

Wilcox, M.H., Fawley, W.N., 2000. Hospital disinfectants and spore formation by *Clostridium difficile*. Lancet 356, 1324.

Witte, W., Cuny, C., Klare, I., Nubel, U., Strommenger, B., Werner, G., 2008. Emergence and spread of antibiotic-resistant Gram-positive bacterial pathogens. Int. J. Med. Microbiol. 298, 365–377.

World Health Organization, 2017. Integrated surveillance of antimicrobial resistance in foodborne bacteria: application of a one health approach: guidance from the WHO Advisory Group on Integrated Surveillanec of Antimicrobial Resistance (AGISAR). World Health Organization. ISBN No. 9789241512411, pp. 1–88. https://apps.who.int/iris/handle/10665/255747. License: CC BY-NC-SA 3.0 IGO.

Yashoda, K., Sachindra, N., Sakhare, P., Rao, D.N., 2001. Microbiological quality of broiler chicken carcasses processed hygienically in a small scale poultry processing unit. J. Food Qual. 24, 249–259.

Yohannes, A., Habtamu, M., Abreha, M., Endale, B., Habtamu, T., 2014. Presence of egg contaminant bacteria and egg deformities (external and internal qualities) which potentially affect public health and egg marketability in Mekelle, Ethiopia. J. Vet. Sci. Med. Diagn. 3 (2), 2.

CHAPTER

9

Combinatorial approach to combat drug resistance in human pathogenic fungi

Manzoor Ahmad Mir, Hafsa Qadri, Shariqa Aisha and Abdul Haseeb Shah

Department of Bioresources, School of Biological Sciences, University of Kashmir, Srinagar, Jammu and Kashmir, India

9.1 Introduction

All over history, human beings have been afflicted by various deadly infectious diseases, and the continuing COVID-19 (Coronavirus disease-2019) pandemic is an intimidating reminder that such susceptibility persists in the current world. Communicable diseases overall represent a major leading cause of mortality around the globe (Kainz et al., 2020). It is unfortunate that such microbial-associated diseases/infections are generally overlooked and ignored by various healthcare professionals, even though yearly the health of millions of individuals is at risk around the whole world. Diseases/Infections associated with different fungal pathogens belong to such overlooked growing microbial infections/diseases, which account for millions of deaths per year (Bongomin et al., 2017).

Kingdom fungi has been estimated to possess about 6,000,000 species which are broadly distributed in the surroundings (Taylor et al., 2014). Among the various fungal species, many are pathogenic to humans in different ways (Govorushko et al., 2019).Around 625 species of fungi are found responsible for the occurrence of fungal disease/infection in the case of vertebrates. Interestingly, kingdom fungi consist of 167 orders, out of which around 40 species have been continuously reported in different medical research studies (de Hoog et al., 2018). These pathogenic fungi are known to cause a broad range of superficial (hair, skin, nails, or mucosal surface infections) as well as severe invasive infections if left undiagnosed and not exactly treated (Konopka et al., 2019). Moreover, to cause invasive fungal infections in human hosts, fungal organisms should satisfy the following four basic criteria:

Human Pathogenic Microbes
https://doi.org/10.1016/B978-0-323-96127-1.00011-5

- Firstly, these fungal organisms possess the capacity to develop at a body temperature near/above the mammalian hosts.
- Secondly, these fungal organisms have the capacity to influence the internal tissues via piercing/escaping the host barriers.
- Thirdly, these fungal organisms possess the capacity to break the host tissues and assimilate the host components as well.
- Fourthly, these fungal organisms have the capacity to escape the entire immune defense mechanism of the host (Kohler et al., 2017).

From the last few years, fungal pathogens have evolved as a growing human health issue globally. In general, fungal organisms are opportunistic in a way that they are always in search of opportunities for hosts with a compromised immune system because of poor health conditions like AIDS, cancer, and other related cases. These organisms represent the causal agents of a broad spectrum of diseases varying from superficial to invasive ailments influencing about millions of persons worldwide (Brown et al., 2012b; Spitzer et al., 2017b). Per year around 1.7 million deaths occur because of these increasing fungal infections/diseases. Unfortunately, with the rising social and medical establishments the cases associated with the growing spread of fungal infections are increasing rapidly (Brown et al., 2012a). Socioeconomic and geoecological features, as well as the exposure to a large fungal inoculum in their natural surroundings, represent other essential factors that impact the overall frequency and prevalence of fungal infections/disease around the globe (Bongomin et al., 2017).

In terms of disease and death rates, IFIs (Invasive Fungal Infections) which are mostly caused by the pathogenic strains of *Cryptococcus*, *Candida*, and *Aspergillus* represent a severe clinical health burden (Arendrup et al., 2014; Pappas et al., 2016). The primary risk elements connected with the acquirement of IFIs are given in Table 9.1 (Firacative, 2020). Consisting of more than 30 species, genus *Cryptococcus* harbors two essential ones including *Cryptococcus neoformans* and *Cryptococcus Gattii,* which are well known for producing various human fungal related-diseases. On the other hand, *Aspergillus* genera harbor around 340 species,

TABLE 9.1 Primary risk elements connected with the acquirement of IFIs (Invasive Fungal Infections) (Firacative, 2020).

S.No	Risk element	Specified state	Concerned and most frequent pathogenic
1.	Biological/Health intercession	**1.** Use of catheters and other such medical equipment. **2.** Utilization of a wide range of antibiotics.	**1.** Different *Candida* species like *C. albicans.* **2.** Different *Candida* species like *C. albicans.*
2.	Coinfection	**1.** TB	**2.** *C. neoformans, C. albicans, A. fumigatus,* etc.
3.	Immunosuppression caused by a disease	**1.** HIV/AIDS, etc.	**1.** *C. neoformans* etc.
4.	Immunosuppression caused by a treatment	**1.** Various biological agents. **2.** Solid-organ transplantation.	**1.** Different pathogenic species of Candida, Aspergillus, and Cryptococcus genus. **2.** *C. neoformans, C. albicans, A. fumigatus* etc.

among which 40 species are correlated with different issues connected to health (Samson et al., 2014; Thakur et al., 2015). In persons particularly in immunocompromised ones, Aspergillosis presents a critical clinical health problem (Chowdhary et al., 2014). Among different *Aspergillus* species, *Aspergillus fumigatus* represents the most commonly present pathogenic mold in the case of human beings, causing serious hazards in susceptible/vulnerable individuals (Szalewski et al., 2018).

Among different *Candida* species, *Candida albicans* represent the main causative species of multiple bloodstream infections around the world, which accounts for about 43% to 56% of all cases. *C. albicans* is followed by other major essential species (Pfaller et al., 2011b). Additional *Candida* species like *C. guilliermondii*, *C. lusitaniae*, *C. kefyr*, *C. dubliniensis*, etc., are the lesser commonly occurring members of the *Candida* genus (Sanguinetti et al., 2015). The pathogenic species belonging to the genus *Candida* have gained much significance due to the intensity of their infection as well as their capacity to withstand the impact of multiple antifungals via the development of the process of resistance (Prasad and Kapoor, 2005). Table 9.2, displays some of the specific medically important species of the *Candida* genus (Maurya et al., 2020).

C. albicans represent the most frequent causative agent of human fungal diseases, which can survive in a wide range of environments, leading to a variety of infections like superficial, oral, vaginal, and multiple severe systematic fungal diseases (Calderone and Gow, 2002; Brown et al., 2006; Rodaki et al., 2009; Prasad et al., 2014; Kühbacher et al., 2017; Cortegiani et al., 2018a).

C. albicans represent the fourth highest prevalent cause of bloodstream ailments in the United States, and it can cause both surfaces associated and systemic infections (Gudlaugsson et al., 2003; Wisplinghoff et al., 2004). For a very extended period, *C. albicans* was assumed to be an asexual organism, but various researches have shown that it also possesses the ability of mating sexually (generating tetrasomic-cells)/parasexually (by concerted chromosome damage, reverting cells to the diploid phase, since true meiosis fails to take place) (Wang et al., 2018). It has been reported that thrush is caused by *C. albicans* in domestic poultry, aquatic fowls, etc., with the upper gut flora of small birds being the most affected (Dadar et al., 2018). Biofilms of *C. albicans* can build on any medical equipment. Vascular and urinary devices, heart valves, prosthetic vascular bypass devices, pacemakers, etc., are among the most frequently involved medical equipment (Bryers and Ratner, 2004; Kojic and Darouiche, 2004).

TABLE 9.2 Showing some of the specific medically important species of Candida genus (Maurya et al., 2020).

	Concerned pathogenic species	Abundance (%)
1.	*C. albicans*	Fifty to seventy (approx.)
2.	*C. parapsilosis*	Ten to twenty (approx.)
3.	*C. glabrata*	Fifteen to thirty (approx.)
4.	*C. tropicalis*	Six to twelve (approx.)

The biotic-surface biofilms of *C. albicans* are made up of two kinds of cells: yeast and filamentous cells (Finkel and Mitchell, 2011; Ramage et al., 2006). One of *C. albican's* most critical virulence factors is its capability to switch from yeast to hyphal shape (Han et al., 2011).

Candida glabrata is a common microbial pathogen that can be found in the mouth cavity, digestive tract (GIT), and vaginal system. *C. glabrata* infections account for around 29% of all *Candida* bloodstream ailments (Pfaller et al., 2011a; Montagna et al., 2014). *C. glabrata* is becoming highly significant with the growing phenomenon of resistance to commonly employed antifungal agents as well as to the host immune defense process (Musa et al., 2018). *C. parapsilosis* is another prominent fungal pathogen that is presently one of the major causes of invasive candidiasis (Trofa et al., 2008). Tóth et al. indicated that *C. parapsilosis* is the secondor third highest frequently detected *Candida* species globally. People with weakened immune systems, such as those struggling with HIV/AIDS or those receiving surgeries, especially gastrointestinal surgical procedures, are more susceptible to *Parapsilosis*-associated ailments (Trofa et al., 2008).

Remarkably, *Candida auris* is rapidly gaining popularity among non-albican pathogens. Infectious diseases produced by *C. auris* have recently been documented all over the world, and they are frequently associated with higher morbidity and mortality rates. *C. auris* was discovered as a novel fluconazole-resistant strain in an outer ear sampling of a Japanese person in 2009 (Spivak and Hanson, 2018). Ever since the pathogen has been related to hospital-acquired infections and illnesses throughout five distinct nations (Cortegiani et al., 2018b). *C. auris*, which causes a variety of serious diseases, is a new MDR fungal pathogen that has been related to higher mortality rates (Morales-López et al., 2017). The potential of *C. auris* to confer resistance to routinely available antifungal drugs may account for its increasing death rates (Cortegiani et al., 2018b; Navalkele et al., 2017). *C. auris* was responsible for 5.6% of nosocomial candidemia patients in geographically distinct medical care settings (Chakrabarti et al., 2015). *C. auris* spread is thought to be caused by contact with infected facilities and spread from medical workers, according to preliminary findings. Hand transmission and surface contamination have been linked to prolonged breakouts (Cortegiani et al., 2018b; Biswal et al., 2017). The overall deaths for *C. auris*-associated candidemia are estimated to be between 30% and 72% (Cortegiani et al., 2018b). *C. auris* has an unusual capacity to transmit across persons in healthcare settings, which is likely due to its propensity to evade skin surface and remain for longer durations (Escandón et al., 2019). When compared to several pathogens, along with its close relatives, *C. auris* has unique characteristics. The organism could thrive at temperatures exceeding 420°C (Chatterjee et al., 2015). *C. auris* cells have an oval structure under the microscope, with no evidence of pseudohyphae development. Under different culture environments, *C. auris* can have a variety of structural forms, involving round-to-round, extended, and pseudohyphal forms (Borman et al., 2016). *C. auris*'s propensity to survive and spread, combined with its highly proliferative clonal life events, has resulted in an increasing number of nosocomial diseases around the globe (Calvo et al., 2016; Morales-López et al., 2017). The organism cannot propagate, make hyphae, or generate chlamydospores; however, it can develop biofilms, which are finer than those produced by *C. albicans*. Many investigations have demonstrated that *C. auris* possess a substantially lower capability to express numerous pathogenicity determinants than *C. albicans* (Larkin et al., 2017). Adherence, germination, and biofilm development, as well as the synthesis of phospholipase and proteinase, are all pathogenicity-related elements (Larkin et al., 2017).

Regardless of the possibility that numerous remedies are presently being used to combat the rising number of human fungal diseases, only minimal progress has been obtained (Redhu et al., 2018). *Candida* pathogens are susceptible to the most commonly utilized antifungal agents in varying degrees (Sanguinetti et al., 2015b). Azoles, polyenes, echinocandins, and pyrimidine analogs are the only antifungal drugs currently used in clinical trials (Ksiezopolska and Gabaldón, 2018). The increased usage of antifungals, particularly azoles, has led in the enhancement in the process of drug resistance, and despite effective antifungal treatments, the disease progresses (Shah et al., 2015).

The fungal pathogenic organisms employ multiple resistance mechanisms (Fig. 9.1) to overcome the consequences of distinct antifungal classes (Qadri et al., 2021c).

Among *Candida* species, the mechanism of MDR has been more intensively studied in *C. albicans* (Gulshan and Moye-Rowley, 2007). A significant knowledge regarding different mechanisms and biological elements contributing toward the establishment of drug resistance in these pathogens is very essential for the establishment of proper treatment and control strategies (Cowen et al., 2015). The restricted availability of antifungal drugs is the main hurdle in the way of the development of proper treatment and control strategies against different fungal diseases (Vandeputte et al., 2012). It is furthermore intensified by the aspect that the development of novel antifungal agents has lagged in comparison to the rate of the origination of different fungal diseases. The fungal cell wall components like mannans, glucans, and chitins, and some ergosterol biosynthetic pathway enzymes that are peculiar to fungal cells are frequently targeted for the evolution of different antifungal drugs (Vandeputte et al., 2012). The antifungals azoles and echinocandins are often considered

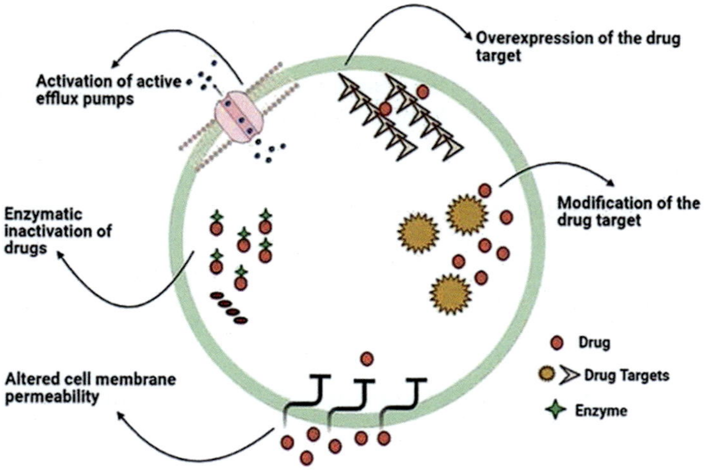

FIGURE 9.1 Drug-resistance mechanisms including modification of the drug target, active efflux pump activation, enzymatic degradation of drugs, etc., adopted by fungal pathogens to overcome the effect of different antifungals.

as the basis for the cure of IFIs (Arendrup et al., 2014; Pappas et al., 2016). Moreover, the continuous usage of different antifungal agents has contributed to the development of various drug-resistant fungal organisms, including the emerging and exceptionally virulent human fungal pathogen *C. auris* (Brown et al., 2012b). The robust usage of a variety of antifungal drug classes has ultimately led to the development of resistant fungal infections. The fungal pathogenic organisms display various resistance processes to eminently nullify the impact of multiple antifungal drugs. Furthermore, in order to successfully colonize as well as for the development of infection inside the host, different pathogenic fungal species undergo metabolic flexibility for the modulation of specific nutrient uptake processes and to regulate their metabolism properly (Qadri et al., 2021b). Following the rise in the occurrence of globally spreading fungal diseases and the development of various antifungal drug–resistance mechanisms outcompeting the establishment of innovative antifungal agents, it is essential to comprehend different aspects of fungal diseases, as well as to recognize the fundamental drug-resistance processes that govern the advancement of such resistance issues.

Due to the restricted availability of antifungal agents for clinical usage and the growing issue of multidrug resistance fungal infections, a continuous demand for the establishment of new extended antifungal agents having greater potency is required. Nowadays, combination therapy represents a successful option and a possible treatment of choice against a wide range of different infectious diseases. Moreover, the application of drugs in combinations for the proper treatment and control of fungal pathogenic organisms has gained tremendous attention throughout many years. In this chapter, we are presenting a brief account of the currently available antifungal classes and their possible mechanism of action along with a detailed account of the possible benefits and other important aspects of drug combinatorial strategies for the proper cure and control of growing fungal diseases. A variety of novel combined treatments, procedures, and regimes have been tried beside each other and have proven to improve patient populations as the system offers novel combination solutions. Drug combination therapy is often used not to cure but to relieve symptoms and increase life expectancy. As technology progresses and the number of permitted entities grow, delivering safe and long-lasting solutions for patients will become a top priority for all future research and establishments. On the Indian subcontinent, combination drug treatment and therapy are commonly used. Owing to the extensive rise of MDR-associated pathogens, and increasing drug-resistant human microbial infections drug combination therapy is being employed rapidly and continuously. Tuberculosis, cancer, HIV/AIDS, etc., are among the diseases treated by combination therapy. Drug combinations minimize the emergence of the phenomenon of drug resistance because a pathogenic organism is less inclined to produce resistance to many compounds at the same time.

9.2 Major classes of antifungal drugs

The development of antifungal drugs for the cure and control of systemic fungal diseases/ infections is increasing recently and presently consists of the following major antifungal agent categories (Prasad et al., 2016).

9.2.1 Azoles

The antifungal drugs azoles principally affect the ergosterol biosynthetic pathway by targeting the enzyme 14′- lanosterol demethylase (*CYP51*) encoded by the *ERG11* gene leading to the suppression of cytochrome P450-dependent transformation of lanosterol to ergosterol. Ergosterol depletion restricts the major ergosterol functions (ergosterol is an important constituent of the fungal cell membrane) and thereby impacts the overall growth and proliferation of the fungal organism (Shapiro et al., 2011). Moreover, the inhibition of 14′-demethylase leads to the cumulation of toxic methylated sterols which ultimately causes membrane tension (Shapiro et al., 2011).

In general azole drugs contain two major drug classes viz. Class 1—Imidazoles (miconazole, oxiconazole, econazole, ketoconazole, tioconazole, and clotrimazole) Class 2—Triazoles (fluconazole, posaconazole, itraconazole, terconazole, and voriconazole). Fluconazole drug displays major antifungal action in the case of several pathogenic organisms of *Aspergillus*, *Cryptococcus*, *Candida,* and other related pathogenic organisms (Pardasani, 2000; Pappas et al., 2004). Similarly, the drug itraconazole possesses better antifungal activity in the case of different *Aspergillus* species and many additional yeast pathogens as well (Denning and Hope, 2010).

In the 1990s, in *C. albicans,* resistance to azoles was identified as a serious issue, when among the 90% (approx.) of the AIDS patients having oral candidiasis, majority received a lifelong azole remedy and few patient populations evolved resistance (White et al., 1998). For many years, in pathogenic fungi, various molecular mechanisms governing azole resistance have been elucidated through various research studies. Some important azole drug-resistance processes/mechanisms (Fig. 9.2) have been outlined below. All of these mechanisms are frequently reported in various drug-resistant clinical strains of *Candida* (Bhattacharya et al., 2020).

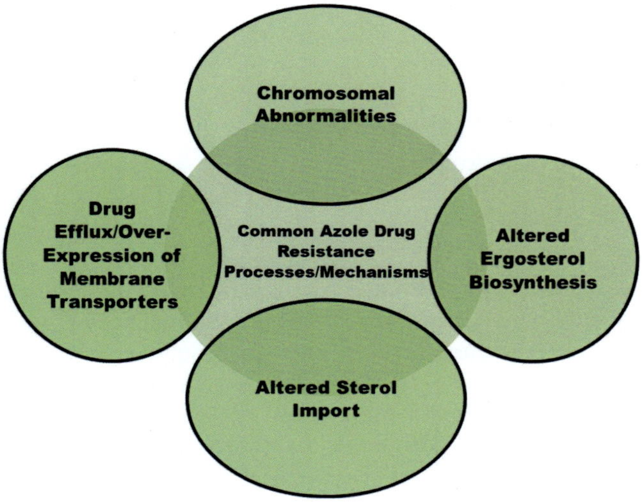

FIGURE 9.2 Diagrammatic representation of various molecular mechanisms/processes governing azole drug resistance phenomenon in different human pathogenic fungi.

9.2.1.1 *Drug efflux/overexpression of membrane transporters*

Activation of membrane-linked extrusion proteins, recognizing a wide range of chemical compounds, represents a remarkable drug resistance process enabling the MDR phenomenon. In general, in fungi, two distinct drug extrusion systems, viz. the ATP-binding cassette (ABC) superfamily and the major facilitator superfamily (MFS) regulate azole drug resistance. ABC-proteins represent the primary active transporters consisting of two transmembrane spanning domains (TMDs) and two cytoplasmic nucleotide-binding domains (NBDs) catalyzing ATP hydrolysis (Cowen et al., 2015). *C. albicans* is predicted to harbor 28 ABC proteins (Prasad and Goffeau, 2012). On the other hand, major facilitator transporters, which are secondary active transporters, use a proton gradient generated across the plasma membrane as an energy source for drug efflux. These transporters do not have the NBDs and consist of 12 (DHA1) or 14 (DHA2) trans-membrane segments (Shah et al., 2015; Prasad et al., 2014). The *C. albicans* genome sequence is predicted to harbor 95 MFS proteins distributed across 17 families (Gaur et al., 2008).

Increased expression of these membrane transporters has been found correlating with the phenomenon of azole drug resistance in fungal pathogenic organisms. The overexpression of *CaCDR1*, *CaCDR2*, and *CaMDR1* is frequently reported in azole-resistant oral, systemic, and vaginal clinical isolates of *C. albicans* (Bhattacharya et al., 2016; White et al., 2002). Moreover, the enhanced expressions of *CaCDR1*, *CaCDR2*, and *CaMDR1* have been reported in the matching pair resistant isolates on comparing with their susceptible members (Bhattacharya et al., 2016). In azole-resistant strains, in addition to *CaCDR1*, *CaCDR2*, and *CaMDR1*, increased expression of *CaFLU1* (an MFS-transporter) and *CaPDR16* (an ABC-transporter) has also been reported (Calabrese et al., 2000). In the clinical isolates of *C. glabrata*, the upregulation of the ABC-transporter proteins, viz. *CgCDR1*, *CgSNQ2*, and *CgPDH1* has been found contributing toward azole drug resistance (Sanguinetti et al., 2005; Vermitsky et al., 2004). Similarly, in other non-albican *Candida* species, like *C. kruseiABC1*, and other pathogenic non-*Candida* species like *C. neoformans*, *AFR1* represents the multidrug transporters having a significant part in the advancement of the phenomenon of drug resistance in such pathogenic fungal organisms (Sanguinetti et al., 2006; Prasad et al., 2016).

9.2.1.2 *Altered ergosterol biosynthesis*

The process of altered ergosterol biosynthesis usually occurs via mutation and/or overexpression of different ergosterol biosynthesis pathway genes (Bhattacharya et al., 2020). *Erg11p* is a cytochrome p450 containing enzyme regulating a rate-limiting phase in the ergosterol biosynthetic pathway (Veen et al., 2003). *ERG11* overexpression was found correlated with azole drug resistance in different pathogenic fungi. Multiple azole-resistant clinical isolates of *C. albicans* display enhanced *CaERG11* expression (Bhattacharya et al., 2016; White et al., 2002). Moreover, *CgERG11* overexpression was documented in azole-resistant isolates of *C. glabrata* (Pam et al., 2012). *ERG11* overactivation was also found in *C. auris* showing enhanced azole drug resistance (Bhattacharya et al., 2019). In *C. albicans*, several point mutations in the *ERG11* gene connected with azole drug resistance have been reported in resistant clinical isolates (Xiang et al., 2013). It has also been found that in *C. albicans*, the deletion of the *ERG3* gene causes enhanced resistance against azoles (Sanglard et al., 2003). In *C. glabrata* isolates, Q139A in Erg3p has been found linked with the phenomenon of azole drug resistance

(Yoo et al., 2010). Moreover, a remarkable resistance against azoles has been witnessed in the heterozygous *ERG6* deletion in *C. albicans* (Xu et al., 2007). In *C. glabrata ERG6* contributes to azole resistance because of several base pair variations causing missense mutations (Vandeputte et al., 2007).

9.2.1.3 Altered sterol import

Azole drugs decrease the cell's ergosterol levels, which are compensated by the process of exogenous sterol import. In *S. cerevisiae*, *C. albicans*, and *C. glabrata* the mechanism of sterol import has been very well marked (Zavrel et al., 2013). Both *S. cerevisiae* and *C. glabrata* import sterols under anaerobic/microaerophilic conditions utilizing the sterol importers *Aus1p* and *Pdr11p* (Kuo et al., 2010), and displaying enhanced resistance against azoles (Zavrel et al., 2013). It has been reported that *C. albicans,* unlike *C. glabrata* and *S. cerevisiae,* imports sterols aerobically, developing resistance against azole drugs in the presence of both serum and cholesterol (Zavrel et al., 2013). By importing cholesterol and serum from the blood, *C. albicans* could evolve drug resistance. Thus, sterol import acts as an efficient candidate of azole drug resistance in pathogenic fungi (Bhattacharya et al., 2020).

9.2.1.4 Chromosomal abnormalities

Various genetic modifications like loss of heterozygosity (LOH) and aneuploidy are found associated with the azole drug resistance phenomenon. In the clinical isolates of *C. albicans,* LOH is reported in the regions consisting of azole-resistance determinants viz. *CaTAC1*, *CaERG11*, and *CaMRR1*, connected with enhanced drug resistance process (Ford et al., 2015). It has been found that clinical strains of *C. albicans* may also involve segmental aneuploidy, where two copies of the left arm of chromosome 5 containing *CaERG11* and *CaTAC1* form an isochromosome correlating with the azole drug resistance phenomenon (Selmecki et al., 2006). The evaluation of sequential *C. albicans* strains which developed resistance in persons demonstrated that mutations in these genes often develop in the heterozygous state and become homozygous by the loss of heterozygosity (LOH) (Selmecki et al., 2010).

Furthermore, chromosomal variations are found linked with drug resistance in *C. glabrata* and *C. neoformans*. In *C. glabrata* azole resistance is evolved by enhancing the copy number of the ERG11 gene (Marichal et al., 1997). Moreover, in the azole-resistant isolates of *C. glabrata*, the emergence of segmental aneuploidies and novel chromosome configurations has been reported (Poláková et al., 2009). In *C. neoformans*, azole drug resistance is found linked with particular chromosomal alterations, mainly disomies of chromosomes (1 and 4) (Sionov et al., 2010).

9.2.2 Polyenes

The antifungal drug polyenes represent the amphipathic organic natural molecules referred to as macrolides (Vandeputte et al., 2012). These drugs directly attach to the important constituent of the cell membrane, i.e., ergosterol of the fungal organism resulting in the development of pores in the cell membrane, thereby causing ionic imbalance, disruption of membrane stability, and ultimately cell destruction (Sanglard et al., 2009). These antifungals chiefly consist of amphotericin B, nystatin, and natamycin. The antifungal amphotericin-B shows major potential in the case of systemic IFIs as well as against different pathogenic strains of *Candida, Cryptococcus,* and *Aspergillus* (Sanglard et al., 2009; Lemke et al., 2005). On the other hand, the antifungals natamycin and nystatin because of their low absorption property are being successfully applied in case of various topical infections (Vandeputte et al., 2012).

9.2.3 Pyrimidine analogs

The antifungal drugs pyrimidine analogs, comprising 5-FC and 5-FU represent the synthetic structural analogs of nucleotide cytosine. 5-fluorocytosine is converted to 5-fluorouracil by the enzyme cytosine deaminase, which after transformation to downstream products incorporate into the nucleic acids thereby altering the entire DNA replication/protein synthesis processes. These antifungal agents have been known to possess good antifungal properties in the case of various pathogenic species of *Candida* and *Cryptococcus* genus (Sanglard et al., 2009; Lemke et al., 2005).

9.2.4 Echinocandins

There are three major echinocandin antifungal drugs (micafungin, anidulafungin, and Caspofungin) that generally target the cell wall of different fungal species (Sucher et al., 2009; Mukherjee et al., 2011). These drugs display both in vitro and in vivo fungicidal characteristics in the case of various pathogenic strains of *Candida* as well as fungi-static characteristics in the case of various *Aspergillus* pathogens. The antifungal caspofungin is widely used in the case of invasive aspergillosis while micafungin and anidulafungin are utilized in the case of invasive and esophageal candidiasis (Akins, 2005; Gershkovich et al., 2009).

9.3 Overview of combination therapy

Drug combinations have emerged as novel treatment strategies (Prasad et al., 2016). Given the scarcity of efficient antimicrobial drugs and the scarcity of antimicrobial techniques, there is an ongoing search for new and more potent antimicrobial agents to battle different microbial diseases/infections. Because the establishment of innovative drug alternatives has not kept up with the rate of resistance to presently available treatments, it is critical to look for new ways to tackle different pathogenic yeast diseases. For example, the transcription factor *Upc2*, which controls *Erg* genes expression, is identified as a possible target for antifungal drug establishment. In the organisms like *S. cerevisiae* and *C. glabrata*, several minor chemicals identified from commercially available compounds are found to prevent the azole-regulated activation of *Upc2* and its target genes. Nevertheless, before the *Upc2* transcription factor can be effectively used to enhance antifungal tactics, its entire potential must be studied (Gallo-Ebert et al., 2014). Glucan synthesis, microtubule mechanics, translational extension, etc., are just a few of the novel antifungal strategies that have been discovered (Roemer and Krysan, 2014). Furthermore, combination therapies, in which multiple drugs are given together to address fungal diseases, are one of the most popular approaches. Combination therapeutic approaches provide several advantages, including a wide variety of treatment strategies, the synergy of actions among different medications, reduced drug dosages, and a reduced risk of developing drug resistance. It has been reported that a drugs' synergistic action is mostly attributed to additive impacts on both the cell wall and cell membrane constituents of the concerned pathogenic organism. Impaired cell walls caused by one of the antifungal constituents, for example, enhance the efficacy of drugs that target cellular membrane constituents. Drug permeability across cellular membranes toward intracellular targets could be aided by the reduced cell wall integrity. The synergistic actions of the antifungals viz. azoles and allylamines are because of the inhibition of the similar route at various phases (Tobudic et al., 2010; Rodrigues et al., 2016). Combinatorial strategy, in particular, necessitates a thorough examination of the antagonistic and agonist characteristics of various drugs when used together (Lewis et al., 2001).

Combination therapy combines two or more therapeutic agents to get better outcomes in comparison to conventional monotherapy. Combination therapy usually has one or more of the following goals (Malik et al., 2020):

(1) Lowering the rate at which acquired resistance develops by integrating agents having negligible cross-resistance, requiring the formation of numerous mutations in fast succession, an unusual situation.
(2) Reducing the dosage of combining agents with similar remedial profiles and nonoverlapping toxic effects to achieve efficiency with minimal side effects.
(3) Sensitizing cells to the activity of a drug by using another drug (chemosensitization) or radiation (radiosensitization), generally by changing cell-cycle phase or growth parameters (cytokinetic development).
(4) Leveraging additivity, or even more, greater-than-additive impacts, in the metabolic actions of two drugs to achieve increased efficiency (Malik et al., 2020).

The research studies involving different drug combination therapies are generally based on the following principles:

(1) Mechanisms of action, drug combination with complementary targets in the fungal cells.
(2) Range of activity.
(3) Strengthened pharmacokinetic/pharmacodynamic properties.

Among the above-mentioned three major principles, the majority of the research studies associated with drug combination therapies mainly involve those combinatorial drugs having interrelated mechanisms and sites of action (Fig. 9.3) (Mukherjee et al., 2005).

Moreover, drug combination therapy in the case of the presently available antifungal agents has proven to be very beneficial against different disease-causing fungal pathogens. The combination of two different drugs displays an intensifying killing impact, minimizing the population size of the concerned pathogen and the likeliness of obtaining resistance mutations (Anderson, 2005; Hill and Cowen, 2015a; Spitzer et al., 2017). The continuous usage of a wide range of chemical compounds has identified several potential inhibitory drug combinations against different fungal pathogens, e.g., the leading human fungal pathogen *C. albicans*, and others like *C. neoformans*, etc. (Polvi et al., 2016; Robbins et al., 2015; Spitzer et al., 2011).

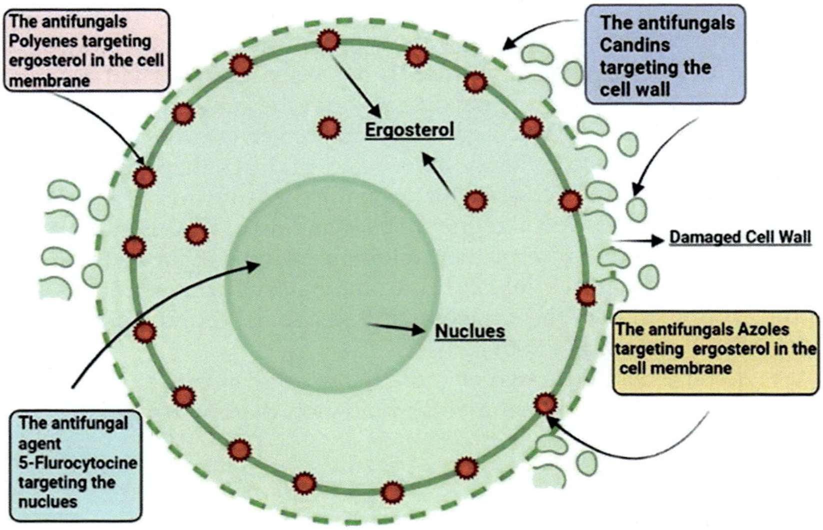

FIGURE 9.3 Diagrammatic demonstration of the sites of action of the most commonly employed antifungal drugs. In general, the antifungal agents, candins, target the fungal cell wall, cause its destruction and eventually enabling other antifungal agents to act. Moreover, the antifungals, viz. polyenes and azoles target the important cell membrane constituent, i.e., ergosterol, causing cell destruction and making way for 5-fluorocytosine to target the nucleus and block the DNA/RNA synthesis (Mukherjee et al., 2005).

Decades ago, for the treatment of cryptococcal meningitis, an investigation performed by Bennett et al. evaluated the antifungal potential of amphotericin B alone and also in combination with 5-fluorocytosine, highlighting the significance of combination therapy using different drug combinations. However, the application of the combinatorial approach in the case of invasive fungal diseases takes place at a slower rate and is restricted only to some antifungal drug classes viz. (a) Amphotericin B in association with 5-fluorocytosine (b) 5FC in association with FLC. With the acceptance of the thirdgeneration azole (i.e., voriconazole) as well as candins (caspofungin)) the interest in drug combination therapy has increased, mainly because these agents share different mechanisms of action (Mukherjee et al., 2005).

The research studies concerning associations of antifungal drugs in comparison to antibacterial/antiviral drugs are in their initial phases of examination and are therefore extremely dynamic (Mukherjee et al., 2005). Various fungal stress response inhibitors such as Hsp90 inhibitors, etc., are known to show different therapeutic benefits when utilized in association with additional antifungals (LaFayette et al., 2010a; Shekhar-Guturja et al., 2016a, 2016b). The usefulness of different antifungal drug combos in the case of various fungal pathogenic organisms has been evaluated in vitro, in vivo, and in different clinical operations (Mukherjee et al., 2005; Vazquez, 2008; Hatipoglu and Hatipoglu, 2013). The important application of the in vitro studies is the proficiency of the strategy and evaluation, which makes it credible to evaluate different drug combos against various pathogenic fungal organisms (Steinbach, 2006). The clinical potency of a drug combo is contributed by various elements viz. fungal virulence and resistance, immune status of the host, and their interplay with the therapeutic compounds. Moreover, the in vitro outcomes in relation to the action of peculiar drug combinations in the case of a particular fungal organism usually vary among different research labs (Johnson et al., 2004).

The antifungal agent fluconazole (a triazole) represents one of the major potent drugs utilized for the proper control and treatment of fungal-associated diseases. Unfortunately, the rise of drug-resistant fungal species has made this first-generation antifungal along with some other antifungal drugs worthless in case of the treatment of growing fungal diseases. Combination therapy has been utilized in the cure of fungal diseases resistant to traditional antimicrobial agents in order to improve the potency of different antifungal drugs. In *C. albicans*, a combination of FLC and dexamethasone successfully blocked the process of replication (Sun et al., 2017). Moreover, the study reported that the administration of both of these antifungal agents produced an additive impact that ultimately resulted in the increase in their antifungal potential in the case of drug-resistant strains in the mammalian cells and animals in culture (Sun et al., 2017).

It has been reported that in the case of the *Fusarium* species, which represent the main plant fungal pathogens in the agricultural sector, the antifungal agent "gentamycin" (an aminoglycoside) showed poor antifungal potency (Lu et al., 2018). The agent was employed in association with FLC to increase the overall potency in order to restrict the proliferation of drug-resistant *C. albicans* cells. At the same time, utilization of each of these agents remarkably increased the antifungal potency and eventually resulted in a pronounced decrease in the fungal load (Lu et al., 2018). Moreover, following coadministration of both of the antimicrobial agents, the investigation was performed in an animal model of *Galleria mellonella* in the same fashion. The results of this study showed that gentamycin and FLC worked together to combat drug-resistant *C. albicans* infections by inhibiting the extrusion pump action that reduces overall drug uptake and potency (Lu et al., 2018).

FLC and 2-isopropyl-5-methylphenol (thymol) were tested in the case of *Candida* species resistant clinical strains in an independent investigation (Sharifzadeh et al., 2018). The use of both antifungal drugs in combination had a cumulative impact against all *C. albicans* strains. Furthermore, the drug association was found to be more effective against the pathogenic strains *C. krusei* and *C. glabrata*. The positive outcomes of the study suggested that using thymol and FLC together could be a useful method for combating multiple drug-resistant pathogenic *Candida* infections (Sharifzadeh et al., 2018). One more study investigated the antifungal action following treating resistant sessile and planktonic *C. albicans* strains using a cocktail of various antifungal drugs (Touil et al., 2018). AMP-B-caspofungin and AMP-B-voriconazole were the drug combos tested in the study. The MIC-values were significantly reduced by both drug combinations in the case of sessile and planktonic *C. albican* strains. Interestingly, a significant reduction in fungal burden was seen, indicating that a cocktail of AMP-B with either CAS/VORI might be used as a remedial strategy to treat both planktonic and sessile *C. albicans* strains (Touil et al., 2018).

9.4 Mechanism of drug interactions

The treatment of growing fungal diseases is usually indicated by various factors such as greater toxicity, lower endurability, etc. Such issues have paved the way to evaluate the potency of drug combination therapy for the proper control and treatment of growing fungal diseases. For increasing the durability of different antimicrobial drugs, the drug combinatorial approach nowadays has been found to be a successful treatment and control strategy. For the successful reduction in the progression of drug resistance, prudent evaluation of the utilization of drug combos should be adopted ideally (Hill and Cowen, 2015a).

Many attempts were made to categorize different mechanisms of the process of drug interactions (Johnson et al., 2004; Jia et al., 2009; Hamdani et al., 2020). Yet, the majority of the drug combinations possess various mechanisms and fail to fit into any particular class. The association between two agents could be synergistic as one of the agents will be potent in case of the disease itself while the second agent will enhance the beneficial effect/concentration of the first agent at a particular target spot (Toews and Bylund, 2005). Here the second agent/drug fails to act directly but might impact the pharmacokinetic properties of the first drug via enhancing the degree of absorption/diffusion. It further directs the first agent to the intended site of action and also helps to reduce the metabolism or removal of the first agent. Such kind of drug association/interaction is further depicted as the bioavailability model (Cokol et al., 2011). In general, synergy among different antifungal agents is regarded either as a positive interaction or a negative interaction. In a positive association, two or more drugs could develop a cumulative impact; on the other hand, antagonism will be linked to a negative association (Cuenca-Estrella, 2004).

Various techniques are established to report the in-vitro association among different antifungal drug combos. The absence of the normalization of such techniques represents the major issue, although a majority of such methods are based on CLSI and EUCAST guidelines (Cuenca-Estrella, 2004; Wayne, 2008). Furthermore, commercialized techniques such as diffusion and Etest techniques have been carried out in certain labs in order to have a proper representation of the in vitro antifungal property of the combining drugs (Cantón et al., 2005; Serena et al., 2005).

Antifungals viz. azoles and AMP-B targeting the cell membrane of the fungal organism describe the process of synergism between these two antifungal agents with flucytosine (Yamamoto et al., 1997). On the other hand, two antifungal agents targeting distinct steps of a similar biological pathway/protein could also enhance the pharmaceutical potency upon the combination, usually in an additive way. Because of the minimum single doses, the drug combinations allow maximum beneficial impacts and minimum toxicities. The antifungal drugs, azoles and terbinafine, are known to inhibit the fungal ergosterol biosynthetic pathway and disrupt the proper functioning of the fungal cell membrane. However, upon the combination of the two antifungal agents the drug potency can be enhanced (Perea et al., 2002).

Furthermore, the combinatorial approach with the antifungal agents which inhibit different biological pathways and converge on similar important biological activity could also result in synergistic drug associations (Spitzer et al., 2017b). It has been reported that the drug association network of about 21 antibiotic agents has displayed successful synergistic drug interactions among specific categories of antibiotic agents which successfully target distinct biologically important roles and activities (Yeh et al., 2006). The drug combinations between the antifungal agents, viz. echinocandins, azoles, and polyenes display successful synergistic properties and at once target the cell wall and the cell membrane of the concerned fungal organism (Shalit et al., 2003).

The choice of agents used for combination therapies could be achieved via multiple approaches, on the basis of either the previous understanding of the resistance processes or on the systematized assessment for potent drug combinations. Inhibitors of TOR signaling (Cowen et al., 2009), protein kinase C (LaFayette et al., 2010b), Hsp90 (Shekhar-Guturja et al., 2016c), and calcineurin (Cruz et al., 2002), etc., represent some of the important stress response inhibitors which in the presence of many presently available antifungal agents display tremendous remedial advantages (Robbins et al., 2017). Of late, Nishikawa et al. (2016), recognized that in *C. glabrata* a little molecule "iKIX1"revoke the mechanism of azole drug resistance in vivo by damaging the association among the mediator complex and the Pdr5 efflux pump transcriptional activator *Pdr1*, essential for azole drug resistance (Nishikawa et al., 2016). Additionally, multiple research investigations have proposed that targeting the development of aneuploidies connected with the process of drug resistance in the leading pathogenic fungi *C. albicans* might offer a feasible remedial approach. By framing an evolutionary device that could target both the genotypic diversity as well as health, the phenomenon of azole drug resistance was relapsed in *C. albicans* bearing the i(5L) aneuploidy (Chen et al., 2015), pointing to the capability of targeting the modulators of the process of drug resistance in association with different antifungal drugs for the better cure and control of various systemic diseases of fungi (Robbins et al., 2017).

Many effective antifungal drug combos are reported by various in vitro/in vivo and clinical studies. The combo of 5-FC with AMP-B or flucytosine is the most frequently employed antifungal drug combination in the case of cryptococcosis caused by the pathogenic species belonging to *Cryptococcus* genus. The newly recommended antifungal agents such as voriconazole and caspofungin were also employed for combination therapy. The combinations of the novel drugs having clinical usefulness for the cure and control of diseases caused by various pathogenic *Candida* species involve a combo of voriconazole and micafungin, a combo of 5-fluorocytosine and CAS, a combo of CAS and amphotericin B, or a combination of fluconazole and terbinafine. On the other hand, the combinations with clinical usefulness for the cure and control of filamentous fungal diseases specifically caused by *Aspergillus* species involve a combo of VORI and CAS, VORI, and liposomal AMP-B, VORI and TERB, and 5-fluorocytosine and CAS (Mukherjee et al., 2005). One limitation is that because antifungal drug-drug associations are strain-specific, there is no way to choose a single interaction that would work for the entire representatives of a species. This involves the evaluation of the best mixture for each clinical strain. Clearly, it adds a significant amount of time and price to the process, and it should not be done on a regular basis. As a result, in the therapeutic situation, antifungal combination therapies must be used only in individuals who have failed to respond to monotherapies and have high MICs for individual antifungals (Mukherjee et al., 2005).

Multiple categories of antifungal agents are also being used in triple combinations/combos. On the basis of the synergistic action of the interaction of CAS, 5-FC, and AMB in the case of *Aspergillus* species, combos of three or more antifungal agents have also been explored and utilized in the clinical context (Dannaoui et al., 2004). There was also a synergistic effect in double combinations in such an event. Due to the lack of a truly synergic impact in dual interactions, some triple combinations, such as CAS, 5FC, and VRZ, showed contradictory antagonistic and synergistic effects against the species of the *Aspergillus* genus (Dannaoui et al., 2004). The findings of interactions among voriconazole, amphotericin B, and caspofungin in the case of *A. fumigatus*, *A. flavus*, and *A. terreus* come to the realization that in vitro antifungal dosage determines the drug combination's strength and prospective efficiency (Meletiadis et al., 2007). Following this, the synergistic effect was seen at lower concentration of all these three drugs, and on other hand raised antifungal concentrations displayed antagonistic effect. It has also been reported that the availability of a third antifungal agent impacts the dual interaction of two others (Meletiadis et al., 2007).

Combinations between antifungal agents and nonantifungal agents are also being successfully demonstrated in various research investigations. Antifungal agents are being frequently blended with different nonantimicrobial compounds, and various research studies concerning human recombinant antibodies, calcineurin blockers, proton pump blockers, microbial metabolites, etc., are accessible (Finquelievich et al., 2014). The exposure of *C. albicans* cells to fluconazole results in calcineurin suppression which proves fatal to the concerned organism (Cruz et al., 2002). The suggested method involves fluconazole inhibiting ergosterol biosynthesis, which promotes calcineurin blocker entry or prevents calcineurin formation. The interaction of fluconazole and cyclosporine produced a fungicidal synergistic effect in the case of *C. albicans* having an unknown mechanism of activity not based on the multidrug efflux pumps coded by *CDR1*, *CDR2*, and other related genes (Marchetti et al., 2003).

In different *Candida* species, "Mycograb," a human recombinant antibody against HSP90 of *Candida* spp., has been found to display an intrinsic antifungal action and also an additive impact on combination with AMP-B or their lipidic/liposomal formulations and caspofungin, both in vitro as well as in vivo and in the case of humans for the cure of invasive candidiasis (Finquelievich et al., 2014). Presently "Mycograb" has been evaluated in various clinical trials in persons with invasive candidiasis (*C. albicans*, *C. krusei*, and *C. glabrata*) receiving AMP-B, with FLC and CAS for the cure of infections associated with *C. neoformans* (Finquelievich et al., 2014; Rowlands et al., 2006).

One more important suggested strategy is the interaction of different antifungal agents with various compounds of known antibacterial properties (Liu et al., 2014). Data is provided for antibacterials that have a mechanism of action relevant to targets found in both cell models of the concerned organisms and a joint impact. Rifampicin/rifabutin inhibits the transcriptional process via targeting the enzyme RNA polymerase. The antifungal agents AMP-B/nystatin with a mechanism of activity linked at the cell membrane site have been found to increase the passage of active concentrations of rifabutin/rifampicin acting on Rna polymerase. The mechanism has been found in different pathogenic species of *Cryptococcus*, *Candida*, *Aspergillus*, etc. (Cuenca-Estrella, 2004).

9.5 Benefits of using drug combinations/applications of combinatorial approach

Drug combination therapy involving different antifungal agents represents an important currently available treatment and control strategy in case of the growing invasive fungal infections which present the medical practitioners with effectual devices. The possible merits of the drug combinatorial approach are that it aids in the improvement of the antifungal remedial efforts used against different human fungal infections and infections caused by various drug-resistant human fungal pathogenic organisms, as drug combination therapy aims at multiple fungal targets simultaneously. But, the actual benefits are more likely achieved with some specific drug combinations and in the case of some specific fungal diseases and some particularly affected individuals (Finquelievich et al., 2014).

In comparison to monotherapies, combination therapies provide numerous benefits (Fig. 9.4) (Qadri et al., 2021c).

The possible advantages of the utilization of combination therapy involve the following:

(a) Wide range of potency.
(b) Higher potential of the drugs used in combination than in comparison to each one utilized in case of monotherapy.
(c) Better security and endurability.
(d) Decrease in the number of concerned resistant microbial species (Lewis et al., 2001).

The increased inhibitory impacts of drug combinations make it possible to use reduced single drug doses and decreased treatment length, thereby diminishing the toxicity of the host (Spitzer et al., 2017a; Hill and Cowen, 2015b). Fortunately, particular drug associations/combinations display the ability to inverse the process of drug resistance via the phenomenon referred to as *selection inversion* (Baym et al., 2016), which had been recently demonstrated utilizing the tetracycline-resistant *E.coli* (Stone et al., 2016). In the current times, drug combination remedies represent a high degree of supervision and management in the case of cure and control of various highly complicated infectious ailments including AIDS and malaria; yet, the drug combinations stay comparatively unmarked as antifungal therapies.

It has been reported that the synergistic approach of drug combination therapy is a sound and empirical method to identify drugs having new mechanisms of action. It could also possibly minimize the dosage of single drug usage having enhanced drug potency, and eventually lesser drug toxicity. The application of targeting two or more drug targets at the same

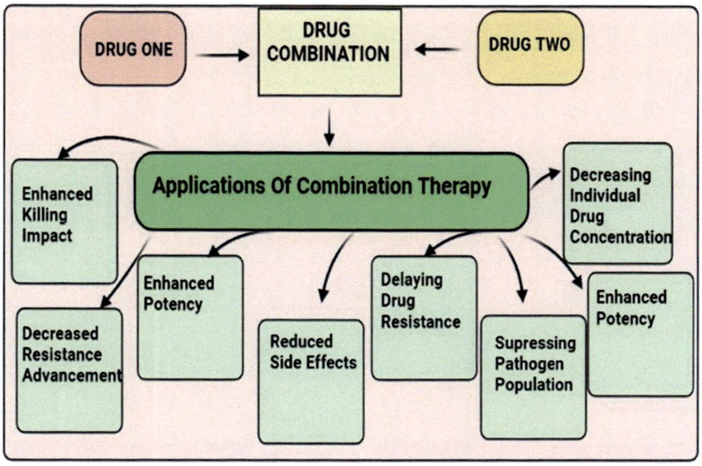

FIGURE 9.4 Applications of drug combination therapy. In comparison to monotherapies, combination therapies provide numerous benefits (Qadri et al., 2021).

TABLE 9.3 Representing the most frequently utilized antifungal drug combinatorial remedies (Malik et al., 2020).

S.No	Drug combo	Concerned fungal infection/Disease
1.	• Combination of amphotericin-B and flucytosine. • Combination of amphotericin-B and azole. • Combination of echinocandin and azole.	Invasive candidiasis (caused by various pathogenic species of *Candida*).
2.	• Combination of azole/amphotericin-B and echinocandin. • Combination of voriconazole and anidulafungin.	Invasive aspergillosis (caused by various pathogenic species of *Candida*).

time is constant with the notion that a particular infection/disease is a systematized and complex result brought about by multiple impacts (Cui et al., 2015). Table 9.3, represents the most frequently utilized antifungal drug combinatorial remedies (Malik et al., 2020).

Moreover, the establishment of the process of drug resistance could be minimized to a large extent with the aid of multiple target approaches (Cui et al., 2015). In general, there are three distinct stages (Fig. 9.5) for the system of synergistic antifungal drug combinations viz:

(1) Stage I: In-vitro evaluation
(2) Stage II: In-vivo animal model confirmations, and
(3) Stage III: The clinical examinations (Cui et al., 2015).

Fora very long period, the clinical utilization of two or more antifungal drugs for the proper cure and control of serious invasive fungal diseases has been successfully reported.

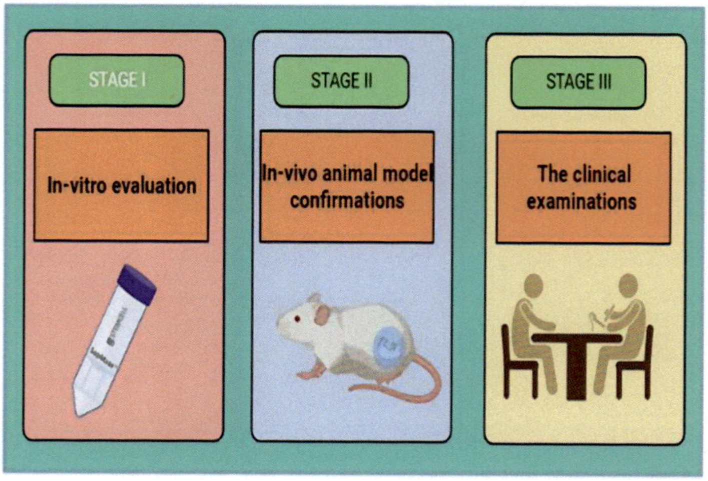

FIGURE 9.5 Representation of the three distinct stages of the system of synergistic antifungal drug combinations.

The very first utilization of the synergistic therapy in case of invasive candidiasis involves the two antifungal agents, viz. FLC and Amp-B. Drug monotherapy involving flucytosine very often causes drug resistance and unpredicted lethal consequences, while amphotericin B decreased such issues (Cui et al., 2015; Hatipoglu and Hatipoglu, 2013). Such beneficial drug association had been approved by the "Infectious Diseases Society of America"recommendations for the proper control and cure of candidiasis in individuals with specific conditions, together with other severe *Candida* infections mainly the endovascular diseases, central neuron system infections, and also the serious intraabdominal candidiasis (Cui et al., 2015).

There are few in vivo and clinical research studies involving the utilization of antifungal drug combinations in the case of some of the most serious and challenging invasive fungal diseases. Further investigations are needed before the combination of azoles (FLC) and polyenes (Amp B) can be used as standard anti-*Candida* therapy. Prior to the addition of a second antifungal agent to the treatment regimen in the case of different antifungal drug-resistant and invasive fungal diseases, scientists must examine numerous combos of antifungal drug-drug associations. Moreover, in a person who is already immunosuppressed and in a critical situation, mismatched antifungal drug combos, unfortunately, might cause antagonistic action, low efficiency, and increasing unpleasant effects (Maurya et al., 2020). Table 9.4 represents some of the specific drug combos, displaying a synergistic or antagonistic action (Maurya et al., 2020).

There has been an inconsistency among the rate of antifungal resistance phenomenon and the establishment of novel antifungal agents. Therefore, the concept of mixing different drugs/compounds with antifungal properties for the proper cure and control of invasive fungal diseases seems attractive for various medical practitioners, especially after various in-vitro studies displayed synergism among different antifungal drugs. A number of randomized controlled experiments have been reported concerning the potency and security of different antifungal drug combinations; however, there are certain factors like high cost, minimum cases, and multiple contradictory elements which at times may result in negative outcomes. Moreover, the absence of proper consent in various clinical settings emphasizes the significance of the requirement for additional studies regarding antifungal drug combination therapies (Campitelli et al., 2017).

TABLE 9.4 Specific drug combos display a synergistic or antagonistic action (Maurya et al., 2020).

S.No	Drug combo	Feasible impact	Feasible activity mechanism
1.	Combination of amphotericin B with flucytosine	Synergistic/Semi antagonistic	Destruction of the fungal cell wall and enhanced uptake of flucytosine.
2.	Combination of amphotericin-B with azole	Antagonistic	Alterations in the cell wall.
3.	Combination of azole with flucytosine	Synergistic	Destruction of the fungal cell wall and enhanced uptake of flucytosine.
4.	Combination of flucytosine with azole	Synergistic	Destruction of the fungal cell wall and uptake of flucytosine.

9.6 Future perspectives

Infectious diseases are constantly posing a serious threat to world health and finances, necessitating regular exploration, in the concerned field (Sheikh et al., 2021). For the appropriate treatment of the growing infectious diseases, there is a need for the establishment of various molecular, genetic, and immunological procedures (MIR et al., Mir, Mir, 2015, Mir & Al-baradie, 2013, Mir and Albaradie, 2014, Mir et al., 2021). Moreover, human fungal infections are rising at a rapid rate mainly because of the growing population of individuals with a compromised immune system. The constant utilization of multiple antifungal drugs has ultimately led to the formation of resistant organisms (Qadri et al., 2021a). The phenomenon of MDR associated with the growing human fungal diseases is without any doubt creating a big risk to the worldwide health system, and if nothing is done to stop the development of MDR, worldwide mortality rates could increase. Therefore, it becomes very essential to switch to the latest techniques and methodologies to master the issue of the rapidly growing antifungal drug resistance phenomenon. The phenomenon of MDR has already wrought havoc on diseases like TB, leprosy, cancers, malaria, and AIDS. In the United States only, there are over 10,000 active clinical studies investigating various drug combinations for diseases like cancer, infectious illnesses, metabolic, cardiovascular, autoimmune, and neurological problems. Now, research in combination therapy with synergistic efficacies has grown. Combination treatment, which involves the use of two or more agents at the same time to treat a particular disease condition, has shown excellent potential among numerous techniques. Combination remedies have the benefits of minimizing drug resistance since a pathogen is less inclined to produce resistance to many agents at the same time (Buchbinder and Desai, 2016). The idea of combination treatment is to employ drugs that function through different mechanisms, reducing the chances of the establishment of resistance. When several agents having distinct properties are coupled, each agent is used at its optimum dosage without causing unacceptably severe symptoms (Hanahan and Weinberg, 2011). Long-term survivability for a larger patient population has been proved to be an advantage of combination therapy.

In a nutshell, the novel drug combinatorial approach covered in this chapter offers a powerful support system to overcome the globally arising antifungal drug resistance issue as well as guides toward the proper recognition of additional potent and latest approaches which could be utilized in an associative way for the complete eradication of this serious problem. The chapter focusing on various aspects of drug combination therapies might help in the advancement of the antifungal remedial efforts in the case of fungal infections and diseases resistant to presently available treatment therapies via aiming for various fungal targets simultaneously. It will further aid in upgrading our understanding regarding different fungal pathogens and fungal diseases to stimulate much better and quality research and also increase our abilities to overcome such human fungal infections/diseases.

References

Akins, R., 2005. An update on antifungal targets and mechanisms of resistance in Candida albicans. Med. Mycol. 43, 285–318.

Anderson, J.B., 2005. Evolution of antifungal-drug resistance: mechanisms and pathogen fitness. Nat. Rev. Microbiol. 3, 547–556.

Arendrup, M.C., Boekhout, T., Akova, M., et al., 2014. ESCMID and ECMM joint clinical guidelines for the diagnosis and management of rare invasive yeast infections. Clin. Microbiol. Infect. 20, 76–98.

Baym, M., Stone, L.K., Kishony, R., 2016. Multidrug evolutionary strategies to reverse antibiotic resistance. Science 351.

Bhattacharya, S., Holowka, T., Orner, E.P., Fries, B.C., 2019. Gene duplication associated with increased fluconazole tolerance in Candida auris cells of advanced generational age. Sci. Rep. 9, 1–13.

Bhattacharya, S., Sae-Tia, S., Fries, B.C., 2020. Candidiasis and mechanisms of antifungal resistance. Antibiotics 9, 312.

Bhattacharya, S., Sobel, J.D., White, T.C., 2016. A combination fluorescence assay demonstrates increased efflux pump activity as a resistance mechanism in azole-resistant vaginal Candida albicans isolates. Antimicrob. Agents Chemother. 60, 5858–5866.

Biswal, M., Rudramurthy, S., Jain, N., et al., 2017. Controlling a possible outbreak of Candida auris infection: lessons learnt from multiple interventions. J. Hosp. Infect. 97, 363–370.

Bongomin, F., Gago, S., Oladele, R.O., Denning, D.W., 2017. Global and multi-national prevalence of fungal diseases—estimate precision. J. Fungi 3, 57.

Borman, A.M., Szekely, A., Johnson, E.M., 2016. Comparative pathogenicity of United Kingdom isolates of the emerging pathogen Candida auris and other key pathogenic Candida species. MSphere 1 e00189-16.

Brown, G.D., Denning, D.W., Gow, N.A., Levitz, S.M., Netea, M.G., White, T.C., 2012a. Hidden killers: human fungal infections. Sci. Transl. Med. 4, 165rv13-rv13.

Brown, G.D., Denning, D.W., Levitz, S.M., 2012b. Tackling human fungal infections. In: American Association for the Advancement of Science.

Brown, V., Sexton, J.A., Johnston, M., 2006. A glucose sensor in Candida albicans. Eukaryot. Cell 5, 1726–1737.

Bryers, J.D., Ratner, B.D., 2004. Bioinspired implant materials befuddle bacteria. ASM 70, 232.

Buchbinder, E.I., Desai, A., 2016. CTLA-4 and PD-1 pathways: similarities, differences, and implications of their inhibition. Am. J. Clin. Oncol. 39, 98.

Calabrese, D., Bille, J., Sanglard, D., 2000. A novel multidrug efflux transporter gene of the major facilitator superfamily from Candida albicans (FLU1) conferring resistance to fluconazole. Microbiology 146, 2743–2754.

Calderone, R., Gow, N.A., 2002. Host Recognition by Candida Species. Candida and Candidiasis. ASM Press, Washington, DC, pp. 67–86.

Calvo, B., Melo, A.S., Perozo-Mena, A., et al., 2016. First report of Candida auris in America: clinical and microbiological aspects of 18 episodes of candidemia. J. Infect. 73, 369–374.

Campitelli, M., Zeineddine, N., Samaha, G., Maslak, S., 2017. Combination antifungal therapy: a review of current data. J. Clin. Med. Res. 9, 451.

Cantón, E., Pemán, J., Gobernado, M., Viudes, A., Espinel-Ingroff, A., 2005. Synergistic activities of fluconazole and voriconazole with terbinafine against four Candida species determined by checkerboard, time-kill, and Etest methods. Antimicrob. Agents Chemother. 49, 1593–1596.

Chakrabarti, A., Sood, P., Rudramurthy, S.M., et al., 2015. Incidence, characteristics and outcome of ICU-acquired candidemia in India. Intensive Care Med. 41, 285–295.

Chatterjee, S., Alampalli, S.V., Nageshan, R.K., Chettiar, S.T., Joshi, S., Tatu, U.S., 2015. Draft genome of a commonly misdiagnosed multidrug resistant pathogen Candida auris. BMC Genet. 16, 1–16.

Chen, G., Mulla, W.A., Kucharavy, A., et al., 2015. Targeting the adaptability of heterogeneous aneuploids. Cell 160, 771–784.

Chowdhary, A., Sharma, C., Kathuria, S., Hagen, F., Meis, J.F., 2014. Azole-resistant Aspergillus fumigatus with the environmental TR46/Y121F/T289A mutation in India. J. Antimicrob. Chemother. 69, 555–557.

Cokol, M., Chua, H.N., Tasan, M., et al., 2011. Systematic Exploration of Synergistic Drug Pairs, vol. 7, p. 544.

Cortegiani, A., Misseri, G., Fasciana, T., Giammanco, A., Giarratano, A., Chowdhary, A., 2018a. Epidemiology, clinical characteristics, resistance, and treatment of infections by Candida auris. J. Intensive Care 6, 69.

Cortegiani, A., Misseri, G., Fasciana, T., Giammanco, A., Giarratano, A., Chowdhary, A., 2018b. Epidemiology, clinical characteristics, resistance, and treatment of infections by Candida auris. J. Intensive Care 6, 1–13.

Cowen, L.E., Sanglard, D., Howard, S.J., Rogers, P.D., Perlin, D., 2015. Mechanisms of antifungal drug resistance. Cold Spring Harb. Perspect. Med. 5, a019752.

Cowen, L.E., Singh, S.D., Köhler, J.R., et al., 2009. Harnessing Hsp90 function as a powerful, broadly effective therapeutic strategy for fungal infectious disease. Proc. Natl. Acad. Sci. 106, 2818–2823.

Cruz, M.C., Goldstein, A.L., Blankenship, J.R., et al., 2002. Calcineurin is essential for survival during membrane stress in Candida albicans. EMBO Rep. 21, 546–559.

Cuenca-Estrella, M., 2004. Combinations of antifungal agents in therapy—what value are they? J. Antimicrob. Chemother. 54, 854–869.

Cui, J., Ren, B., Tong, Y., Dai, H., Zhang, L., 2015. Synergistic combinations of antifungals and anti-virulence agents to fight against Candida albicans. Virulence 6, 362–371.

Dadar, M., Tiwari, R., Karthik, K., Chakraborty, S., Shahali, Y., Dhama, K., 2018. Candida albicans-Biology, molecular characterization, pathogenicity, and advances in diagnosis and control—An update. Microb. Pathog. 117, 128–138.

Dannaoui, E., Lortholary, O., F, D., 2004. In vitro evaluation of double and triple combinations of antifungal drugs against Aspergillus fumigatus and Aspergillus terreus. Antimicrob. Agents Chemother. 48, 970–978.

De Hoog, G.S., Ahmed, S.A., Danesi, P., Guillot, J., Gräser, Y., 2018. Distribution of pathogens and outbreak fungi in the fungal kingdom. In: Emerging and Epizootic Fungal Infections in Animals. Springer, pp. 3–16.

Denning, D.W., Hope, W., 2010. Therapy for fungal diseases: opportunities and priorities. Trends Microbiol. 18, 195–204.

Escandón, P., Chow, N.A., Caceres, D.H., et al., 2019. Molecular epidemiology of Candida auris in Colombia reveals a highly related, countrywide colonization with regional patterns in amphotericin B resistance. Clin. Infect. Dis. 68, 15–21.

Finkel, J.S., Mitchell, A.P., 2011. Genetic Control of Candida Albicans Biofilm Development, vol. 9, pp. 109–118.

Finquelievich, J., Tur-Tur, C., Eraso, E., Jauregizar, N., Quindós, G., Giusiano, G., 2014. Combination antifungal therapy: a strategy for the management of invasive fungal infections. Rev Esp Quimioter. 27, 141–158.

Firacative, C., 2020. Invasive fungal disease in humans: are we aware of the real impact? Mem. Inst. Oswaldo Cruz 115.

Ford, C.B., Funt, J.M., Abbey, D., et al., 2015. The evolution of drug resistance in clinical isolates of Candida albicans. elife 4, e00662.

Gallo-Ebert, C., Donigan, M., Stroke, I.L., et al., 2014. Novel antifungal drug discovery based on targeting pathways regulating the fungus-conserved Upc2 transcription factor. Antimicrob. Agents Chemother. 58, 258–266.

Gaur, M., Puri, N., Manoharlal, R., et al., 2008. MFS transportome of the human pathogenic yeast Candida albicans. BMC Genom. 9, 1–12.

Gershkovich, P., Wasan, E.K., Lin, M., et al., 2009. Pharmacokinetics and biodistribution of amphotericin B in rats following oral administration in a novel lipid-based formulation. J. Antimicrob. Chemother. 64, 101–108.

Govorushko, S., Rezaee, R., Dumanov, J., Tsatsakis, A., 2019. Poisoning associated with the use of mushrooms: a review of the global pattern and main characteristics. Food Chem. Toxicol. 128, 267–279.

Gudlaugsson, O., Gillespie, S., Lee, K., et al., 2003. Attributable mortality of nosocomial candidemia, revisited. Clin. Infect. Dis. 37, 1172–1177.

Gulshan, K., Moye-Rowley, W., 2007. Multidrug resistance in fungi. Eukaryot. Cell 6, 1933–1942.

Hamdani, S.S., Bhat, B.A., Tariq, L., et al., 2020. Antibiotic resistance: the future disaster. Int. J. Res. Appl. Sci. Biotechnol. 7.

Han, T.-L., Cannon, R.D., Villas-Bôas, S., 2011. The metabolic basis of Candida albicans morphogenesis and quorum sensing. Fungal Genet. Biol. 48, 747–763.

Hanahan, D., Weinberg, R., 2011. Hallmarks of cancer: the next generation. Cell 144, 646–674.

Hatipoglu, N., Hatipoglu, H., 2013. Combination antifungal therapy for invasive fungal infections in children and adults. Expert Rev. Anti Infect. Ther. 11, 523–535.

Hill, J.A., Cowen, L.E., 2015. Using combination therapy to thwart drug resistance. Future Microbiol. 10, 1719–1726.

Jia, J., Zhu, F., Ma, X., Cao, Z.W., Li, Y.X., Chen, Y.Z., 2009. Mechanisms of drug combinations: interaction and network perspectives. Nat. Rev. Drug Discov. 8, 111–128.

Johnson, M.D., Macdougall, C., Ostrosky-Zeichner, L., Perfect, J.R., Rex, J., 2004. Combination antifungal therapy. Antimicrob. Agents Chemother. 48, 693–715.

Kainz, K., Bauer, M.A., Madeo, F., Carmona-Gutierrez, D., 2020. Fungal infections in humans: the silent crisis. Microb. Cell 7, 143.

Kohler, J., Hube, B., Puccia, R., Casadevall, A., Perfect, J., 2017. Fungi that infect humans. Microbiol. Spectr. https://doi.org/10.1128/microbiolspec. In.: FUNK-0014-2016.

Kojic, E.M., Darouiche, R., 2004. Candida infections of medical devices. Clin. Microbiol. Rev. 17, 255–267.

Konopka, J.B., Casadevall, A., Taylor, J.W., Heitman, J., Cowen, L., 2019. One Health: Fungal Pathogens of Humans, Animals, and Plants.
Ksiezopolska, E., Gabaldón, T., 2018. Evolutionary emergence of drug resistance in Candida opportunistic pathogens. Genes 9, 461.
Kühbacher, A., Burger-Kentischer, A., Rupp, S., 2017. Interaction of Candida species with the skin. Microorganisms 5, 32.
Kuo, D., Tan, K., Zinman, G., Ravasi, T., Bar-Joseph, Z., Ideker, T., 2010. Evolutionary divergence in the fungal response to fluconazole revealed by soft clustering. Genome Biol. 11, 1–12.
Lafayette, S.L., Collins, C., Zaas, A.K., et al., 2010. PKC signaling regulates drug resistance of the fungal pathogen Candida albicans via circuitry comprised of Mkc1, calcineurin, and Hsp90. PLoS Pathog. 6, e1001069.
Larkin, E., Hager, C., Chandra, J., et al., 2017. The emerging pathogen Candida auris: growth phenotype, virulence factors, activity of antifungals, and effect of SCY-078, a novel glucan synthesis inhibitor, on growth morphology and biofilm formation. Antimicrob. Agents Chemother. 61 e02396-16.
Lemke, A., Kiderlen, A., Kayser, O., 2005. Amphotericin B. Appl. Microbiol. Biotechnol. 151–162.
Lewis, R.E., Kontoyiannis, D.P., 2001. Rationale for combination antifungal therapy. Pharmacotherapy 21, 149S–164S.
Liu, S., Hou, Y., Chen, X., Gao, Y., Li, H., Sun, S., 2014. Combination of fluconazole with non-antifungal agents: a promising approach to cope with resistant Candida albicans infections and insight into new antifungal agent discovery. Int. J. Antimicrob. Agents 43, 395–402.
Lu, M., Yu, C., Cui, X., Shi, J., Yuan, L., Sun, S., 2018. Gentamicin synergises with azoles against drug-resistant Candida albicans. Int. J. Antimicrob. Agents 51, 107–114.
Malik, M.A., Wani, M.Y., Hashmi, A.A., 2020. Combination therapy: current status and future perspectives. In: Combination Therapy against Multidrug Resistance. Elsevier, pp. 1–38.
Marchetti, O., Moreillon, P., Entenza, J.M., et al., 2003. Fungicidal synergism of fluconazole and cyclosporine in Candida albicans is not dependent on multidrug efflux transporters encoded by the CDR1, CDR2, CaMDR1, and FLU1 genes. Antimicrob. Agents Chemother. 47, 1565–1570.
Marichal, P., Vanden Bossche, H., Odds, F.C., et al., 1997. Molecular biological characterization of an azole-resistant Candida glabrata isolate. Antimicrob. Agents Chemother. 41, 2229–2237.
Maurya, I.K., Semwal, R.B., Semwal, D.K., 2020. Combination therapy against human infections caused by Candida species. In: Combination Therapy against Multidrug Resistance. Elsevier, pp. 81–94.
Meletiadis, J., Stergiopoulou, T., O'shaughnessy, E.M., Peter, J., Walsh, T., 2007. Concentration-dependent synergy and antagonism within a triple antifungal drug combination against Aspergillus species: analysis by a new response surface model. Antimicrob. Agents Chemother. 51, 2053–2064.
Mir M, Albaradeh R, Agrewala J. Innate–Effector Immune Response Elicitation against Tuberculosis through Anti-B7-1 (Cd80) And Anti-B7-2 (Cd86) Signaling in Macrophages.
Mir MA. Costimulation and Costimulatory Molecules.
Mir, M.A., 2015. Developing Costimulatory Molecules for Immunotherapy of Diseases. Academic Press.
Mir, M.A., Al-Baradie, R., 2013. Tuberculosis time bomb-A global emergency: need for alternative vaccines, 1, 77–82.
Mir, M.A., Albaradie, R., 2014. Inflammatory mechanisms as potential therapeutic targets in stroke. Adv. Neuroimmune Biol. 5, 199–216.
Mir, M.A., Bhat, B.A., Sheikh, B.A., Rather, G.A., Mehraj, S., Mir, W.R., 2021. Nanomedicine in human health therapeutics and drug delivery: nanobiotechnology and nanobiomedicine. In: Applications of Nanomaterials in Agriculture, Food Science, and Medicine. IGI Global, pp. 229–251.
Montagna, M., Lovero, G., Borghi, E., et al., 2014. Candidemia in Intensive Care Unit: A Nationwide Prospective Observational Survey (GISIA-3 Study) and Review of the European Literature from 2000 through 2013.
Morales-López, S.E., Parra-Giraldo, C.M., Ceballos-Garzón, A., et al., 2017. Invasive infections with multidrug-resistant yeast Candida auris. Colombia 23, 162.
Mukherjee, P., Sheehan, D., Puzniak, L., Schlamm, H., Ghannoum, M., 2011. Echinocandins: are they all the same? J. Chemother. 23, 319–325.
Mukherjee, P.K., Sheehan, D.J., Hitchcock, C.A., Ghannoum, M., 2005. Combination treatment of invasive fungal infections. Clin. Microbiol. Rev. 18, 163–194.
Musa, K., Ahmed, M.A., Shahpudin, S.N., et al., 2018. Resistance of Candida Glabrata to Drugs and the Host Immune System.

Navalkele, B.D., Revankar, S., Chandrasekar, P., 2017. Candida auris: a worrisome, globally emerging pathogen. Expert Rev. Anti Infect. Ther. 15, 819–827.

Nishikawa, J.L., Boeszoermenyi, A., Vale-Silva, L.A., et al., 2016. Inhibiting fungal multidrug resistance by disrupting an activator–mediator interaction. Nature 530, 485–489.

Pam, V.K., Akpan, J.U., Oduyebo, O.O., et al., 2012. Fluconazole susceptibility and ERG11 gene expression in vaginal Candida species isolated from Lagos Nigeria. Int. J. Mol. Epidemiology Genet. 3, 84.

Pappas, P.G., Kauffman, C.A., Andes, D.R., et al., 2016. Clinical practice guideline for the management of candidiasis: 2016 update by the Infectious Diseases Society of America. Clin. Infect. Dis. 62, e1–e50.

Pappas, P.G., Rex, J.H., Sobel, J.D., et al., 2004. Guidelines for treatment of candidiasis. Clin. Infect. Dis. 38, 161–189.

Pardasani, A., 2000. Oral antifungal agents used in dermatology. Curr. Probl. Dermatol. 12, 270–275.

Perea, S., Gonzalez, G., Fothergill, A.W., Sutton, D.A., Rinaldi, M., 2002. In vitro activities of terbinafine in combination with fluconazole, itraconazole, voriconazole, and posaconazole against clinical isolates of Candida glabrata with decreased susceptibility to azoles. J. Clin. Microbiol. 40, 1831–1833.

Pfaller, M.A., Messer, S.A., Moet, G.J., Jones, R.N., Castanheira, M., 2011a. Candida bloodstream infections: comparison of species distribution and resistance to echinocandin and azole antifungal agents in Intensive Care Unit (ICU) and non-ICU settings in the SENTRY Antimicrobial Surveillance Program (2008–2009). Int. J. Antimicrob. Agents 38, 65–69.

Pfaller, M.A., Moet, G.J., Messer, S.A., Jones, R.N., Castanheira, M., 2011b. Geographic variations in species distribution and echinocandin and azole antifungal resistance rates among Candida bloodstream infection isolates: report from the SENTRY Antimicrobial Surveillance Program (2008 to 2009). J. Clin. Microbiol. 49, 396–399.

Poláková, S., Blume, C., Zárate, J.Á., et al., 2009. Formation of new chromosomes as a virulence mechanism in yeast Candida glabrata. Proc. Natl. Acad. Sci. 106, 2688–2693.

Polvi, E.J., Averette, A.F., Lee, S.C., et al., 2016. Metal chelation as a powerful strategy to probe cellular circuitry governing fungal drug resistance and morphogenesis. PLoS Genet. 12, e1006350.

Prasad, R., Goffeau, A., 2012. Yeast ATP-binding cassette transporters conferring multidrug resistance. Annu. Rev. Microbiol. 66, 39–63.

Prasad, R., Kapoor, K., 2005. Multidrug resistance in yeast Candida. Int. Rev. Cytol. 242, 215–248.

Prasad, R., Shah, A.H., Dhamgaye, S., 2014. Mechanisms of drug resistance in fungi and their significance in biofilms. In: Antibiofilm Agents. Springer, pp. 45–65.

Prasad, R., Shah, A.H., Rawal, M.K., 2016. Antifungals: mechanism of action and drug resistance. Adv. Exp. Med. Biol. 892, 327–349.

Qadri, H., Haseeb, A., Mir, M., 2021a. Novel strategies to combat the emerging drug resistance in human pathogenic Microbes. Curr. Drug Targets 22, 1–13.

Qadri, H., Qureshi, M.F., Mir, M.A., Shah, A.H., 2021c. Glucose-The X Factor for the survival of human fungal pathogens and disease progression in the host. Microbiol. Res. 126725.

Ramage, G., Martínez, J.P., López-Ribot, J.L., 2006. Candida biofilms on implanted biomaterials: a clinically significant problem. FEMS Yeast Res. 6, 979–986.

Redhu, A.K., Banerjee, A., Shah, A.H., et al., 2018. Molecular basis of substrate polyspecificity of the Candida albicans Mdr1p Multidrug/H+ antiporter. J. Mol. Biol. 430, 682–694.

Robbins, N., Caplan, T., Cowen, L.E., 2017. Molecular evolution of antifungal drug resistance. Annu. Rev. Microbiol. 71, 753–775.

Robbins, N., Spitzer, M., Yu, T., et al., 2015. An antifungal combination matrix identifies a rich pool of adjuvant molecules that enhance drug activity against diverse fungal pathogens. Cell Rep. 13, 1481–1492.

Rodaki, A., Bohovych, I.M., Enjalbert, B., et al., 2009. Glucose promotes stress resistance in the fungal pathogen Candida albicans. Mol. Biol. Cell 20, 4845–4855.

Rodrigues, M.E., Silva, S., Azeredo, J., Henriques, M., 2016. Novel strategies to fight Candida species infection. Crit. Rev. Microbiol. 42, 594–606.

Roemer, T., Krysan, D., 2014. Antifungal drug development: challenges, unmet clinical needs, and new approaches. Cold Spring Harb. Perspect. Med. 4, a019703.

Rowlands, H.E., Morris, K., Graham, C., 2006. Human recombinant antibody against Candida. J. Pediatr. Infect. Dis. 25, 959–960.

Samson, R.A., Visagie, C.M., Houbraken, J., et al., 2014. Phylogeny, identification and nomenclature of the genus Aspergillus. Stud. Mycol. 78, 141–173.

Sanglard, D., Coste, A., Ferrari, S., 2009. Antifungal drug resistance mechanisms in fungal pathogens from the perspective of transcriptional gene regulation. FEMS Yeast Res. 9, 1029–1050.

Sanglard, D., Ischer, F., Parkinson, T., Falconer, D., Bille, J., 2003. Candida albicans mutations in the ergosterol biosynthetic pathway and resistance to several antifungal agents. Antimicrob. Agents Chemother. 47, 2404–2412.

Sanguinetti, M., Posteraro, B., Fiori, B., et al., 2005. Mechanisms of azole resistance in clinical isolates of Candida glabrata collected during a hospital survey of antifungal resistance. Antimicrob. Agents Chemother. 49, 668–679.

Sanguinetti, M., Posteraro, B., La Sorda, M., et al., 2006. Role of AFR1, an ABC transporter-encoding gene, in the in vivo response to fluconazole and virulence of Cryptococcus neoformans. Infect. Immun. 74, 1352–1359.

Sanguinetti, M., Posteraro, B., Lass-Flörl, C., 2015. Antifungal drug resistance among Candida species: mechanisms and clinical impact. Mycoses 58, 2–13.

Selmecki, A., Forche, A., Berman, J., 2010. Genomic plasticity of the human fungal pathogen Candida albicans. Eukaryot. Cell 9, 991–1008.

Selmecki, A., Forche, A., Berman, J., 2006. Aneuploidy and isochromosome formation in drug-resistant Candida albicans. Science 313, 367–370.

Serena, C., Pastor, F.J., Gilgado, F., Mayayo, E., Guarro, J., 2005. Efficacy of micafungin in combination with other drugs in a murine model of disseminated trichosporonosis. Antimicrob. Agents Chemother. 49, 497–502.

Shah, A.H., Rawal, M.K., Dhamgaye, S., Komath, S.S., Saxena, A.K., Prasad, R., 2015. Mutational analysis of intracellular loops identify cross talk with nucleotide binding domains of yeast ABC transporter Cdr1p. Sci. Rep. 5, 1–17.

Shalit, I., Shadkchan, Y., Samra, Z., Osherov, N., 2003. In vitro synergy of caspofungin and itraconazole against Aspergillus spp.: MIC versus minimal effective concentration end points. Antimicrob. Agents Chemother. 47, 1416–1418.

Shapiro, R.S., Robbins, N., Cowen, L.E., 2011. Regulatory circuitry governing fungal development, drug resistance, and disease. Microbiol. Mol. Biol. Rev. 75, 213–267.

Sharifzadeh, A., Khosravi, A., Shokri, H., Shirzadi, H., 2018. Potential effect of 2-isopropyl-5-methylphenol (thymol) alone and in combination with fluconazole against clinical isolates of Candida albicans, C. glabrata and C. krusei. J. Mycol. Med. 28, 294–299.

Sheikh, B.A., Bhat, B.A., Mehraj, U., Mir, W., Hamadani, S., Mir, M., 2021. Development of new therapeutics to meet the current challenge of drug resistant tuberculosis. Curr. Pharm. Biotechnol. 22, 480–500.

Shekhar-Guturja, T., Gunaherath, G.K.B., Wijeratne, E.K., et al., 2016a. Dual action antifungal small molecule modulates multidrug efflux and TOR signaling. Nat. Chem. Biol. 12, 867–875.

Shekhar-Guturja, T., Tebung, W.A., Mount, H., et al., 2016b. Beauvericin potentiates azole activity via inhibition of multidrug efflux, blocks Candida albicans morphogenesis, and is effluxed via Yor1 and circuitry controlled by Zcf29. Antimicrob. Agents Chemother. 60, 7468–7480.

Sionov, E., Lee, H., Chang, Y.C., Kwon-Chung, K., 2010. Cryptococcus neoformans overcomes stress of azole drugs by formation of disomy in specific multiple chromosomes. PLoS Pathog. 6, e1000848.

Spitzer, M., Griffiths, E., Blakely, K.M., et al., 2011. Cross-species discovery of syncretic drug combinations that potentiate the antifungal fluconazole. Mol. Syst. Biol. 7, 499.

Spitzer, M., Robbins, N., Wright, G.D., 2017. Combinatorial strategies for combating invasive fungal infections. Virulence 8, 169–185.

Spivak, E.S., Hanson, K., 2018. Candida auris: an emerging fungal pathogen. J. Clin. Microbiol. 56 e01588-17.

Steinbach, W., 2006. Combination antifungal therapy for invasive aspergillosis—Is it indicated? Med. Mycol. 44, S373–S382.

Stone, L.K., Baym, M., Lieberman, T.D., Chait, R., Clardy, J., Kishony, R., 2016. Compounds that select against the tetracycline-resistance efflux pump. Nat. Chem. Biol. 12, 902–904.

Sucher, A.J., Chahine, E.B., Balcer, H., 2009. Echinocandins: the newest class of antifungals. Ann. Pharmacother. 43, 1647–1657.

Sun, W., Wang, D., Yu, C., Huang, X., Li, X., Sun, S., 2017. Strong synergism of dexamethasone in combination with fluconazole against resistant Candida albicans mediated by inhibiting drug efflux and reducing virulence. Int. J. Antimicrob. Agents 50, 399–405.

Szalewski, D.A., Hinrichs, V.S., Zinniel, D.K., Barletta, R.G., 2018. The pathogenicity of Aspergillus fumigatus, drug resistance, and nanoparticle delivery. Can. J. Microbiol. 64, 439–453.

Taylor, D.L., Hollingsworth, T.N., Mcfarland, J.W., Lennon, N.J., Nusbaum, C., Ruess, R., 2014. A first comprehensive census of fungi in soil reveals both hyperdiversity and fine-scale niche partitioning. Ecol. Monogr. 84, 3–20.

Thakur, R., Anand, R., Tiwari, S., Singh, A.P., Tiwary, B.N., Shankar, J., 2015. Cytokines induce effector T-helper cells during invasive aspergillosis; what we have learned about T-helper cells? Front. Microbiol. 6, 429.

Tobudic, S., Kratzer, C., Lassnigg, A., Graninger, W., Presterl, E., 2010. In vitro activity of antifungal combinations against Candida albicans biofilms. J. Antimicrob. Chemother. 65, 271–274.

Toews, M.L., Bylund, D., 2005. Pharmacologic principles for combination therapy. Proc. Am. Thorac. Soc. 2, 282–289.

Touil, H., Boucherit-Otmani, Z., Boucherit, K., 2018. In vitro activity of antifungal combinations against planktonic and sessile cells of Candida albicans isolated from medical devices in an intensive care department. J. Mycol. Med. 28, 414–418.

Trofa, D., Gácser, A., Nosanchuk, J., 2008. Candida parapsilosis, an emerging fungal pathogen. Clin. Microbiol. Rev. 21, 606–625.

Vandeputte, P., Ferrari, S., Coste, A., 2012. Antifungal resistance and new strategies to control fungal infections. Int. J. Microbiol. 713687, 2012.

Vandeputte, P., Tronchin, G., Bergès, T., et al., 2007. Reduced susceptibility to polyenes associated with a missense mutation in the ERG6 gene in a clinical isolate of Candida glabrata with pseudohyphal growth. Antimicrob. Agents Chemother. 51, 982–990.

Vazquez, J., 2008. Combination antifungal therapy for mold infections: much ado about nothing? Clin. Infect. Dis. 46, 1889.

Veen, M., Stahl, U., Lang, C., 2003. Combined overexpression of genes of the ergosterol biosynthetic pathway leads to accumulation of sterols in *Saccharomyces cerevisiae*. FEMS Yeast Res. 4, 87–95.

Vermitsky, J.-P., Edlind, T., 2004. Azole resistance in Candida glabrata: coordinate upregulation of multidrug transporters and evidence for a Pdr1-like transcription factor. Antimicrob. Agents Chemother. 48, 3773–3781.

Wang, J.M., Bennett, R.J., Anderson, M., 2018. The genome of the human pathogen Candida albicans is shaped by mutation and cryptic sexual recombination. MBio 9, e01205–e01218.

Wayne, P., 2008. Clinical and Laboratory Standards Institute. Reference Method for Broth Dilution Antifungal Susceptibility Testing of Filamentous Fungi. CLSI Document M38-A2. Clinical and Laboratory Standards Institute, 2008. In.

White, T.C., Holleman, S., Dy, F., Mirels, L.F., Stevens, D., 2002. Resistance mechanisms in clinical isolates of Candida albicans. Antimicrob. Agents Chemother. 46, 1704–1713.

White, T.C., Marr, K.A., Bowden, R., 1998. Clinical, cellular, and molecular factors that contribute to antifungal drug resistance. Clin. Microbiol. Rev. 11, 382–402.

Wisplinghoff, H., Bischoff, T., Tallent, S.M., Seifert, H., Wenzel, R.P., Edmond, M., 2004. Nosocomial bloodstream infections in US hospitals: analysis of 24,179 cases from a prospective nationwide surveillance study. Clin. Infect. Dis. 39, 309–317.

Xiang, M.-J., Liu, J.-Y., Ni, P.-H., et al., 2013. Erg11 mutations associated with azole resistance in clinical isolates of Candida albicans. FEMS Yeast Res. 13, 386–393.

Xu, D., Jiang, B., Ketela, T., et al., 2007. Genome-wide fitness test and mechanism-of-action studies of inhibitory compounds in Candida albicans. PLoS Pathog. 3, e92.

Yamamoto, Y., Maesaki, S., Kakeya, H., et al., 1997. Combination therapy with fluconazole and flucytosine for pulmonary cryptococcosis. Chemotherapy 43, 436–441.

Yeh, P., Tschumi, A.I., Kishony, R., 2006. Functional classification of drugs by properties of their pairwise interactions. Nat. Genet. 38, 489–494.

Yoo, J.I., Choi, C.W., Lee, K.M., Lee, Y., 2010. Gene expression and identification related to fluconazole resistance of Candida glabrata strains. Osong Public Health Res. Perspect. 1, 36–41.

Zavrel, M., Hoot, S.J., White, T., 2013. Comparison of sterol import under aerobic and anaerobic conditions in three fungal species, Candida albicans, Candida glabrata, and *Saccharomyces cerevisiae*. Eukaryot. Cell 12, 725–738.

CHAPTER

10

Recent trends in the development of bacterial and fungal vaccines

Manzoor Ahmad Mir, Muhammad Usman, Hafsa Qadri and Shariqa Aisha

Department of Bioresources, School of Biological Sciences, University of Kashmir, Srinagar, Jammu and Kashmir, India

10.1 Introduction

Disease can be defined as a composite of symptoms and signs. Disease and illness are relatively seen to be similar (Tikkinen et al., 2012). The outcome of a disease, comprising of specific signs and symptoms, leads to an overall harm to the body of an organism. In most of the cases it has been seen to be fatal when left untreated. Diseases are caused by many factors which include external ways like infectious diseases (tuberculosis, Jaundice, malaria, Influenza, etc.), internal disorders and unhealthy behaviors such as acute respiratory infections, asthma and autoimmune diseases. Apart from these internal and external factors, it has been reported that many of the diseases are also caused by microbes or microorganism (Tikkinen et al., 2012; Scully, 2004). Microbes are very oldest, tiny, single cell organisms. Being small in size, microbes are unable to be seen with the naked eye. Microbes are found on every part of the Earth like in air, water, soil, and rock. Apart from this, they have been seen to be present in plants, animals, and human body. Microbes are very diversely found in nature and environment. They include prokaryotes (bacteria, archae), eukaryotes including protozoa, fungi, algae, and animals like rotifers and planarians (Singh et al., 2014) When the body of an organism gets colonized by the microbes, it results into the infectious disease. Many similar states of the disease arise due to the different causes, e.g., pneumonia is caused by many pathogenic bacteria, fungi, and other causes (Scully, 2004).

Microbes like bacteria, fungi, viruses, and protozoa have the ability to become pathogenic and cause severe infections. These pathogens have the strength to even cause death to the host. The immune system of the host has the excellent ability to prevent these pathogens from entering the host. The innate and adaptive immune system of the host comes into action

Human Pathogenic Microbes
https://doi.org/10.1016/B978-0-323-96127-1.00003-6

and starts the killing of these pathogens (Hayward et al., 2018). These pathogenic microbes possess the ability of various levels of invasion which occur from mild to life-threatening. Bacteria are the most common organisms. Humans have a common experience of getting infected because of the bacteria residing on the external surfaces which also include the skin, gut, and lungs. On a daily routine, humans are continuously exposed to bacteria. Due to the strength of the defense system of the host, many of these bacteria are harmless. It has been seen that only a little number of the bacteria are harmful and cause infection which leads to the lethal diseases, even death. In compromised patients, whose defense system is weak, these bacteria enter the blood stream after the surgeries and other treatment methods resulting in a number of infectious diseases (Doron and Gorbach, 2008).

Different kinds of transmission factors are responsible for the bacteria to causes infections and diseases. To spread and cause infections a good number of bacteria need to have the potential to survive in the environment and finally reach to the host and cause infections. Different bacteria live, survive, and adapt in water, soil, and other places. Many bacteria do not directly infect the host. They initially infect the vectors like animals and insects which in turn reach and infect the host. Because of the increase in the individuals with compromised immune system due to various reasons like AIDS and organ transplantations, this has resulted in an increase in number of individuals who are getting infected due to these bacterial pathogens (Doron and Gorbach, 2008).

There are various causes and factors for the bacteria to cause infection and finally the disease. The infection of an organism is the first important cause that will totally determine the number of individuals that will get affected and infected. Pathogenicity of an organism having the ability to cause infection is an important factor for the pathogenic organism to infect and cause the disease. There are many attributes for pathogenic bacteria which it effectively brings in order to invade the body of an organism, its immune strength, and defense. These attributes lead to the cause of the disease in the organism. These include the aggressive invasion, the release of the toxins. On the other hand, there are various attributes and factors of an organism that decide whether the infection will transmit and lead to the disease. After the infection by the pathogen, there are various factors of the host which will provide an idea of transmission of the disease. Those comprise of the immune system response, age, genetic makeup of the organism, nutrition, the time of the exposure to the organism, and cooccurring illnesses. Harmful chemicals and contaminants in the environment also lead to the susceptibility to the infection (Doron and Gorbach, 2008).

Despite the research and development in the strategies of protection and treatment, infectious diseases continue to create havoc worldwide. Every year millions of deaths occur due to these microbial infections in the world and developing countries in particular. In the meantime there is the discovery of the occurrence of the new species and their new variants. One of the pathogenic bacteria called *Mycobacterium tuberculosis* has been seen to be the main cause of the concerning disease called tuberculosis (TB). It has been reported that there are nearly nine million new cases per year, which comprise of almost 0.1% of the world population. Through these figures there is an estimate of 1.5–2 million deaths per year. These reports also give a data of 450,000 people who are infected with multidrug-resistant tuberculosis per year (Sarmah et al., 2018; Qadri et al., 2021).

There is a rising concern regarding the continuous and large-scale use of drugs against these deadly microbes. This large-scale use of antimicrobial drugs creates an opportunity for microbes to develop a resistant mechanism called antimicrobial drug resistance. This leads to the difficulty in the treatment ways and hence causing severe infections which are difficult to treat (Al-Mehdar and Al-Akydy, 2017). An estimate of World Health Organization (WHO) is that nearly about one-third of the population of the world possesses a latent kind of *M. tuberculosis* infection. Accordingly, in future there will be a pool for tuberculosis cases. Another report of the WHO (2014) is about the 22 African countries saying that, more than 70% of untimely mortalities have been calculated due to infectious diseases, with coinfections like HIV and helminthes. These kinds of reports suggest that there is a need to develop knowledge technology and research-based activities so that these human microbial pathogens can be efficiently dealt (Soares et al., 2016).

Another estimate by the WHO (2007) gives a data that there are a total of 9.27 million active diseases, with an occurrence of 139 out of 100,000 populations throughout the world (Zhu et al., 2016). In Asia there were the highest figures (55%) of these TB cases followed by Africa (31%). In Africa there was the highest occurrence due to the high HIV infection (Ahmad and Immunology, 2011). In Asia there has been an estimation of more than 50% of tuberculosis cases worldwide. These countries include China, India, Indonesia, Pakistan, Bangladesh, and Philippines. One of the main modes of transmission of TB is that the healthy individual gets infected by it from an infected host due to the inhalation of aerosol droplets which are highly infectious (Sarmah et al., 2018). *M. tuberculosis* benefits fully from the defense system of the host. One of the worrying things about the disease is that the patient can get an initial infection after the attack or the disease can remain hidden in the body of the host. These pathogens enter the lungs and then the alveoli. The immune system of the host considers them as outsiders and then eliminates them by the attack of the macrophages (Singh et al., 2015).

Various strains of TB are more virulent and they get transmitted at a greater rate. Apart from the usage of attenuated live vaccines and antibiotics, the cases of TB have been on a rise. This gives a reflection of the more research, more efficient diagnosis technologies, and developments of the new vaccine and medicines (Smith, 2003). This microbial disease has been challenging, as it is more fatal of all the other infectious diseases. In 1920 Bacillus Calmette-Guerin [BCG] has been the first kind of vaccine developed against TB. This vaccine is administered as a part of immunization to newly born and children (Sarmah et al., 2018).

Another bacterium named *Helicobacter pylori* is microaerophilic, that is, it needs oxygen (Mehmood et al., 2010). *H. pylori* cause chronic infection (Leone et al., 2003). It gets developed in childhood and mostly among the low income class (Mendall et al., 1992; Queiroz et al., 2012). This bacterium causes various disorders and diseases like gastroduodenal diseases, peptic ulcer disease (Hopkins et al., 1996), gastric carcinoma, and gastric MALT lymphoma (Wotherspoon et al., 1993; Mosa et al., 2013). One of the main factors for its transmission is the low economic status and the hygiene status. These are the factors which lead to its infection in developing countries. One of the reports of the Global Systemic Review of 2015 provides an estimation of nearly 4.4 billion individuals being positive for *H. pylori* (Graham et al., 1993). The estimates are high in Africa, Latin America, Caribbean, and Asia with 79.1%, 63.4%, and 54.7%, respectively. One of the effective drugs against this bacterium is clarithromycin. It has been successful to treat 40% of the infection (Lind et al., 1999; Peterson et al., 1993) (Table 10.1).

TABLE 10.1 Pathogens with the ways of transmission and the sites of infection (Sarmah et al., 2017).

Pathogen	Ways of infection	Infection sites
Mtb	Air	Lung
Helicobacter pylori	Mouth and feces	Viscous lining of the stomach
Hep.	Feces, blood donations, and infected body discharges	Liver

Apart from these, various other diseases are also caused by bacterial pathogens like inflammatory bowel disease (IBD). As the name suggests, it is an array of inflammatory forms of colon and small intestine which results in many types of diseases like Crohn's and ulcerative colitis. IBD falls under autoimmune diseases in which the immune system of the body behaves abnormally and starts attacking the digestive system (Friswell et al., 2010; Katz et al., 2011). Another bacterial disease in women has been reported, known as bacterial vaginosis. Bacterial vaginosis is a vaginal infection seen in the child-bearing age in women. One of the main reasons for the occurrence of this disease is women having multiple partners. It develops after sexual intercourse by a woman with a new partner (Prescott, 2002; Mir, 2015).

Apart from bacteria, there are also many fungal pathogens which cause many life-threatening diseases. Fungi are ubiquitous and highly diversified. Out of 1.5 million species, over 300 fungi have been estimated to be the main cause for infections like in humans and become responsible for various types of infections beginning their effect on outer surfaces like superficial infections of the skin, hair, nails and mucosal surfaces to the infection of the internal organs (invasive) (Kim, 2016). Superficial infections are common and are easy to cure while invasive infections have a low occurrence rate and cause life-threatening diseases. The diseases occur especially in immune compromised patients with HIV/AIDS, diabetics, oral and genital tract infections, vulvo genital candidiasis in 70%–75% women or autoimmune diseases like in individuals undergoing anticancer chemotherapy and persons who have undergone organ transplantation (Li and Nielsen, 2017; Kim, 2016). A report gives an estimate of 1.5 million people dying every year because of the human pathogenic fungi causing invasive functions, thus making it a concern worldwide (Brown et al., 2012). The research regarding the human fungal pathogen has not shown any further progress resulting in the decline in the development of antifungal drugs and techniques that could diagnose and treat the invasive fungal diseases.

Human fungal pathogen attributed deaths occur because of many fungal species like *Candida, Aspergillus,* and *Cryptococcus* initially affecting immune compromised patients. The fungal pathogens which are the major cause for infections include *Candida albicans, Cryptococcus neoformans* and *Aspergillus fumigatus.* Among them there are environmental fungi like *C. neoformans* and *A. fumigatus* commonly occurring in soil and other places. Inhalation of the airborne spores leads to exposure to these pathogens (Ikeh et al., 2017). Cryptococcal meningitis disease is due to the main causative agent that is *C. neoformans.* This has been a highly occurring incidence in HIV-1/AIDS populations. It leads to 180,000 deaths each year (Brown et al., 2012).

In immunocompromised individuals, Invasive aspergillosis is another fungal disease (Kosmidis and Denning, 2015). In persons suffering from this disease, the growth of the *Aspergillus* in the lungs results in damage to the tissue, and sometimes hemoptysis (Bongomin et al., 2017). The kinds of diseases which are at larger risk to invasive aspergillosis include chronic granulomatous disease, HSCT, prolonged neutropenia, and organ transplantations like in heart, lung, and pancreas (Gavaldà et al., 2014). Every year there is an estimation of more than 200,000 cases of invasive aspergillosis, and it is because of *A. fumigatus* in those individuals with compromised immune system. *A. fumigatus* also causes pulmonary aspergillosis (lung disease) affecting three million people worldwide. Change in morphology is linked with the fungal pathogens and hence associated with virulence (Brown et al., 2012).

Globally there are various estimates and reports of the fungal diseases like there are nearly 223,100 cases of the disease cryptococcal meningitis complicating HIV/AIDs, invasive candidiasis with nearly 700,000 cases, 500,000 cases of *Pneumocystis jirovecii* pneumonia. Similarly there are 100,000 cases of disseminated histoplasmosis, more than 10,000,000 cases of fungal asthma, and finally ~1,000,000 cases of fungal keratitis in a year (Bongomin et al., 2017). Among the various fungal pathogens, *Candida, Aspergillus* and *Cryptococcus* species, *Pneumocystis jirovecii*, endemic dimorphic fungi such as *Histoplasma capsulatum* and *Mucormycetes* have been recognized to be the main fungal pathogens causing diseases which kill millions of people annually in the world. For mucosal infections and diseases, *Candida albicans* has been found to be responsible. For most of the allergic fungal diseases, *Aspergillus fumigatus* has been seen to be the main fungal pathogen responsible. Similarly various *Trichophyton* spp., especially *T. rubrum* are seen to be responsible for many infections of the skin (Boral et al., 2018).

Candida albicans has been one of the main fungal species which displays its sign and effects in both mucocutaneous and disseminated infections. At the same time there is an another aspect of nonalbicans *Candida* (NAC). The disease Candidiasis is caused because of NAC spp. It reports are increasing (Sardi et al., 2013). There are many factors which are associated with the disease like severe immunosuppression or illness, prematurity, the wide spectrum use of antibiotics and also the use of antimycotic drugs. Because of these Non-*Candida* species, the clinical signs of infection caused have been unrecognized. Various NAC species show and acquire a great resistance to the various antifungal drugs called antifungal drug resistance (Sullivan et al., 1996). As it now seen that *C. albicans* are said be one of the main important cause of many candidal bloodstream infections (BSIs), the study how the disease spreads and how it is to be controlled has changed now. There are many research studies which claim that a number of BSIs are caused by the by non-albican *Candida* species. Those non-albicans include *Candida glabrata, Candida krusei, Candida parapsilosis,* and *Candida tropicalis,* which consist of one-half of candidamia and data related to the infections is limited (Pfaller et al., 1998).

In adult women the vulvovaginal candidiasis has been seen as the most frequently occurring fungal infection. The symptoms of this include the frequent itching, irritation in the vulva, soreness, loss of comfort with the sexual [dyspareunia], and a white discharge of vagina. In this vulvovaginal candidiasis disease there are various species like *C. glabrata, C. krusei, C. tropicalis, C. parapsilosis,* and *C. dubliniensis* which have been the main factor of its occurrence. Among them *C. glabrata* alone has been the main contributor of 50%–67% of vulvovaginal candidiasis (Makanjuola et al., 2018).

Another fungal disease called the oral candidiasis has been seen to occur by an unusual growth of the fungus in the mouth. There are two forms called the white oral candidiasis (hyperplastic candidiasis, pseudomembranous candidiasis) and another one is erythematous oral candidiasis (median rhomboid glossitis, atrophic candidiasis, linear gingival erythema, and angular cheilitis) (Brown et al., 2012). Non-albicans species reported to cause oral candidiasis have been identified as *Candida dubliniensis, Candida glabrata , Candida kefy, Candida parapsilosis, Candida stellatoidea, and Candida tropicalis* (Millsop and Fazel, 2016). Despite the use of new antifungals, the most common occurrence of invasive candidiasis has been related with the candidemia resulting in greater number of mortality rates (>40%) (Arendrup and J, 2013; Kullberg and Arendrup, 2015). In UK there has been reported the occurrence of invasive candidiasis with 5142 cases in 2016 (Pegorie et al., 2017). In Intensive care units, there was an estimation of the 60% of the cases which were followed by cancer and transplantation units (13%). In Pakistan candidemia occurrence was reported as highest (38,795 cases) that comprise a number of 21 cases per 100,000. This record was followed by Brazil (28,991 cases) which comprise to a number of 14.9 cases per 100,000 and finally Russia (11,840 cases) with the given numbers of 8.29 cases per 100,000 (Arendrup, 2010).

So far due to these bacterial and fungal diseases millions of deaths have occurred, creating havoc worldwide. But at the same time, parallelly research, development, and treatment strategies have been studied also. Among these developmental strategies, vaccination has also been a great source to protect and treat individuals from these dreadful infections and diseases. These vaccines have prevented from diseases caused by bacterial and fungal pathogens. At the end of the 19th century, the vaccines developed against bacterial pathogens were among the successful human applications of microbiology. The most successful vaccines were against the tetanus and diphtheria with efficiency over 95%. Increased understanding of the research identification of new determinants and analysis of the pathogenic processes by bacterial pathogens has led us toward a hope that vaccines will be developed against all these diseases with more affectivity (Strugnell et al., 1997).

Th1 mediated immune response that is the Th1 release of the cytokines has been associated with the protection of the bacterial infections which are caused by the intracellular pathogenic bacteria. This emerges into an efficient way of an effective acceleration in the killing of the phagocytes which are harboring pathogens (Kaufmann, 1995). A *Mycobacterium bovis* strain called Bacillus Calmette_Guerin (BCG) has been used mostly. It is modifiable in its effectiveness and efficiency because of the independent strains of the vaccine with multiple stages of with attenuation. There is no answer yet for the mutations which were meant for attenuation in BCG. DNA vaccines could have an impact in overcoming the effect of the trouble by the infections by the pathogenic bacteria where T cells produce IFN-gamma which in turn help in mediating the disease resistance and cytotoxic T lymphocyte responses in the protection (Strugnell et al., 1997). Two research studies were carried out. In first it was seen that *M. tuberculosis* protein encoded mice with naked immunized DNA were saved from the virulence of the bacterium (Lowrie et al., 1994). One of the studies has also reported that mice can be protected from *M. tuberculosis* by the DNA vaccine through the expression of a stress protein of a bacteria that is the GroEL (Tascon et al., 1996; Mir, 2015) (Table 10.2).

Related to fungal infections, there has also been research and development regarding the drugs and vaccines to protect and minimize the threat created by these fungal diseases. In immune compromised individuals, it has been seen that live attenuated vaccines can provide a long term and stronger immune response. Vaccinologists have provided many of the

TABLE 10.2 During infection bacteria produce molecules like PAMPs and DAMPs, which in turn activate multiprotein complexes and detectors (Hayward et al., 2018).

Pathogen	Multiprotein complex detector with in cell
Mycobacterium bovis	AIM2 sensor in hematopoetic cells (dsDNA)
Mycobacterium tuberculosis	AIM2 sensor in hematopoetic cells (dsDNA), NLRP3 detects peptidoglycan
Mycoplasma pneumonia	NLRP3

AIM, Absent in melanoma; *NLRP*; Nucleotide binding oligomerization domain.

outcomes of live attenuated vaccine strategies which have been seen to be very much effective regarding the highly infectious diseases. These product outcomes and strategies have been very workable and effective in containing and decreasing the effect of the highly infectious disease. One important finding is the Heat-killed *Saccharomyces cerevisiae* (HKS) vaccine has been the product with the capability and effective role in the preservation of different fungal infections. This heat killed vaccine has provided the efficient result against the different virulent forms of the fungus *C. albicans*, *C. posadasii*, and *A. fumigatus*. Many *Candida* species secrete a highly expressed factor of virulence called secreted aspartyl proteinase-2 [Sap-2].This factor has the ability to provide protective roles in opposition to the disease vaginal candidiasis. This role is established in a virosome-based format of the vaccine. This vaccine has been applied in the vaginal candidiasis of the rat. Apart from this there was a trial which gave effective outputs (Nami et al., 2019; Mir, 2015) Fig. 10.1.

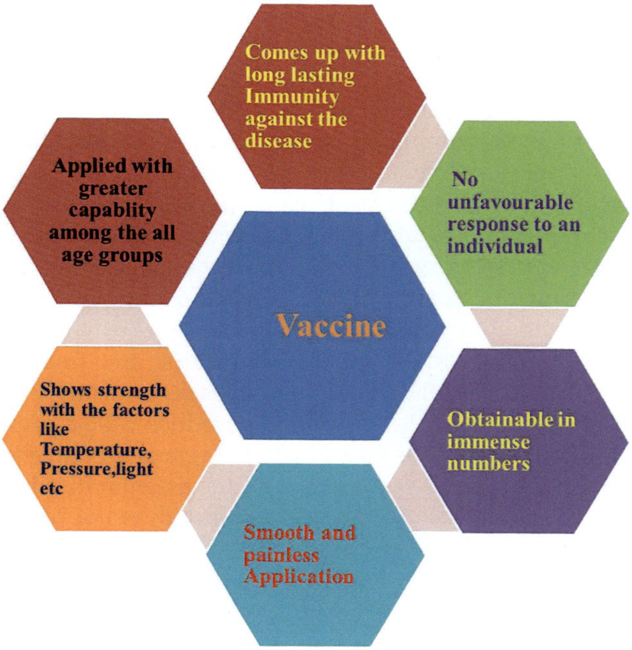

FIGURE 10.1 The various characteristics of an ideal vaccine which make its application successful at all levels.

10.2 Understanding host–pathogen interaction

The specificity of the pathogen toward a host is known as the potential of the pathogen. This potential is used to invade and infect the host (Kirzinger and Stavrinides, 2012). It depends on the specificity and different pathogens have wide and different levels of specificity. Many kinds of bacteria possess different and a wide range of the host specificity. Those kinds of pathogenic bacteria infect insects, humans, and other wild animals. However a low level of specificity is shown by many bacteria. In between humans and the other kinds of hosts, there is a variation in the pathogenicity caused by these pathogens. An example to understand this is *Salmonella typhimurium*. After orally ingesting, it causes gastroenteritis in humans, but differently it causes typhoid fever in mice (Bäumler and Fang, 2013).

Molecular-based attachments between the host and pathogen determine the specificity of pathogens toward host. Due to the different molecular, cellular, and complex compositions, the specificity of the host toward the bacterial pathogens is not fully understood. However, the recent developments have made this host specificity toward the bacterial pathogens understandable to a greater extent. However, the extent to which it is difficult to understand is due to more complex molecular compositions and cellular structures (Pan et al., 2014).

Pathogens are totally different from normal flora. When the immune system of the host is weak or the extent when it gets exposed to a sterile part of the body (when the gut flora enter the peritoneal cavity of the abdomen by bowel perforation, it causes peritonitis), then the problem by the normal microbial inhabitants occurs. On the other hand without being immunocompromised, pathogens attack a normal host. Pathogens enter the cellular boundaries, create panic in the body mechanism, and evoke expected response from the host and this results in its better survival and further work. There are various attributes regarding a pathogen which it needs to have for survival and multiplication within a host. They include invading the host, creating a better place for its survival inside the host, managing to get a place with nutritional richness, avoiding the host's response, destroying the innate and adaptive immune system of the host, creating a system for its effective multiplication by using the resources of the host, and finally colonizing, invading, another host. Through these attributes, pathogens have gained the full potential to use the resources of the host in all possible ways. By these ways, they enable the effective host cell response Fig. 10.2.

By using all the above-mentioned attributes, pathogens are able to get inside the body of the host, multiply, and replicate inside the region of the target. Pathogens possess various virulence factors which aid in causing disease. The more the virulence is, the more easily the disease will occur. There are various factors like the level of virulence, the kind of pathogenic effects, various environmentally controlled factors, and finally the host's immune system. In a general way, we see pathogenic microbes as an army of invaders that attack the body of the host in order to live simply and reproduce themselves. One of their ways is that they do all these processes at the cost of a host organism in a very smart and effective strategic move. Because the host body provides it all the essentialities to live like the temperature and environment, they are able to reproduce (Sarmah et al., 2018).

The infections and diseases because of pathogens cause stress on all living organisms. It is like an armed race between host and pathogen. Host tries its ways to make its defense system strong, evolves various defensive approaches to pathogens, and avoids their brutal attacks

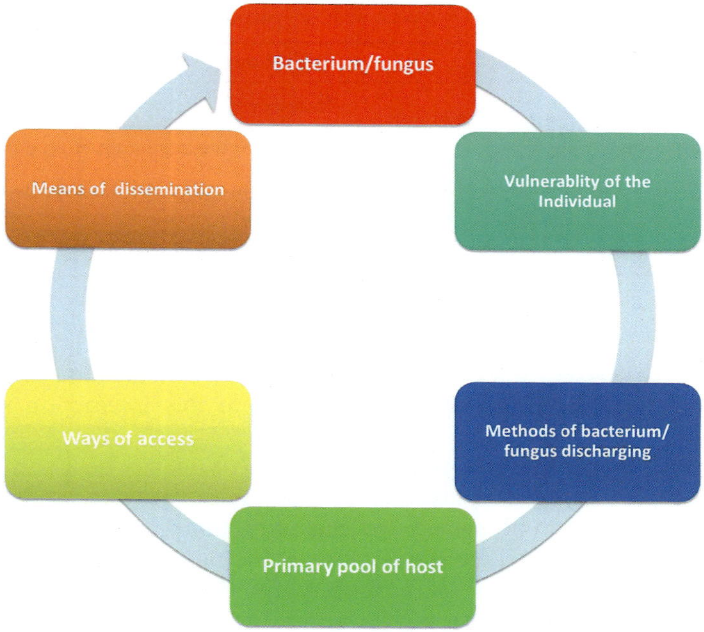

FIGURE 10.2 The important host–pathogen interaction—its methods and its means.

while pathogens also make their ways of infection more and more efficient and minimize and get away from the host measures to control it. It means that there is a defensive and counter-defensive approach. These approaches are at some level genetically controlled (Sironi et al., 2015). Red Queen hypothesis states that the organisms need to adapt and evolve the strategies to survive from pathogens. Allele frequency fluctuations in the pathogen and host populations are determined by the frequency-dependent selection. In this way the rare allele is benefited by the selection that means pathogen may be adapted to the most occurring genotype of the host and fails to infect the hosts which contain the rare allele. There can also be the benefit of the diversity maintenance. This selection results in the speedy increase in the phenotypes of both the host and the pathogen in terms of resistance and virulence. The host is totally under stress to minimize the strength of the interaction. In turn pathogen evolves by decreasing the distance of the phenotype. In case of complex genetic basis (polygenic) of the host-pathogen interactions, there is the occurrence of the Chase; there is the reduction in the genetic diversity within the population and the determination of the selective sweeps (Benton, 2009).

The interactions of the host and the pathogen have been described on different levels. They include the organismal levels (that is the condition where the pathogens infecting the host), population level (virus is infecting in a human population), molecular level (pathogen protein binding to a receptor on human cell). Host–pathogen interaction research has been done on various attributes like environmental factors (temperature, light, season and populations of pathogen and host). *C. elegans* and *Drosophila melanogaster* have been declared to be the model

TABLE 10.3 Values and importance of *Drosophila* in host–pathogen interactions (Younes et al., 2020).

Uses	Utilizations
Little reproduction time duration	To research about the function of gene
Require low cost for rearing	To analyze the beginning of an outbreak
Various genes in human disease outbreaks possess comparisons in *Drosophila*	To perform analysis with in the living organism
Low moral and social worries	Antidoping concealing

organisms in order to efficiently study the behavior of the pathogen on humans. Also some of the proteins have been identified for having a part in this interaction (Pan et al., 2014) (Table 10.3).

There are various substrates and pathways through which a host can be invaded by a pathogen. Human body has the ability to deal easily with many common pathogens while resisting them with the immune system. At the same time pathogens have the capability to get attached to host and decrease the defense of the host. But these observations can be varied for each pathogen. There are various ways through which the host–pathogen interactions can be seen like by disturbing the natural pathways, proteins, and genes, biofilm formation, and inhibiting the macrophage activity (Sen et al., 2016).

10.2.1 Host–pathogen bacterial interaction

Protein secretion is one of the important ways by which a pathogen attacks the host. Among them Gram-negative bacteria are important. They cause pathogenesis in host. These Gram-negative bacteria comprise of secretion systems. Proteins called effectors are released by these secretion systems during the contact with the host. Six specialized protein systems have been seen in Gram-negative bacteria. They are Type I, Type II, Type III, Type IV, Type V, and Type VI. These are the noticeable ones based on their mechanism of host infection. In bacterial pathogenesis, various secreted proteins are vital like toxins, urease, and multivalent adhesion. Due to various pathways, plants and animals synthesize the toxins that are very harmful and poisonous to humans. In humans huge numbers of toxins that create problems often come from germs like bacteria. They can be small molecules, peptides, or proteins. These toxins have the great ability to cause a disease when they come in contact with the host. These toxins also cause disease, when they get absorbed by the tissues of the body when they interact with the macromolecules enzymes or cellular receptors. Once there is the entrance activity of the toxin in the host, they start to interfere in the normal functioning system of the host's metabolism. On the occasions of the attack, when there is less expression of the toxin then there is a less impact on the TCR signaling pathway of the host than with higher toxin expression. Host cells get damaged because of the molecules which are secreted by these Gram-negative pathogens. From the growing bacteria enclosure, the vesicle which gets released functions as the holder for the proteins and lipids of Gram-negative bacteria. This seeks the value of the vesicle-mediated toxin delivery of infection in host (Sen et al., 2016).

In *Mycobacterium*–host interaction, enzyme urease plays an essential role (Clemens et al., 1995). Urease is present in various species of *Mycobacterium*. The presence and absence of urease is mostly used in the speciation of *Mycobacteria*. For many of the microorganisms possessing pathogenic behavior, urease is considered to be a virulence factor. Because of urease, that is, the urinary pathogens which generate ammonia, play their part in pathogenesis, the reason being the toxicity to renal epithelium, participation in complement inactivation, and promotion of urinary stone formation. In the stomach, microenvironment of the bacteria gets alkalinized by the urease. This happens in the pathogenic bacteria *H. pylori*. It is also toxic to the epithelium of the stomach (Sen et al., 2016).

Both intracellular and extracellular sides in the host, urea is available to the bacteria. During the initial stages of the infection, one of the molecules called Multivalent Adhesion Molecule (MAM) has been reported for creating a binding affinity with host cells (Krachler et al., 2011). With the host, MAM7 connects through protein–lipid and protein–protein interactions. For virulence, MAM7 has been seen to be a contributor of virulence. MAM7 is found on the outer membrane of the Gram-negative pathogens. For bacterial pathogens to survive inside the host, they need to get away from the defense system of the host. *M. tuberculosis* qualitatively shows that it transcribes various genes which normally take part in fortification and runaway from host system (Rachman et al., 2006). Because of glycolipids and host system interaction, signaling for antiinflammation is considered as a way to avoid host system. It can also play a role in saving the much inflammatory response (Torrelles and Schlesinger, 2010).

Another way to evade the defense system of the host is that bacterial pathogens interfere with the important pathways of the host, like NF-κB, a family of transcription factors. It helps in the development of the APC (Antigen Presenting Cell) and the lymphocyte (Tato et al., 2002). Whenever the host is compromised, NF-κB gets activated. Due to the accelerated replication, pathogens enable themselves to grow and survive inside the host. In the beginning of this process, there is a need of few genes and proteins in order to survive in the host. At the same time various genes and proteins are also required so that they survive outside the host (Sarmah et al., 2018). *Mycobacterium tuberculosis* genome consists of a large family of Ser/Thr protein kinases (STPKs). They have been seen to possess a great role in cell division and cell envelope biosynthesis (Molle and Kremer, 2010). The interaction between the host and the pathogen is facilitated by the outer membrane of the bacteria (Kuehn et al., 2005).

In order to preserve the occurrence of the infection and the disease, the body of the host evokes immune response against the invasion of the pathogen. The genes of the host play an essential role in its immune response. Mutated β-catenin homolog bar 1 or homeobox gene egl-5 of *C. elegans* has resulted in defective response and hypersensitivity to *Staphylococcus aureus* (Irazoqui et al., 2008). Autophagy is seen to be another mechanism for the hosts to defend against the pathogen. It can be used in the elimination of the Mtb (Vergne et al., 2006). LRG-47 has been reported to start autophagy (Singh et al., 2006). IRGM (Immunity-related GTPase family M protein) also seems to play an important role in autophagy and degradation of intracellular bacillary load (Mir, 2015).

Dendritic cells (DCs) have been seen to play an important role in the immune system activation (Rescigno and Borrow, 2001). After bacterial infection, DCs are brought to the lamina propria of the small intestine. Chemokines facilitate the bringing up of the DCs and cytokines activate the DCs. These solubility factors are stimulated by the infection.

Pathogens, viruses, and their components can activate DCs directly. Dcs possess a great ability to migrate. This characteristic can be both useful and harmful. Like in various infections, its migration and change in location can lead to the activation of T cells as seen in migration from periphery to lymph nodes.; whereas, it is harmful when it is responsible for the spread of infection (Sen et al., 2016).

10.2.2 Understanding host—pathogen fungal interaction

Human fungal pathogens are very challenging to the world and cause diseases. These infections lead to an estimate of almost 1.5 million deaths annually worldwide. Despite their challenging diseases and reports, these fungal diseases receive little attention as compared with malaria and tuberculosis (Brown et al., 2012; Bongomin et al., 2017). Deaths over 70% occurring due to these fungal infections can be related to fungi of the genera *Aspergillus, Candida, Cryptococcus,* and *Histoplasma* (Brown et al., 2012). These kinds of opportunistic fungal pathogens are the normal commensal of the human microbiota or they live in the environment. This feature results in their exposure to humans. In immunocompromised host individuals, there is a large spread of superficial fungal diseases,; skin fungal infection being the most common and vulvovaginal candidiasis (VVC) having an estimate of 70% effect in women (Gonçalves et al., 2016). In order to cope with these fungal pathogens, there is an essential requirement of an efficient immune system. It totally depends on the innate immune system which comprises of cells such as macrophages, monocytes, neutrophils, natural killer (NK) cells, and dendritic cells (DCs) and the adaptive immune system, which mainly consists of T-helper cell responses. At the same time these types of cells are ineffective and do not provide any defense when they are evident or absent. In an Immunocompromised host, various immunosuppressive therapies expose them easily to invasive candidiasis (Lionakis, 2014) and also aspergillosis (Ullmann et al., 2012), whereas the effective innate immune system provides a defense against cryptococcosis (Byrnes et al., 2010). Also due to the infections of the HIV patients with immunocompromised adaptive immune system are mainly exposed (Warkentien et al., 2010). While as fungal species like *Aspergillus* and *Cryptococcus* species, *Histoplasma* mainly are cause of infections in healthy individuals (Köhler et al., 2017).

The resident-tissue macrophages and monocyte-derived macrophages play a major role against the disease invasive candidiasis (Austermeier et al., 2020). Alveolar macrophages (AMs) play a significant role in elimination of fungi like *Aspergillus, Cryptococcus,* or *Histoplasma* species to which host gets exposed when they enter host body (Whitaker and Newman, 2005). There is an essential role of a process called neutropenia in antifungal host defense against aspergillosis and invasive candidiasis (Herbrecht et al., 2000). Dendritic cells (DCs) have been playing a great role for the adaptive immune system activation. There is as a factor that is the reduction in the CD4+ T cell function in AIDS patients which accelerates the susceptibility toward the diseases like *C. albicans, A. fumigatus, C. neoformans* or *H. capsulatum* (van de Veerdonk and Netea, 2010). NK cells have also been to create antifungal effects antifungal effects (Schmidt et al., 2017). While as late restoration of NK (e.g., after allogeneic stem cell transplantation) can be a large risk of invasive aspergillosis (Weiss et al., 2020).

Fungal pathogens have evolved effective strategies to evade successfully the defense system of the host and colonize the places of the human body. Efficiently going away from the recognition of the host immune system (for example, by binding negative regulators of the complement cascade to inhibit complement activation), and therefore constitute part of the normal microbial flora of the gut, oral and vaginal cavity in healthy individuals (Ikeh et al., 2017). There is another way for the fungal pathogens. It involves hiding the cell wall pathogen-associated molecular patterns (PAMPs). Extracellular polysaccharide capsule of *Cryptococcus neoformans*, it prevents the recognition and uptake to the host cells by covering the cell (Rajasingham et al., 2017). Other pathogens mask underlying PAMPs, including β-glucan by molecules that are not recognized by PRRs, such as the outer cell wall α-glucan layer of *Histoplasma capsulatum* or the external hydrophobin layer of *Aspergillus fumigatus*-resting conidia (Brown et al., 2012). The hyphae of *Candida albicans* can also hide this by morphogenesis that does not display β-glucan on the surface (Noble and Johnson, 2007).

One of the important aspects to the innate immune response involves macrophages and PMNs. These macrophages and PMN's recognize pathogen-associated molecular patterns (PAMPs) which are in the cell wall of the fungi through pattern recognition receptors (PRRs) present on the phagocytic cell membrane, endosomes and cytoplasm (Homann et al., 2009). Receptors engage and empower the phagocyte to engulf and destroy fungal cells within the phagolysosome taking into account various oxidative and nonoxidative mechanisms which include the production of toxic reactive oxygen and nitrogen species (ROS and RNS).There is also the expression of many antimicrobial peptides and the activity of hydrolytic enzymes (Sardi et al., 2013). *C. albicans* encode a catalase and six superoxide dismutases; unusually, three Sod enzymes (Sod4–6) are secreted and detoxify extracellular ROS produced by macrophages (Kao et al., 1999).

By PAMP-PRR interaction, indirect killing of the Candida is achieved with the initiation of proinflammatory cytokines and chemotactic factors. This functions to activate other areas of the immune system and helps in the cleansing of body from *Candida*. Various studies have identified tolllike receptors (TLRs), C-type lectin receptors (CLRs), and Nodlike receptors (NLRs).These receptors are thought to be the major PRRs involved in *C. albicans* PAMP recognition. Also studies report that factors like composition of the fungal cell wall, morphogenesis, and species are responsible for the phagocytic process in macrophages and PMN's (Pappas et al., 2003).

10.3 Immunity and vaccination

Since 1960s the application of vaccines has created a great change in terms of the public health. This has been able due to the programs of immunization applied nation wise. The greater these national programs, the greater success have been achieved and many diseases have been tackled. Due to these nationwide vaccination programs especially in children, there has been a great success in tackling diseases (Chan et al., 2013). The World health organization (WHO) reports that 2–3 million are saved every year due to the very effective vaccination programs. These programs have greatly reduced the deaths of the children less than 5 years of age from 93 deaths in 1000 live births in the year 1990 to 39 deaths in 1000 live births in the year 2018 (WHO report). While applying the vaccines, they use the potential of the immune system of the host in order to attack and keep the memory of the attacks which occur one after the other.

Vaccine is a biologically based output which can be applied efficiently and safely to evoke an immune response against the pathogen. A vaccine has the ability to give protection against the infections and diseases. It protects against the repeated attacks of the pathogen. Vaccine comprises of the antigens which have been taken from the pathogen or they are formed synthetically to be able to possess the pathogen components. Vaccine consists of one or more protein antigens. These protein antigens evoke an immune response to give protection. However, there can also be polysaccharide antigens which evoke an immune response to give protection. These antigens have been specially developed against various bacterial infections like in pneumonia which is caused by the *Streptococcus pneumoniae* and meningitis which is caused by the *Neisseria* (Robbins et al., 1989).

Vaccines have been categorized as live and nonlive in order to differentiate the kind of vaccines possessing the diminishing cloning variants of the infectious among those possessing the constituents of a germ or fully killed organisms. Apart from these, various other forms are manufactured including viral transmitters, molecular-based RNA and DNA injections, and virus resembling fragments (Pollard and Bijker, 2021). There is a difference between live and nonlive vaccines. Live vaccines possess the ability to show replication by an unrestricted way like in individuals with compromised system like patients having HIV infection, usage of immunosuppressive drugs, and children possessing the initial immune deficiencies. They are being limited in their use (Rubin et al., 2014), while nonlive do not have any kind of risk in individuals with compromised immune system. At the same time there may not be any kind of protection in individuals with B cell and combined immunodeficiency (Fig. 10.3).

Live vaccines have been developed to provide an efficient immune response by showing replication in an immunocompetent host. Replication does not exceed the limit so as to create disease manifestations like the immunizations developed against the diseases like rubella, rotavirus, measles, polio, and vaccines against the tuberculosis like BCG and also for the influenza, etc. (Pollard and Bijker, 2021).

There is a commutation among the replication of the pathogen in vaccine so as to provide a defensive immune system reaction and an efficient attenuation of the bacterium in order to decrease and nullify the disease. That is mainly one of the reasons that various live attenuated vaccines need to be given in multiple doses and enhance the short-lived immunity like the live attenuated typhoid vaccine, Ty21a (Anwar et al., 2014). At the same time some may create mild diseases like a percentage of 5% of children have developed rashes and an estimate of 15% fever after measles vaccination (mondiale de la Santé and hebdomadaire, 2017). While the nonlive vaccines can be killed as a complete organism by an antigenic part of it (like complete -cell pertussis vaccine and inactivated polio vaccine), proteins filtered from the organism (like acellular pertussis vaccine), recombinant proteins like hepatitis B virus (HBV vaccine), or polysaccharides like the pneumococcal vaccine (against *S. pneumoniae*). Toxoid vaccines like tetanus and diphtheria are formaldehyde-inactivated protein toxins which have been purified from the pathogen. Nonlive vaccines have been combined with an adjuvant to increase its potential to induce an effective immune response that is the immunogenicity (Pollard and Bijker, 2021). There are reported evidences regarding the immune response and

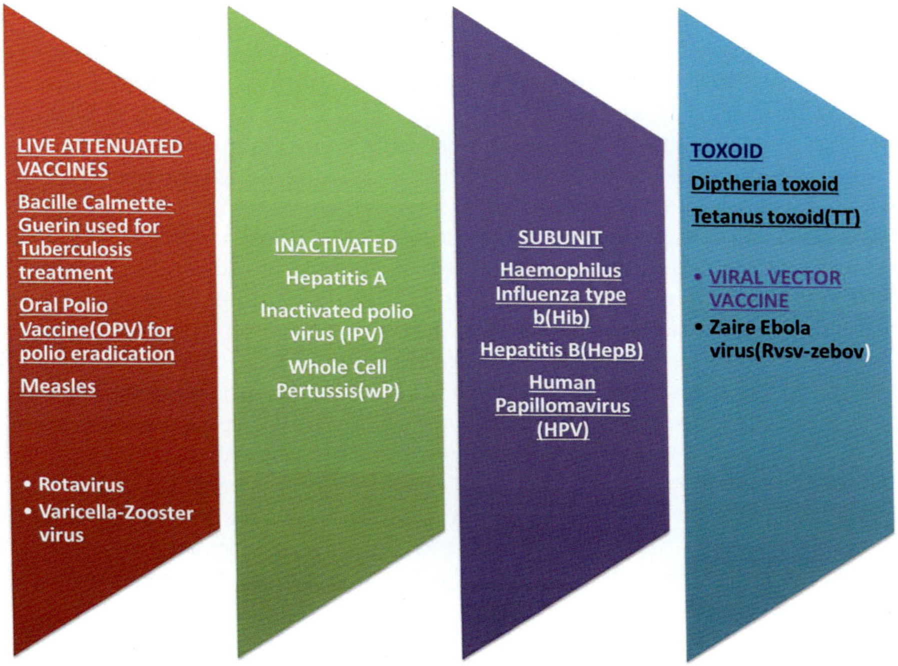

FIGURE 10.3 Different types of vaccines and the examples against them which are available for the diseases.

protection enhancement by the addition of newer adjuvants which can provide danger signals to the innate immune system like oil-in-water emulsion MF59, effectively used in some influenza vaccines (Wilkins et al., 2017). Vaccines possess various constituents which can serve as preservatives, emulsifiers (like polysorbate 80), or stabilizers (for example, gelatin or sorbitol).

One of the main aims of the vaccination is to create immunity against the diseases and infections. In addition to this, it provides herd immunity by decreasing the spread chances of the disease to the exposed individuals. Immune system of the body protects an individual from the foreign attack. It is defined as the resistance mechanism exhibited by the immune cells of the body against any invading antigen like microorganisms, viruses, proteins, and carbohydrates. This resistance mechanism provided by the body against these antigens plays a main role in preventing large number of infections and spread of the disease. Immunity can be innate or acquired. Immunogenicity is the potential to evoke a humoral or cell-mediated immune response. Antigen induces a kind of special immune response. Antigenicity is the potential to combine with antibodies or cell surface receptors. Innate or inborn immunity is seen to consist of barriers which act as block to the pathogens like anatomic barriers; physiologic, phagocytic, inflammatory skin also serves as an anatomic barrier to the pathogen. The microbes are blocked in the nasal passage. The secretions of the mucus of respiratory tract also act as a trapping mechanism. Through this cilia traps the particles and from the pharynx, they are taken out through coughing. Acidic pH of the gastric juice acts as an inhibition to microorganisms. Acidic pH of the stomach also functions as a destroyer to the microorganisms. The normal body temperature of the body also functions as a defense against the pathogens. Increase in the body temperature of the organism also responds to the growth of invading pathogens (Sahoo et al., 2021).

Acquired immunity is the defense system of the immune system acquired during the life of an individual. This type of immunity has the potential to recognize foreign antigens. Among the acquired immunity, natural immunity is acquired by an organism upon exposure with the pathogens having the ability to cause infections. This type of acquired immunity is having long lasting effect. Like the individuals with the recovery from the smallpox have been seen to develop this natural immunity. While as the artificial type of acquired immunity is developed with the vaccination (Sahoo et al., 2021).

The kind of reaction performed by the host immune defense system to the pathogens is known as an immune response. It can be Humoral or Antibody-mediated response and cell-mediated immune response. Humoral type of immune response gets started with the proliferation undergone by the B cells followed by the differentiation into the plasma cells which in turn form and secrete antibodies. B cell consists of receptors consisting of various immunoglobulins. Due to the evoking of the immune response by the antigen, all B cells do not form plasma cells. A number of cells convert into memory cells through which they recognize and remember the subsequent attacks of the pathogen. It is because of the memory cells that the secondary immune response is generated at a greater speed. In contrast, the cell-mediated immunity there is a display of the peptides through the MHC proteins. These peptides are checked by the T cells which possess the receptors of the antigen on their plasmalemma. T-cell antigen receptors know the peptide and the MHC complex. They are fit for Class I or Class II MHC proteins. T lymphocytes have been subcategorized into "helper" Th cells and "cytotoxic" Tc cells. T lymphocytes possess a CD4 surface protein and they recognize the peptides which are attached to the Class II MHC molecules, usually on myeloid APCs. Apart from this, they also possess the ability to recognize the peptides which are shown by B cells. Cell-mediated immunity possesses a characteristic to clear some of the tumors and many of the cells infected with the virus before they can infect other type of cells (Sahoo et al., 2021).

The adaptive immunity is acquired by the B cells which secrete antibodies and also by the T cells. All types of vaccines have been reported to give protection with the antibodies initiation. There is an exception to this which is the BCG vaccine which has been reported to initiate T-cell response and prevents the disease. There are mainly three factors which have given us evidence that different types of functional antibodies are essential in the protection by the vaccine. The three factors are studies of passive protection, immunodeficiency states, and immunological data (Pollard and Bijker, 2021).

Many evidences suggest that antibodies play a main role in the immunity, which has been induced because of the vaccination. Various vaccines also cause T-cell responses. However, there are very less characteristic of T cell in terms of protection. Although they possess a role in B-cell development and the production of antibodies in the lymph, there are various evidences where in it is suggested that in individuals with immunodeficiency, there was a decrease in the count of the antibody resulting in the increase in the exposure to the infection. This has led to the failure of the T cells in assuring protection by controlling the pathogen after the infection occurred, for example, the deficiency in T cells has resulted in the fatal

Varicella zoster virus infection in an uncontrolled way. At the same time, individuals with the deficiency in antibody have developed infection but they have recovered as immunocompetent individuals. The relative suppression of T-cell responses that occurs at the end of pregnancy increases the severity of infection with influenza and *Varicella zoster* viruses (Kourtis et al., 2014).

T cells are classified as cytotoxic (killer) T cells and helper T cells. On the basis of cytokine production, their subcategories include T helper 1 (TH1) cells and TH2 cells. These subcategories are essential in providing cellular immunity and humoral immunity. TH1 cells generate IgG antibody subclasses IgG1 and IgG3. The other types also include TH17 cells functioning in the mucosal surfaces and T follicular helper cells present in the secondary lymphoid organs having their function in high affinity antibody generation. Through this B-cell and T-cell mechanisms, vaccination has provided protection. An important feature of the immune system is the immune memory. Vaccines enhance development by protecting the clinical signs of the infection. They also help in protecting the asymptomatic infections of the disease, reduce the obtainment of the pathogen and finally its spread thus enhancing and establishing the herd immunity. This has been seen as the most important feature of the vaccination programs. Many vaccines nonspecifically provide an enhanced strength of response to the functions in the near future (Pollard and Bijker, 2021).

After vaccination, when a pathogen attacks the immune system of an individual, the immune system responds in a faster and efficient way to protect the individual. It is because of the immune memory. When there is the gap of incubation more, this has been an efficient for protection against the pathogens to mount an immune response, like in view of HBV. It has an incubation period of 6 weeks to 6 months. An individual who has been vaccinated gets protection after the vaccination even if the exposure occurs at the same time. Apart from this there is a thought that this immune memory is not sufficient for protection because there are bacterial invasive infections which cause disease within days or hours after the pathogen (Kelly et al., 2005), like in the cases of both *Haemophilus influenzae* type B (Hib) and capsular group C meningococcal infection. In these two infections it has been seen that after the immune memory because of vaccination, disease still developed (McVernon et al., 2003).

The decline in the levels of the antigen shows a variation over age like it is very rapid in infants. Application of multiple doses of vaccine in childhood can be a cure for this like in the case of diphtheria, tetanus, pertussis, and polio vaccines. This has been recommended in order to keep the level of antibodies required above the margin of requirement like there is provision of five or six doses of tetanus or diphtheria vaccine in childhood (Vaccine, 2018). It has been seen to provide a lifelong protection. Directly vaccines cannot save every individual of a population, reasons being that everyone does not get vaccinated and many among them despite being vaccinated do not evoke immune responses against the pathogen. At the same time those who are vaccinated in a population are prevented from the disease and its transmission and infection. Similarly the occurrence rates of the disease can also be controlled resulting in the overall protection of the disease and infection in those who are not susceptible. About 95% of the population must be vaccinated in those pathogens which have the transmission at a higher rate. At the same time those pathogens having a lower rate of transmission, a lower vaccine application rate would be sufficient to tackle (for example, in polio, diphtheria, and rubella the coverage of the vaccine could be less or equal to 86%. Likely the level for herd immunity is highly different for varying seasons. There is also the vaccine variation affectivity through each year 30%–40% of the vaccine coverage has and is set to show effect on the influenza (Pollard and Bijker, 2021).

10.4 Vaccine development against bacterial pathogens

In 21st century, there has been a great health challenge with respect to the protection and prevention of bacterial infections. Many factors like aging of the populations with high susceptibility to the infection, rise in the phenomenon of antimicrobial resistance, increase in the mortality rates worldwide add to the concern (O'neill and Nations, 2014). Bacterial vaccines can have great effects on antimicrobial resistance. Vaccines can prevent these concerns of infections having multidrug-resistance pathogens, healthcare-related infections, prevent and reduce bacterial infections of all age groups (Klugman and Black, 2018; Poland et al., 2009). The WHO has recognized AMR as being the topmost threat to the health of the humans. There is an estimation of 63.5% in the year 2015, which provides the figure of AMR infections having the base of healthcare in Europe (Gasser et al., 2019). The individuals having the age of more than 65 have also been seen as the second most concern. Reports have estimated that annually 700,000 deaths are associated to AMR globally (O'neill and Nations, 2014). In AMR there are some specific pathogens related with the infections. They are known as ESKAPE pathogens: *Enterococcus faecium, S. aureus, K. pneumoniae, Acinetobacter baumannii, P. aeruginosa,* and *Enterobacter* spp11. *Mycobacterium tuberculosis* (Mtb) has been reported as the cause of 1.6 million deaths in 2017. Multidrug resistant (MDR) Mtb strains have been a cause of 3.5% of the new and more cases of tuberculosis with 19% of existing cases. Among them 8.5% were expansively resistant. Worldwide, the cure for tuberculosis treatment with the utilization of antimicrobials being low (55%), and this reflects an essential need of antimicrobials and vaccines having the potential to prevent infection and therefore the prevention of the disease occurrence (Harding, 2020).

Polysaccharide-protein-conjugate vaccines have proven very effective and successful. Surface polysaccharides are linked chemically with the protein T cell dependent. APCs take up the glycoconjugate like as saccharide-specific B-cells. The saccharide specific B cells then show the digested peptides on the surface of the cell. APCs present these peptides to T-helper cells, which in turn accelerate the differentiation of the saccharide-specific B cells into antisaccharide antibody producing plasma cells and B memory cells. Licensed PCV, Hib and meningococcal conjugate vaccines have been reported to be the successful bacterial vaccines. Since the implementation of PCVs in low-income countries by the Gavi Vaccine Alliance, vaccination is estimated to have prevented at least 500,000 childhood deaths, mainly from pneumonia (Sundaram et al., 2017). The techniques like bioconjugation and synthetic/semisynthetic saccharide synthesis have been used in order to enhance and expand the reservoir of the second-generation conjugate vaccines. In glycoconjugates, polysaccharides have the ability to activate T cells when they are shown as the targeted glycopeptides with the peptide part presented in the context of MHCII molecules, and the saccharide part which stimulates the T-cell responses (Sun et al., 2019).

M. tuberculosis is an intracellular bacterium. The structure being lipoidal gives additional chances of the immune attack. Once there is infection, there is the assemblage of monocytes, chemokines which promote the adhesion of the cell and the recruitment of the T cell which base the formation of granuloma the assembled and organized structures proceeding to the long inactive phase of the infection. From 1960s, the Bacille Calmette-Guerin (BCG) vaccine has been in use, to prevent the spread of tuberculosis in children. BCG appears to gear up the Mtb circulation from alveolar macrophages to tissue-recruited or lung recruited macrophages and neutrophils via antigen-specific CD4 T-cells. This activation, differentiation, and the transfer of the phagocytosis is believed to enhance the potential ability to limit the replication of the bacteria through some widespread mechanisms (Delahaye et al., 2019). BCG works to induce the inborn immune response which fastens the potential of innate effector cells to respond to nonspecific stimuli (Joosten et al., 2018; Covián et al., 2019). If the disease gets reactivated, this may side step this transfer option which is necessary. This could reach ta stage to decrease the effect of the BCG to save the pulmonary reactivation in the infected ones. This is one of the main ways through which tuberculosis gets transmitted and the most common cause of tuberculosis-related deaths worldwide (Trunz et al., 2006) (Fig. 10.4).

This protective adaptive immune system is seen to be mainly accelerated by by CD4+ Th1 cells, with additional effects and helps from Th17 and CD8+ T-cells (Andersen and Scriba, 2019).

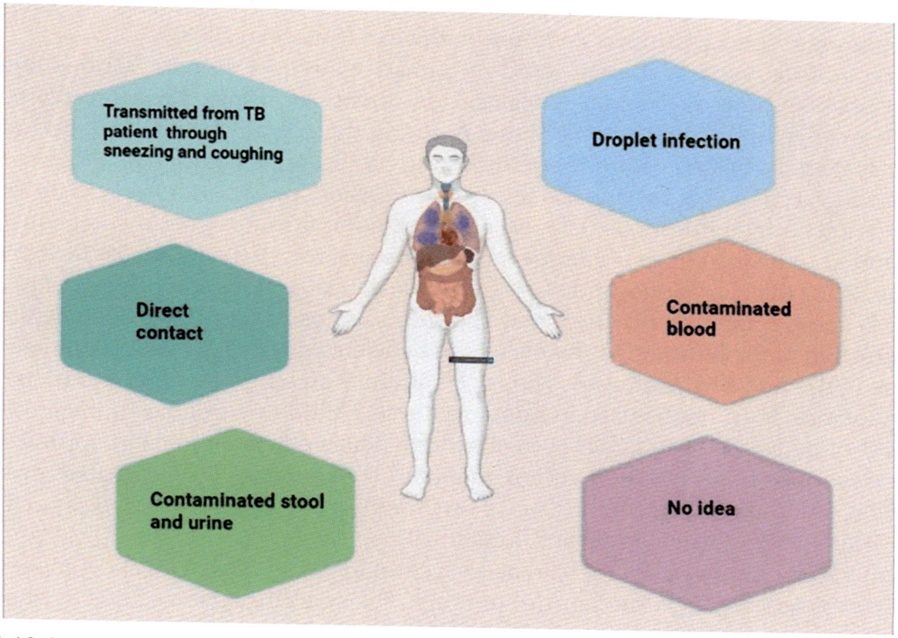

FIGURE 10.4 Various modes of transmission of the deadly tuberculosis, like contact, sneezing, and all other factors which play a role in its infection.

In India the use of BCG revaccination in IGRA+ (IFN-y release assay linked to presence of *Mycobacteria*) gives a remarkable boost up of the Th1/Th17 CD4+ T-cell responses which also include polyfunctional T cells. In South African adolescents with the booster application it prevented the sustained seroconversion (by IFN-γ release assay). With the new development, live attenuated vaccines (MTBVAC; VPM1002) have been confirmed as the adjustments and improvements in the BCG. The CD4+ responses in immunology have been seen to be in major attention, also there has been little attention regarding the bacterial mechanisms. There is a lack of knowledge regarding the Mtb systems of secretion, defenses of bacteria against the reactive oxygen and nitrogen molecules. There are also very much less elaborated mechanisms of the virulence of the bacteria and the protection of the immune system. These are the main reasons why there is no fully effective vaccine regarding the Mtb (Poolman, 2020).

10.5 Understanding vaccine development against fungal pathogens

Vaccines have been effective in preventing six million deaths annually (Leibovitch and Jacobson, 2016). Live attenuated vaccines give a long-term and an efficient immune response against the pathogens. It can be very efficient in case of immunocompetent hosts. Due to the infection of various pathogenic fungi like *Blastomyces dermatitidis* (*B. dermatitidis*), *Pneumocystis carinii (P. carinii*), *Paracoccidioides brasiliensis* (*P. brasiliensis*), *Histoplasma capsulatum (H. capsulatum*) and *C. neoformans*, these kinds of measures are very efficient and workable as they evoke immune response through common pathways (Cassone, 2008).

This type of the vaccine is firstly used in human subjects. There are various types of the studies calculating the effects and efficiencies of attenuated fungi. Probably these vaccines will have an edge in the prevention of fungi in future times with the efficient and full defense immune system. In this way an important finding has been the heat-killed *Saccharomyces cerevisiae* (HKS) vaccine. This vaccine has the application and an important role in the protection against various fungal infections. Hks-based vaccination through the subcutaneous way has been reported to be the efficient vaccine factor in opposition to the strains of the fungus with much virulence like *C. albicans* and *Aspergillus fumigatus* (*A. fumigatus*) (Nami et al., 2019).

There is another method of attenuated vaccine called as H99 g. It has been earlier seen to be in protection of the infection to the mice with deficiency in CD4+ T cells with the induction of murine IFN-γ and Th1 responses. The H99 g strain is an efficient stimulator in the production of cytokines. It cannot be used in human subjects. Another work has estimated that there is an effect of both CD4+ and CD8+ T cells in the perseverance of mice in opposition to the *C. neoformans* infection. The two recent methods (BAD-1 and H99 g) could provide the acceleration in immunizing the CD4+ T cell-deficient subjects, particularly HIV/AIDS patients (Nami et al., 2019; Santos and Levitz, 2014).

Subunit vaccines are the most researched vaccines for fungi. They consist of more than one recombinant proteins or polysaccharides of fungi. The scientific aspect of this technology has been the transference and the expression of the gene which encodes an antigen. Through this there is a sufficient immune response. By these ways, the gene which gets transmitted encodes a region which is related to the pathogenicity and virulence of the organism. Protein antigens are mostly attached with the matchable adjuvant or protein carrier. Among them bacterial toxoids are usually used in order to mount an efficient immune response and hence the sustainable effect of immunization. There are many uses for the recombinant subunit vaccines. Firstly there is the absence of the agent which has the pathogenic potential. That is the reason these have been the safest in their use, particularly among the individuals with compromised immune system. Recombinant subunit vaccines are useful as there is absence of the pathogenic agent, hence its use becomes safe especially in immunocompromised patients (Nami et al., 2019).

A conjugate vaccine is efficiently obtained by the linking of a strong antigen with a poor antigen covalently, most effectively polysaccharide to protein. It is formed in a way to evoke a great immune response against the pathogen (Karch and Burkhard, 2016). While opposing with the B polysaccharide antigens, B cells develop antibody responses without the involvement of the T cells called as T-independent immune response. The epitopes of the polysaccharides are distinguished by the receptors of the B cell. while as for its presentation, they need to attach with the peptides (hapten-carrier system) which in turn needs to be presented by MHC complexes expressed on the APCs. The kind of immunity which is being stimulated by the T cells has been seen to be strong with efficient time. MHC molecules are able to bind proteins and eventually induce the T-cell responses (Furman and Davis, 2015). First kind of conjugated vaccine was formed against the *C. neoformans* which contains glucuronoxylomannan (GXM), a capsular polysaccharide, and tetanus toxoid (TT). The benefit of using this conjugated vaccine is that it is based on its target to polysaccharide epitopes. Those polysaccharide epitopes are usually present in all fungi, especially β-glucans (Nami et al., 2019) Fig. 10.5 Table 10.4.

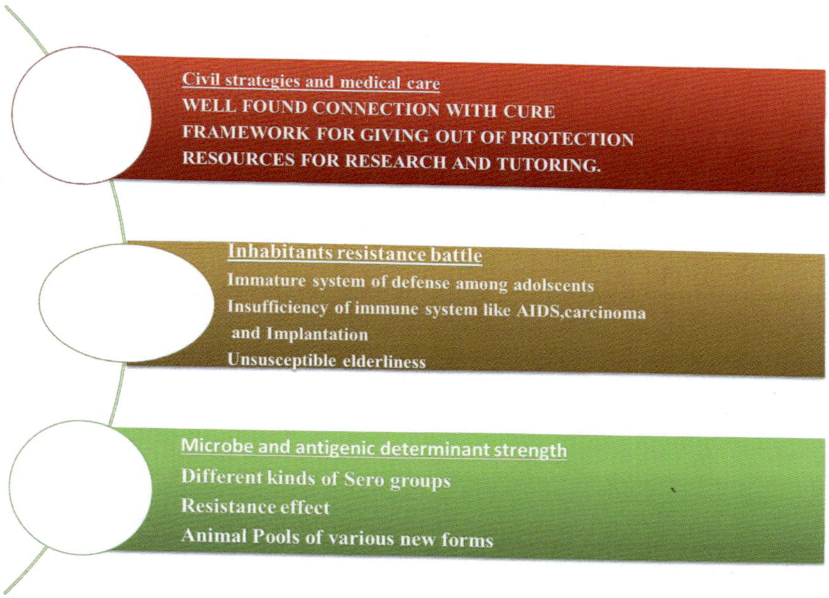

FIGURE 10.5 Illustration of herd immunity as an important process achieved by vaccination—Its advantages, easiness, and the challenges faced.

TABLE 10.4 Individuals with their targets upon (Plotkin, 2005).

Individual stages	Targets
Children	Tetanus, inactive vaccine of polio, Hep b, etc.
Teenagers	Tetanus, human papilloma virus, diphtheria
Elders	Zoster, *Herpes simplex* virus −2
Patients in the treatment centers	Fungal *Candida*
Expecting women	Human orthopneumovirus, group B *Streptococcal* infection
Persons with nonlethal challenges	Abnormal immune system challenges, cancer, addictions in the drug
Persons having the lethal infections	Sexually transmitted HIV, HPV

HIV, human immunodeficiency virus; *HPV*, human papillomavirus; *HSV*, herpes simplex virus.

10.6 Future perspectives

From past many years, it has been a remarkable transformation in the field of vaccinology. Earlier there were uncountable deaths worldwide due to the unavailability of the vaccines and treatment strategies. It is now possible to a greater extent that the most fatal diseases and infections caused by microbes can be efficiently protected and treated through efficient vaccination programs. The field of vaccinology has transformed with the use and application of recombinant DNA technology and other disciplines. In recent times there has been the application of new methods of the discovery and design of the antigens. There has been an efficient application of vaccines globally, leaving a good impact on the global health. At the same time there are various diseases and infections worldwide which have create a great concern because either there are no vaccines available to tackle those diseases or there are the different kinds of strategies adopted by the different pathogens to decrease the resistance effect of the host's immune system.

Due to the addition of the adjuvants to the vaccines, there has been an enhancement in the immune response. In addition to this; it has also decreased the concentration of the antigen and its dosage. Liposomes have been used as delivery systems. They are able to carry both of the hydrophilic and hydrophobic substances. Being nontoxic, they can be conjugated with antibodies. The route through which the vaccines can be delivered is important. Without having this proper channel there are least chances for the engineered vaccine to meet its target. Various traditional vaccine methodologies were carried out. They are limited because of their inconvenience and safety. There are very less number of vaccines that can be used intranasally, orally, and subcutaneously except influenza, polio, and measles-mumps-rubella, respectively.

In recent times there is the use of alternative vaccine administrative strategies which include the transdermal patch, the ballistic delivery to skin (the gene gun), and other intradermal methods. Also there are aerosol, sublingual, rectal, and vaginal mucosal vaccines. These alternative methods have many efficient advantages to enhance the immune response and make it effective at the target site (e.g., replication in the gut area by the oral polio vaccine), convenience (e.g., transdermal patch use), the use of combined vaccines in order to decrease and enhance the schedule of vaccination, and decrease in the application by means of the efficient hypodermic needle. Apart from these alternate strategies there are few vaccines which can be applied and administered through non-IM routes, possibly due to the reasons like the decreased efficiency and less data of safety.

Till now there are the priorities of having vaccination and research strategies and programs for the infectious diseases. Apart from the immunization, there is growing interest to have an effective research and find more and more ways to target and set goals for noninfectious diseases which have been a great concern for the whole world like cancer. Similarly there are future targets to control the allergies with better antigens, with the introduction of IgG apart from IgE antibodies. There are also some goals as how to tolerate auto antigens, contraception by immunization against hormones, combating multiple ailments by immunization against cholesterol, and antibody-mediated clearance of drugs. Furthermore, several efforts have been made in order to develop the humanized mice which have the potential to mimic immune system of humans. Despite this work over the years there has to be an efficient understanding of the host–parasite interaction. The varying immune responses under different conditions have to be studied.

Researches comprising of treatment strategies have to be geared up against challenging diseases which include tuberculosis, AIDS, malaria, and other various infectious diseases. The research and treatment strategies regarding these challenging diseases should revolutionize the world and provide a sigh of relief to the human kind. Also there have been successful developments in the vaccine efficiencies particularly against the allergies, autoimmune diseases, and cancers. Also through these efforts they provide vaccines with great efficiency against the new and emerging diseases. In a Nutshell there has been great work from years regarding the vaccinology. At the same time this work is to be carried forward in an efficient way. More research strategies are to be adopted for the challenges which have been posed by the fatal and challenging infectious diseases due to their changing behaviors toward host defense system.

References

Ahmad, S.J.C., Immunology, D., 2011. Pathogenesis, immunology, and diagnosis of latent *Mycobacterium tuberculosis* infection. Clin. Dev. Immunol 2011.

Al-Mehdar, A.A., Al-Akydy, A.G., 2017. Pattern of antimicrobial prescribing among in-patients of a teaching hospital in yemen: a prospective study. Univ. J. Pharmaceut. Res. 2 (5).

Andersen, P., Scriba, T.J.J., 2019. Moving tuberculosis vaccines from theory to practice. Nat. Rev. Immunol. 19, 550–562.

Anwar, E., Goldberg, E., Fraser, A., Acosta, C.J., Paul, M., Leibovici, L.J.C., 2014. Vaccines for preventing typhoid fever. Coch. Database Syst. Rev. 5 (5).

Arendrup, M.C., 2010. Epidemiology of invasive candidiasis. Curr. Opin. Crit. Care 16, 445–452.

Arendrup, M.C., J, E.D.M., 2013. Candida and candidaemia. Dan. Med. J. 60, B4698.

Austermeier, S., Kasper, L., Westman, J., Gresnigt, M.S.J., 2020. I want to break free—macrophage strategies to recognize and kill Candida albicans, and fungal counter-strategies to escape. Curr. Opin. Microbiol. 58, 15–23.

Bäumler, A., Fang, F.C.J., 2013. Host specificity of bacterial pathogens. Cold Spring Harb. Perspect. Med. 3, a010041.

Benton, M.J.J.S., 2009. The Red Queen and the Court Jester: species diversity and the role of biotic and abiotic factors through time. Science 323, 728–732.

Bongomin, F., Gago, S., Oladele, R.O., Denning, D.W.J., 2017. Global and multi-national prevalence of fungal diseases—estimate precision. J. Fungi. 3, 57.

Boral, H., Metin, B., Döğen, A., Seyedmousavi, S., Ilkit, M.J.F.G., Biology, 2018. Overview of selected virulence attributes in Aspergillus fumigatus, Candida albicans, Cryptococcus neoformans, Trichophyton rubrum, and Exophiala dermatitidis. Fung. Genet. Biol. 111, 92–107.

Brown, G., Denning, D., Gow, N., Levitz, S., Netea, M., White, T., 2012. Hidden killers: human fungal infections. Sci. Transl. Med. 4, 165rv13.

Byrnes Iii, E.J., Li, W., Lewit, Y., et al., 2010. Emergence and pathogenicity of highly virulent Cryptococcus gattii genotypes in the northwest United States. Plos Pathogen. 6, e1000850.

Cassone, A., 2008. Fungal vaccines: real progress from real challenges. Lancet Infect. Dis. 8, 114–124.

Chan, M., Lake, A., Fauci, A., et al., 2013. Global vaccine action plan. Vaccine 31. ELSEVIER SCI LTD THE BOULEVARD, LANGFORD LANE, KIDLINGTON, OXFORD OX5 1GB.

Clemens, D.L., Lee, B.-Y., Horwitz, M.J.J., 1995. Purification, characterization, and genetic analysis of *Mycobacterium tuberculosis* urease, a potentially critical determinant of host-pathogen interaction. J. Bacteriol. 177, 5644–5652.

Covián, C., Fernández-Fierro, A., Retamal-Díaz, A., et al., 2019. BCG-induced cross-protection and development of trained immunity: implication for vaccine design. Front. Immunol. 10, 2806.

Delahaye, J.L., Gern, B.H., Cohen, S.B., et al., 2019. Cutting edge: bacillus calmette–guérin–induced T cells shape *Mycobacterium tuberculosis* infection before reducing the bacterial burden. J. Immunol. 203, 807–812.

Doron, S., Gorbach, S.L., 2008. Bacterial infections: overview. Int. Encyclop. Pub. Health 273.

Friswell, M., Campbell, B., Rhodes, J.J.G., 2010. The role of bacteria in the pathogenesis of inflammatory bowel disease. Gut Liver. 4, 295.

Furman, D., Davis, M.M.J., 2015. New approaches to understanding the immune response to vaccination and infection. Lancet Infect. Dis. 33, 5271–5281.

Gasser, M., Zingg, W., Cassini, A., Kronenberg, A.J.T., 2019. Attributable deaths and disability-adjusted life-years caused by infections with antibiotic-resistant bacteria in Switzerland. J. Bacteriol. 19, 17–18.

Gavaldà, J., Meije, Y., Fortún, J., et al., 2014. Invasive fungal infections in solid organ transplant recipients. Clin. Microbiol. Infect. 20, 27–48.

Gonçalves, B., Ferreira, C., Alves, C.T., Henriques, M., Azeredo, J., Silva, S.J., 2016. Vulvovaginal candidiasis: epidemiology, microbiology and risk factors. Crit. Rev. Microbiol. 42, 905–927.

Graham, D.Y., Opekun, A.R., Klein, P.D., 1993. Clarithromycin for the eradication of *Helicobacter pylori*. J. Clin. Gastroenterol. 16, 292–294.

Harding, E.J., 2020. WHO global progress report on tuberculosis elimination. Lancet Respir. Med. 8, 19.

Hayward, J.A., Mathur, A., Ngo, C., Man, S.M.J., Reviews, M.B., 2018. Cytosolic recognition of microbes and pathogens: inflammasomes in action. Microbiol. Mol. Biol. Rev. 82, e00015–18.

Herbrecht, R., Neuville, S., Letscher-Bru, V., Natarajan-Amé, S., Lortholary, O.J.D., 2000. Fungal infections in patients with neutropenia. Drug. Aging 17, 339–351.

Homann, O.R., Dea, J., Noble, S.M., Johnson, A.D.J., 2009. A phenotypic profile of the Candida albicans regulatory network. Plos Genet. 5, e1000783.

Hopkins, R.J., Girardi, L.S., Turney, E.J.G., 1996. Relationship between *Helicobacter pylori* eradication and reduced duodenal and gastric ulcer recurrence: a review. Gastroenterology 110, 1244–1252.

Ikeh, M., Ahmed, Y., Quinn, J.J.M., 2017. Phosphate acquisition and virulence in human fungal pathogens. Microorganisms 5, 48.

Irazoqui, J.E., Ng, A., Xavier, R.J., Ausubel, F.M.J., 2008. Role for β-catenin and HOX transcription factors in *Caenorhabditis elegans* and mammalian host epithelial-pathogen interactions. Proc. Natl. Acad. Sci. U. S. A. 105, 17469–17474.

Joosten, S.A., Van Meijgaarden, K.E., Arend, S.M., et al., 2018. Mycobacterial growth inhibition is associated with trained innate immunity. J. Clin. Invest. 128, 1837–1851.

Kao, A.S., Brandt, M.E., Pruitt, W.R., et al., 1999. The epidemiology of candidemia in two United States cities: results of a population-based active surveillance. Clin. Infect. Dis. 29, 1164–1170.

Karch, C.P., Burkhard, P.J.B.P., 2016. Vaccine technologies: from whole organisms to rationally designed protein assemblies. Biochem. Pharmacol. 120, 1–14.

Katz, J.A., Melmed, G., Sands, B.E., Colitis Foundation of America NY, 2011. The Facts About Inflammatory Bowel Diseases.

Kaufmann, S.H.J., 1995. Immunity to intracellular microbial pathogens. Annu. Rev. Immunol. 16, 338–342.

Kelly, D.F., Pollard, A.J., Moxon, E.R., 2005. Immunological memory: the role of B cells in long-term protection against invasive bacterial pathogens. JAMA 294, 3019–3023.

Kim, J.-Y., 2016. Human Fungal Pathogens: Why Should We Learn? Springer.

Kirzinger, M.W., Stavrinides, J.J.T.I.M., 2012. Host specificity determinants as a genetic continuum. Trend. Microbiol. 20, 88–93.

Klugman, K.P., Black, S.J.P., 2018. Impact of existing vaccines in reducing antibiotic resistance: primary and secondary effects. Proc. Natl. Acad. Sci. U. S. A. 115, 12896–12901.

Köhler, J.R., Hube, B., Puccia, R., Casadevall, A., Perfect, J.R., 2017. Fungi that infect humans. Microbiol. Spectr. 5, 5.3. 08.

Kosmidis, C., Denning, D.W., 2015. The clinical spectrum of pulmonary aspergillosis. Thorax 70, 270–277.

Kourtis, A.P., Read, J.S., Jamieson, D.J.J., 2014. Pregnancy and infection. N. Engl. J. Med. 370, 2211–2218.

Krachler, A.M., Ham, H., Orth, K.J.P., 2011. Outer membrane adhesion factor multivalent adhesion molecule 7 initiates host cell binding during infection by gram-negative pathogens. Proc. Natl. Acad. Sci. U. S. A. 108, 11614–11619.

Kuehn, M.J., Kesty, N.C., 2005. Bacterial outer membrane vesicles and the host–pathogen interaction. Genes. Dev. 19, 2645–2655.

Kullberg, B.J., Arendrup, M.C., 2015. Invasive candidiasis. N. Engl. J. Med. 373, 1445–1456.

Leibovitch, E.C., Jacobson, S., 2016. Vaccinations for neuroinfectious disease: a global health priority. Neurotherapeutics 13, 562–570.

Leone, N., Pellicano, R., Brunello, F., et al., 2003. *Helicobacter pylori* seroprevalence in patients with cirrhosis of the liver and hepatocellular carcinoma. Canc. Detect. Prev. 27, 494–497.

Li, Z., Nielsen, K.J., 2017. Morphology changes in human fungal pathogens upon interaction with the host. J. Fungi. 3, 66.

Lind, T., Mégraud, F., Unge, P., et al., 1999. The MACH2 study: role of omeprazole in eradication of *Helicobacter pylori* with 1-week triple therapies. Gatroenterology 116, 248–253.

Lionakis, M.S., 2014. New insights into innate immune control of systemic candidiasis. Med. Mycol. 52, 555–564.

Lowrie, D.B., Tascon, R.E., Colston, M.J., Silva, C.L., 1994. Towards a DNA vaccine against tuberculosis. Vaccine 12, 1537–1540.

Makanjuola, O., Bongomin, F., Fayemiwo, S.J., 2018. An update on the roles of non-albicans Candida species in vulvovaginitis. J Fungi. 4, 121.

Mcvernon, J., Johnson, P., Pollard, A., Slack, M., Moxon, E.J., 2003. Immunologic memory in Haemophilus influenzae type b conjugate vaccine failure. Arch. Dis. Child. 88, 379–383.

Mehmood, A., Akram, M., Shahab-Uddin, A.A., et al., 2010. *Helicobacter pylori*: an introduction. Clin. Microbiol. Rev. 1, 1337–1351.

Mendall, M., Goggin, P., Molineaux, N., et al., 1992. Childhood living conditions and *Helicobacter pylori* seropositivity in adult life. Lancet 339, 896–897.

Millsop, J.W., Fazel, N.J.C.I.D., 2016. Oral Candid. 34, 487–494.

Mir, M.A., 2015. Developing Costimulatory Molecules for Immunotherapy of Diseases. Academic Press.

Molle, V., Kremer, L.J.M.M., 2010. Division and cell envelope regulation by Ser/Thr phosphorylation: Mycobacterium shows the way. Mol. Microbiol. 75, 1064–1077.

Mondiale De La Santé, O., Hebdomadaire, W.H., 2017. Measles vaccines: WHO position paper–April 2017–Note de synthèse de l'OMS sur les vaccins contre la rougeole–avril. Wkly. Epidemol. Rec. 92, 205–227.

Mosa, T.E., El-Baz, H.A., Mahmoud, M.S., et al., 2013. *Helicobacter pylori* sero-prevalence in different liver diseases. World J. Gastroent. 5, 414–419.

Nami, S., Aghebati-Maleki, A., Morovati, H., Aghebati-Maleki, L., 2019. Current antifungal drugs and immunotherapeutic approaches as promising strategies to treatment of fungal diseases. Biomed. Pharmacother. 110, 857–868.

Noble, S.M., Johnson, A.D.J., 2007. Genetics of Candida albicans, a diploid human fungal pathogen. Annu. Rev. Genet. 41, 193–211.

O'neill, J.J.T., Nations, W.O., 2014. Antimicrobial Resistance.

Pan, X., Yang, Y., Zhang, J.-R., 2014. Molecular basis of host specificity in human pathogenic bacteria. Emerg. Microbes. Infect. 3, 1–10.

Pappas, P.G., Rex, J.H., Lee, J., et al., 2003. A prospective observational study of candidemia: epidemiology, therapy, and influences on mortality in hospitalized adult and pediatric patients. Clin. Infect. Dis. 37, 634–643.

Pegorie, M., Denning, D.W., Welfare, W.J.J., 2017. Estimating the burden of invasive and serious fungal disease in the United Kingdom. J Infect 74, 60–71.

Peterson, W.L., Graham, D.Y., Marshall, B., et al., 1993. Clarithromycin as monotherapy for eradication of *Helicobacter pylori*. Random. Double-blind Trial. 88, 1860.

Pfaller, M., Jones, R., Messer, S., Edmond, M., Wenzel, R.J.D.M., Disease, I., 1998. National surveillance of nosocomial blood stream infection due to Candida albicans: frequency of occurrence and antifungal susceptibility in the SCOPE Program. Diagon. Microbiol. Infect. Dis. 31, 327–332.

Plotkin, S.J.N.M., 2005. Vaccines: past, present and future. Nat. Med 11, S5–S11.

Poland, G.A., Jacobson, R.M., Ovsyannikova, I.G.J.V., 2009. Trends affecting the future of vaccine development and delivery: the role of demographics, regulatory science, the anti-vaccine movement, and vaccinomics. Vaccine 27, 3240–3244.

Pollard, A.J., Bijker, E.M.J.N.R.I., 2021. A guide to vaccinology: from basic principles to new developments. Nat. Rev. Immunol. 21, 83–100.

Poolman, J.T.J.N.V., 2020. Expanding the role of bacterial vaccines into life-course vaccination strategies and prevention of antimicrobial-resistant infections. NPJ Vaccine 5, 1–12.

Prescott, H., 2002. Laboratory Exercises in Microbiology.

Qadri, H., Haseeb, A., Mir, M., 2021. Novel strategies to combat the emerging drug resistance in human pathogenic microbes. Curr. Drug Target. 22, 1–13.

Queiroz, D.M., Carneiro, J.G., Braga-Neto, M.B., et al., 2012. Natural history of *Helicobacter pylori* infection in childhood: eight-year follow-up cohort study in an urban community in northeast of Brazil. Helicobater 17, 23–29.

Rachman, H., Strong, M., Ulrichs, T., et al., 2006. Unique transcriptome signature of *Mycobacterium tuberculosis* in pulmonary tuberculosis. Infect. Immun. 74, 1233–1242.

Rajasingham, R., Smith, R.M., Park, B.J., et al., 2017. Global burden of disease of HIV-associated cryptococcal meningitis: an updated analysis. Lance Infect. Dis. 17, 873–881.

Rescigno, M., Borrow, P.J.C., 2001. The host-pathogen interaction: new themes from dendritic cell biology. Cell 106, 267–270.

Robbins, J., Schneerson, R., Szu, S., et al., 1989. Prevention of invasive bacterial diseases by immunization with polysaccharide-protein conjugates. Curr. Top. Microbiol. Immunol. 169–180.

Rubin, L.G., Levin, M.J., Ljungman, P., et al., 2014. 2013 IDSA clinical practice guideline for vaccination of the immunocompromised host. Clin. Infect. Dis. 58, e44–e100.

Sahoo, J.P., Nath, S., Ghosh, L., Samal, K.C., 2021. Concepts of immunity and recent immunization programme against COVID-19 in India. Acad. Excell. 3, 103–106.

Santos, E., Levitz, S.M.J., 2014. Fungal vaccines and immunotherapeutics. Cold Spring Harb. Perspect. Med. 4, a019711.

Sardi, J., Scorzoni, L., Bernardi, T., Fusco-Almeida, A., Giannini, M.M.J., 2013. Candida species: current epidemiology, pathogenicity, biofilm formation, natural antifungal products and new therapeutic options. J. Med. Microbiol. 62, 10–24.

Sarmah, P., Dan, M., Adapa, D., Sarangi, T.J.E., 2018. A review on common pathogenic microorganisms and their impact on human health. Elect. J. Biol. 14, 50–58.

Sarmah, P., Dan, M.M., Adapa, D.J.E., 2017. Antimicrobial resistance: a tale of the past becomes a terror for the present. Elect. J. Biol. 13.

Schmidt, S., Tramsen, L., Lehrnbecher, T.J.F., 2017. Natural killer cells in antifungal immunity. Front. Immunol. 8, 1623.

Scully, J.L., 2004. What is a disease? Disease, disability and their definitions. EMBO Rep. 5, 650–653.

Sen, R., Nayak, L., De, R.K., 2016. A review on host–pathogen interactions: classification and prediction. Eur. J. Clin. Microbiol. Infect. Dis. 35, 1581–1599.

Singh, S.B., Davis, A.S., Taylor, G.A., Deretic, V.J.S., 2006. Human IRGM induces autophagy to eliminate intracellular mycobacteria. Science 313, 1438–1441.

Singh, S.D., Masood, T., Sabharwal, R.K., Sharma, N., Nautiyal, S.C., Singh, R.K., 2015. Biochemical and molecular characterization of cerebrospinal fluid for the early and accurate diagnosis of *Mycobacterium tuberculosis*. Biomed. Res. 26, 426–430.

Singh, S.R., Krishnamurthy, N., Mathew, B.B.J.J.E.M., 2014. A review on recent diseases caused by microbes. J. Appl. Environ. Microbiol. 2, 106–115.

Sironi, M., Cagliani, R., Forni, D., Clerici, M.J.N.R.G., 2015. Evolutionary insights into host–pathogen interactions from mammalian sequence data. Nat. Rev. Genet. 16, 224–236.

Smith, I.J.C.M.R., 2003. *Mycobacterium tuberculosis* pathogenesis and molecular determinants of virulence. Clin. Microbiol. Rev. 16, 463–496.

Soares, N.C., Bou, G., Blackburn, J.M.J.F.I.M., 2016. Proteomics of microbial human pathogens. Front. Microbiol. 7, 1742.

Strugnell, R., Drew, D., Mercieca, J., et al., 1997. DNA vaccines for bacterial infections. Exp. Rev. Vaccine. 75, 364–369.

Sullivan, D., Henman, M., Moran, G., et al., 1996. Molecular genetic approaches to identification, epidemiology and taxonomy of non-albicans Candida species. J. Med. Microbiol. 44, 399–408.

Sun, X., Stefanetti, G., Berti, F., Kasper, D.L.J.P.O.T.N.O.S., 2019. Polysaccharide structure dictates mechanism of adaptive immune response to glycoconjugate vaccines. Proc. Nat. Acad. Sci. 116, 193–198.

Sundaram, N., Chen, C., Yoong, J., et al., 2017. Cost-effectiveness of 13-valent pneumococcal conjugate vaccination in Mongolia. Vaccine 35, 1055–1063.

Tascon, R.E., Colston, M.J., Ragno, S., Stavropoulos, E., Gregory, D., Lowrie, D.B.J., 1996. Vaccination against tuberculosis by DNA injection. Nat. Med 2, 888–892.

Tato, C., Hunter, C.J.I., Immunity, 2002. Host-pathogen interactions: subversion and utilization of the NF-κB pathway during infection. Infect Immunol. 70, 3311–3317.

Tikkinen, K.A., Leinonen, J.S., Guyatt, G.H., Ebrahim, S., Järvinen, T.L.J., 2012. What is a disease? Perspectives of the public, health professionals and legislators. BMJ 2, e001632.

Torrelles, J.B., Schlesinger, L.S.J.T., 2010. Diversity in *Mycobacterium tuberculosis* mannosylated cell wall determinants impacts adaptation to the host. Tuberculosis 90, 84–93.

Trunz, B.B., Fine, P., Dye, C.J.T.L., 2006. Effect of BCG vaccination on childhood tuberculous meningitis and miliary tuberculosis worldwide: a meta-analysis and assessment of cost-effectiveness. Lancet. 367, 1173–1180.

Ullmann, A., Akova, M., Herbrecht, R., et al., 2012. ESCMID* guideline for the diagnosis and management of Candida diseases 2012: adults with haematological malignancies and after haematopoietic stem cell transplantation (HCT). Clin. Microbiol. Infect. 18, 53–67.

Vaccine, W.H.O.J., 2018. Tetanus vaccines: WHO position paper, February 2017–recommendations. Tetanus Vaccine.: WHO Post. Pap. 36, 3573–3575.

Van De Veerdonk, F.L., Netea, M.G.J.C.F.I.R., 2010. T-cell subsets and antifungal host defenses. Curr. Fungal. Infect. Rep. 4, 238–243.

Vergne, I., Singh, S., Roberts, E., et al., 2006. Autophagy in immune defense against *Mycobacterium tuberculosis*. Autophagy 2, 175–178.

Warkentien, T., Crum-Cianflone, N.F.J., 2010. An update on Cryptococcus among HIV-infected patients. Int. J. Std. Aids. 21, 679–684.

Weiss, E., Schlegel, J., Terpitz, U., et al., 2020. Reconstituting NK cells after allogeneic stem cell transplantation show impaired response to the fungal pathogen Aspergillus fumigatus. Front. Immunol. 11, 2117.

Whitaker, A., Newman, D.P., 2005. Penetration Testing and Network Defense: Penetration Testing _1. Cisco Press.

Wilkins, A.L., Kazmin, D., Napolitani, G., et al., 2017. AS03-and MF59-adjuvanted influenza vaccines in children. Front. Immunol. 8, 1760.

Wotherspoon, A.C., Diss, T., Pan, L., et al., 1993. Regression of primary low-grade B-cell gastric lymphoma of mucosa-associated lymphoid tissue type after eradication of *Helicobacter pylori*. Lancet 342, 575–577.

Younes, S., Al-Sulaiti, A., Nasser, E.A., Najjar, H., Kamareddine, L.J.F.I.C., Microbiology, I., 2020. Drosophila as a model organism in host–pathogen interaction studies. Front. Cell Infect. Microbiol. 10, 214.

Zhu, C., Liu, S., Zhai, J., Chen, Z., Wu, K., Li, N., 2016. Clinical and pathological features of three types of peritoneal tuberculosis: a single centre in China. Biomed. Res. 1302–1308 (0970-938X) 27.

Index

Note: 'Page numbers followed by "f" indicate figures, "t" indicate tables.'

M

N

O

P

Q

R

Printed in the United States
by Baker & Taylor Publisher Services